D0557232

FIFTY KEY THINKERS
ON THE ENVIRONMENT

Fifty Key Thinkers on the Environment is a unique guide to environmental thinking through the ages. Joy A. Palmer, herself an important and prolific author on environmental matters, has assembled a team of thirty-five expert contributors to summarize and analyse the thinking of fifty diverse and stimulating figures – from all over the world and from ancient times to the present day. Among those included are:

- philosophers such as Rousseau, Spinoza and Heidegger
- activists such as Chico Mendes
- literary giants such as Virgil, Goethe and Wordsworth
- major religious and spiritual figures such as Buddha and St Francis of Assissi

Lucid, scholarly and informative, these fifty essays offer a fascinating overview of mankind's view and understanding of the physical world.

Joy A. Palmer is Professor of Education and Pro-Vice-Chancellor at the University of Durham. She is Director of the Centre for Research on Environmental Thinking and Awareness at the University of Durham, Vice-President of the National Association for Environmental Education, and a member of the IUCN Commission on Education and Communication. She is the author and editor of numerous books and articles on environmental issues and environmental education.

Advisory Editors: David E. Cooper, University of Durham, and Peter Blaze Corcoran, Florida Gulf Coast University.

ROUTLEDGE KEY GUIDES

Ancient History: Key Themes and Approaches
Neville Morley

Cinema Studies: The Key Concepts (second edition)
Susan Hayward

Eastern Philosophy: Key Readings
Oliver Leaman

Fifty Contemporary Choreographers
Edited by Martha Bremser

Fifty Eastern Thinkers
Diané Collinson, Kathryn Plant and Robert Wilkinson

Fifty Key Contemporary Thinkers
John Lechte

Fifty Key Jewish Thinkers
Dan Cohn-Sherbok

Fifty Key Islamic Thinkers
Centre for Middle Eastern and Islamic Studies, University of Durham

Fifty Key Thinkers on History
Marnie Hughes-Warrington

Fifty Key Thinkers in International Relations
Martin Griffiths

Fifty Key Thinkers on the Environment
Edited by Joy A. Palmer with David E. Cooper and Peter Blaze Corcoran

Fifty Major Philosophers
Diané Collinson

Key Concepts in Cultural Theory
Edited by Andrew Edgar and Peter Sedgwick

Key Concepts in Eastern Philosophy
Oliver Leaman

Key Concepts in Language and Linguistics
R.L. Trask

Key Concepts in the Philosophy of Education
Christopher Winch and John Gingell

Key Concepts in Popular Music
Roy Shuker

Post-Colonial Studies: The Key Concepts
Bill Ashcroft, Gareth Griffiths and Helen Tiffin

Social and Cultural Anthropology: The Key Concepts
Nigel Rapport and Joanna Overing

FIFTY KEY THINKERS ON THE ENVIRONMENT

Edited by Joy A. Palmer

Advisory Editors: David E. Cooper and Peter Blaze Corcoran

London and New York

First published 2001
by Routledge
11 New Fetter Lane, London EC4P 4EE

Simultaneously published in the USA and Canada
by Routledge
29 West 35th Street, New York, NY 10001

Routledge is an imprint of the Taylor & Francis Group

© 2001 selection and editorial matter, Joy A. Palmer;
individual entries, the contributors

Typeset in Times by Taylor & Francis Books Ltd
Printed and bound in Great Britain by TJ International Ltd,
Padstow, Cornwall

All rights reserved. No part of this book may be reprinted or
reproduced or utilized in any form or by any electronic,
mechanical, or other means, now known or hereafter
invented, including photocopying and recording, or in any
information storage or retrieval system, without permission in
writing from the publishers.

British Library Cataloguing in Publication Data
A catalogue record for this book is available from the
British Library

Library of Congress Cataloging in Publication Data
A catalog record for this book has been requested

ISBN 0–415–14698–4 (hbk)
ISBN 0–415–14699–2 (pbk)

CHRONOLOGICAL LIST OF CONTENTS

ALPHABETICAL LIST OF CONTENTS

CONTRIBUTORS

Allaby, Michael is an author, based in Argyll, Scotland.

Barry, John is Reader in the School of Politics, the Queen's University of Belfast, Northern Ireland.

Barsam, Ara is D.Phil. student in the Faculty of Theology, Oxford University, England.

Bilimoria, Purushottama is Associate Professor in the School of Social Inquiry, Deakin University, Australia, and continuing Visiting Lecturer in Philosophy at the University of Melbourne.

Browne, Janet is Reader in the History of Biology at the Wellcome Centre for the History of Medicine at University College, London, England.

Callicott, J. Baird is Professor of Philosophy at the University of North Texas, USA.

Casal, Paula is Fellow in Ethics and the Professions at the Kennedy School of Government, Harvard University, USA, and Lecturer in Politics at the School of Politics, International Relations and the Environment at the University of Keele, England.

Clarke, John I. is Professor of Geography Emeritus at the University of Durham, England.

Coletta, W. John is Associate Professor of English at the University of Wisconsin–Stevens Point, USA.

Cooper, David E. is Professor of Philosophy at the University of Durham, England.

Corcoran, Peter Blaze is Professor of Environmental Studies and Environmental Education at the Florida Gulf Coast University, USA.

Dumble, Lynette J. is Senior Research Fellow, History and Philosophy of Science at the University of Melbourne, Australia.

Gates, Phillip J. is Lecturer in Biological Sciences at the University of Durham, England.

Glotfelty, Cheryll is Associate Professor of Literature and Environment in the Department of English at the University of Nevada, Reno, USA.

Griffin, Nicholas is Professor of Philosophy at McMaster University, Ontario, Canada.

Hardie, Philip R. is Reader in Latin Literature at the University of Cambridge, England, and a Fellow of New Hall.

James, Simon P. is Tutor in the Department of Philosophy at the University of Durham, England.

Kumar, Satish is Director of Programme, Schumacher College, Dartington, Devon, England, and Editor of *Resurgence* magazine.

Linzey, Andrew is Senior Research Fellow in Theology and Animal Welfare at Mansfield College, Oxford University, and an Honorary Professor at the University of Birmingham, England.

MacDonald, Paul S. is Lecturer in Philosophy, Murdoch University, Australia.

McCarter, Robert is Director, Professor and Architect in the School of Architecture, University of Florida, USA.

McDowell, Michael is Instructor in the Division of English and Modern Languages at Portland Community College, Oregon, USA.

Moss, Ann is Professor of French at the University of Durham, England, and a Fellow of The British Academy.

Mossley, David J. is freelance philosopher and student of Law at Manchester, England.

Nelson, Michael P. is Associate Professor of Philosophy and of Natural Resources at the University of Wisconsin–Stevens Point, USA.

Palmer, Joy A. is Professor of Education, Pro-Vice-Chancellor and Director of the Centre for Research on Environmental Thinking and Awareness at the University of Durham, England.

Riordan, Colin is Professor of German at the University of New-castle-upon-Tyne, England.

Rolston III, Holmes is University Distinguished Professor and Professor of Philosophy at Colorado State University, USA.

Schnadelbach, R. Terry is Professor of Landscape Architecture at the University of Florida, USA.

Sen Gupta, Kalyan is Professor of Philosophy at Jadavpur University, Calcutta, India.

Simmons, Ian G. is Professor of Geography at the University of Durham, England, and a Fellow of the British Academy.

Smith, Richard is Reader in Education at the University of Durham, England.

Walls, Laura Dassow is Associate Professor in the Department of English, Lafayette College, Easton, Philadelphia, USA.

Weir, Jack is Professor of Philosophy at Morehead State University, Morehead, Kentucky, USA.

Yamauchi, T. is Professor Emeritus of Osaka University of Education and Professor of East Osaka Junior College, Japan.

PREFACE

This book is intended to be a valuable resource for readers with an interest in 'influential lives' relating to critical thinking and action which has influenced the environmental movement, and in the intellectual history of environmental philosophy and related fields.

Each essay follows a common format. An opening quotation sets the scene, then readers are provided with an overview of the subject's work and basic biographical information. Each author then engages in critical reflection which aims to illuminate the influence, importance, and perhaps innovative character, of the subject's thinking and, where appropriate, actions. In other words, authors have moved beyond the purely descriptive and have provided a discussion of the nature of the intellectual or practical impact that the life, thinking and works of each figure made or is making upon our understanding of or attitudes towards environmental matters.

At the end of each essay, I have provided information that will lead interested readers into further and more detailed study. Firstly, there are the references for the notes to which the numbers in the text refer; secondly, there is a cross-referencing with other subjects in the book whose thought or influence relates in some obvious way to that of the subject of the essay; thirdly, there is a list of the subject's major writings (where applicable); and finally, there is a list of references for those who wish to pursue more in-depth reading on the subject.

What a tremendously difficult task it was to decide on the final list of fifty environmentalists to be included in this volume. Inevitably, I and my advisory editors were inundated with suggestions and ideas for influential people, who, for the obvious reason of lack of space, had to be left out. The fifty subjects finally decided upon include the very obvious 'great names' in the environmental world such as Jean-Jacques Rousseau and Rachel Carson, alongside some less well-known yet clearly influential people. Our great environmental thinkers span a very broad timescale, from the fifth century BCE to the present day. They include a number of people who might be described as activists, such as Chico Mendes; alongside philosophers or more traditional 'thinkers and writers' such as John Ruskin and Arne Naess.

Finally, I emphasize that this book is certainly not exhaustive – as already mentioned our choice of subjects proved to be extremely difficult. Furthermore, it certainly does not pretend to be an overview of the lives of the fifty greatest environmental thinkers the world has ever known. We believe that it includes some people who would fall into this category of those who have had arguably the greatest global influence on environmental thought and action; but most importantly, *all* people in the book have made very substantial contributions to environmental thinking in some form or another. It is hoped that some readers will derive great benefit and pleasure from the book because it introduces them to previously unknown lives. As a whole, I hope that this volume will be of interest to all who would like to find out more about the lives of individuals past and present who have influenced thinking about the inter-relationships that exist among people, other species, and the natural world.

Joy A. Palmer

BUDDHA fifth century BCE

> How astonishing it is, that a man should be so evil as to
> break a branch off the tree, after eating his fill.[1]

Born Siddharta Gotama into a royal family in northern India, c. fifth
century BCE, the young prince was overwhelmed by the universality
of suffering, old age, illness and death that he witnessed whenever he
was allowed outside the palace gates. He took early to a life of
contemplation, meditation, austerity and simple living so as to
fathom the riddle of life and death, and to resolve his insufferable
despair over the endless, meaningless cycle of re-death and continual
rebirth, until he attained enlightenment (*nirvana*). The natural
settings surrounding Buddha's whole life appeared to have inspired,
if not Buddha's own thinking directly, the imagery attributed to the
sequence of events leading to his enlightenment. It has been
remarked that 'the Buddha Gotama was born, attained enlightenment, and died under trees'. What textual records we have,
furthermore, testify to 'the importance of forests, not only as an
environment preferred for spiritual practices such as meditation but
also as a place where laity sought instruction'.[2] '[So said the
Buddha] ... Seeking the supreme state of sublime peace, I
wandered ... until ... I saw a delightful forest, so I sat down
thinking, "Indeed, this is an appropriate place to strive for the
ultimate realization of ... Nirvana".'[3] Gotama was likely reacting
to rapid commercial urbanization and the rise of merchant and
artisan classes in his region, and a concomitant agrarian economy
responsible for the deforestation of the Ganges region and
consequent vanishing of animal life from its natural habitat.

In Buddha's collected sermons there are compassionate calls to
show due care and loving kindness towards all sentient creatures.
Birds and animals bear witness to the Buddha's testimony, and they
also become dialogic partners in the ensuing discourses. 'The
Buddha Among the Birds' is only one of the 550 stories from the
Jataka tradition that narrates Buddha's life among animals, and
there are stories that recall Buddha's experiences *as* an animal in his
former births. It would seem that the Buddha was reevaluating the
human–cosmos relationship prevalent in the Indic civilization since
the arrival of the Vedic Aryans with their proclivity towards sacrifice,
exploitation of animals for agriculture and warfare, and subservience
to a brahmanic pan-naturalism, with its ingrained fear of nature.

Buddha succeeded in shifting perception from one of fearful warring nature-forces to that of the benign disposition of nature. The Buddha interacted in deep empathy with people from all stratas of life, including the settled merchant classes and trading groups travelling to the region, and from his reflections developed a form of social ethics which he practised and preached. These teachings were handed down and later recorded in the Pali canons, brought together into 'three baskets'. The coded teachings of the enlightened one (or 'buddha') on a broad ethical paradigm that connected with the path of liberation from suffering, despite their heavy emphasis on ascetic life (i.e. renunciation or withdrawal from society), contain innovative and vital knowledge about Buddha's thinking on the environment. One insight that is nowadays seen as holding a key to the growth of Buddhist ecological consciousness over the course of two millennia and across Asia is that of 'dependent arising' (*pratitya-samutpada*): 'on the arising of this, that arises'. The causal principle of interdependence registers an ecological vision that, as a recent scholar aptly put it, 'integrates all aspects of the ecosphere – particular individuals and general species – in terms of the principle of mutual codependence'.[4] The relational model undermines the sovereignty and presumed autonomy of the self over other beings and creatures (animals or plants). The ideals of *dharma* and virtues developed in accordance with this insight have been topics of intense reflection and debate among Buddhist schools, and have also been implemented at different historical junctures, such as by Emperor Ashoka after his conversion to Buddhism. He institutionalized care and welfare towards animals, as the following edict poignantly records for us: 'Here no animal is to be killed for sacrifice ... the Beloved of the Gods has provided medicines for man and beast ... medicinal plants ... [R]oots and fruits have also been sent where they did not grow and have been planted along the roads for use of man and beast.'[5]

Another side of the causal principle of interdependence is the consequent or karmic continuum, which suggests that every action conditions a being's personal history of suffering, the cessation thereof and subsequent liberation from the karmic continuum: 'on the cessation of this, that ceases'. From the particularity of individual suffering (karmic action-effect), the Buddha was able to generalize to humankind, the animal world and natural environment themselves as distinctive manifestations of the cumulative effect of karmic conditioning. He eschewed any hierarchical dominance of one order of being over the other. A social and ecological ethic (*dharma*) based on undoing the cyclical and all-devouring chain of karmic effects or conditionings was the primary goal of Buddha. His followers applied the teachings in several different directions. This is borne out in the Buddha's expectation that monks and lay Buddhists alike ought to strive always for the 'welfare of the many', 'the happiness of the many', 'compassion for the world'.[6]

Buddha's teachings, however, did not separate out a unidimensional emphasis on environmental ethics from an ontology and ethics of spiritual transformation or sacred-making *dharma* of the human and natural worlds alike. It has been argued that ontological notions such as Buddha-nature or Dharma-nature provide a basis for unifying all existent entities in a common sacred universe, even though the tradition has come to privilege human life *vis-à-vis* spiritual realization.[7] In other words, Buddhism underscores the inherent moral worth and 'considerability', in principle at least, of all beings towards which there are certain mutual and reciprocal obligations. We might not ordinarily consider the humble gurgling stream as having any particular obligation towards human beings, but the small schools of fish might be very appreciative of the sustenance and safe ecosystem that the cool water provides for them. For contemporary Buddhists this qualification has become even more urgent, given that some of the kinds and patterns of disjuncture of human–earth relationships that we face nowadays did not exist and might not even have been foreseen by Gotama in his despondent wanderings through the comparatively less disruptive urban environment of his day.

While Buddha may have realized the diversity and interconnectedness of the biocommunity, his worldview was neither entirely naturalistic nor as biocentric as Buddhism in its different forms has sometimes become. In this context, while the relevance and role of the environment is recognized in the ecology of individual movement towards Nirvana, the blurring of individual autonomy and particularity necessary for an ethic of duties, rights and the legal protection of minorities and endangered species, weakens the empowering strength needed for a balanced ethic of poly-ecoism. The Buddha's refusal to prescribe unqualified vegetarianism, it is often argued, is indicative of such a weak link in the Buddha's otherwise noble and promising prolegomenon for all future environmental ethics.[8] Nevertheless, the Buddha's plea for compassion for all life forms in their mutual interdependency and the aestheticization of nature that undergirds his wisdom-teachings, paved the way for a radical transformation of attitude towards

nature in regions to which Buddhism travelled. For instance, the
Dalai Lama is an ardent advocate of environmental compassion and
an ethic of universal responsibility, which he sees very much lacking
in the present, modern-day hectic world. Again, the Vietnamese
monk Thich Nhat Hanh has evolved another strand of wisdom-
concentration as a necessary ingredient in the development of a
sustainable natural habitat for humans and natural beings alike.
More generally, despite the predominance of non-vegetarianism
in Buddhist communities, the rights and ethical protection of certain
liberties of animals have been recognized in Buddhism. Many
Buddhist monasteries across east Asia have banned the cooking of
animal flesh as this involves the killing of animals, whether or not
the direct intention is the act of consumption at the dinner table.
Buddhist environmentalists are active in modern-day Sri Lanka in
their efforts to preserve the lush beauty of the island state from
despoilment through extensive technological development and the
ravages of an ethnic war. They too can be said to be continuing a
practical environmental ethics fostered centuries ago when Buddhism was brought to Sri Lanka.

Likewise, the arrival of Buddhism in Tibet in the seventh century
engendered a nationwide programme for the preservation of the
heavenly-natural oasis that remained a mysterious land for much of
the outside world. The ruling Lamas proscribed injuring and killing
of animals, big and small. The moral practice of showing respect for
and responsibly using nature became a way of life for the Tibetans.
Even though Tibetan Buddhist metaphysics continued the influential Indian Buddhist doctrine of the absence of self-nature or
intrinsic existence of properties and substances alike, proclaiming
this 'emptiness' of all things has strengthened its moral framework
on three counts.

1 Moral properties such as those of the good, compassion, and
   loving kindness or respect, though by no means absolute, have a
   solid presence (contingently supervenient on 'emptiness'), in as
   much as human interaction or ethical life generally presupposes
   these properties.
2 A pluralistic ontology that has fair regard for members within it,
   without privileging any particular species, easily gets translated
   into a non-anthropocentric respect for biodiversity.
3 The religious-soteriological 'end' requires certain self-motivated
   ethical practices and norms, including restraint on desires,
   meditation on the limits of the ego-self, altruism based on the

moral properties of reverence and deep (but not condescending) compassion for all living and non-sentient beings. In other words, the normative constructs for monks, nuns, lay people, farmers and nomads too, underscore concern for the environment.

The Buddhist ethic of living in harmony with the earth accordingly pervaded all aspects of Tibetan culture. Perched on the 'roof-top' of the world, Tibet's environment was recognized as being crucial to the stability of ecological environs and crop cycles in much of neighbouring Asia. For instance, the ten or so major rivers that wind through Asia feed off the river valleys and smooth glacial icescapes of Tibet, and the monsoon relies on Tibet's abundant natural vegetation and dense forests. Its wildlife and natural animal sanctuaries maintained an equilibrium and contributed in different ways to the enrichment of the environment, providing manure for controlled husbandry and organic re-vegetation, as well as fuel (from yak dung), and so on. However, after the Chinese occupation of Tibet, the situation has dramatically altered: massive deforestation, land erosion, pollution of rivers, depletion of resources, excessive killing of animals, and general degradation of the environment have had an adverse environmental impact on south-east Asia, which has been subjected to uncontrollable flooding from the monsoon deluge each year. Buddhists are also concerned that the construction of the world's largest dam on the Yangtze river in China will add to the eco-disequilibrium being visited upon much of Asia by China's modernist ambitions and imperiousness.

Transcending the human-centric (ego-bounded) perspective is one of the great strengths of Buddha's interdependent or interconnected vision of all things within the natural–human–social matrix. As de Silva puts it: 'The Buddhist environmental philosophy may be described as a shift from an egocentric stance towards an ecocentric orientation'.[9] The key ontological and moral concepts that help ground Buddha's ecological thinking comprise:

pratitya-samutpada	interdependent conditioning
karma (Pali *kamma*)	the law of moral causation
duhkha (Pali *dukka*)	unsatisfactoriness
dharma (Pali *dhamma*)	reciprocity of obligations or rights within the bounds of duties
sila	cultivation of virtues, disciplines, and the overcoming of vices. Among the virtues

5

highlighted are: restraint, simplicity, loving-kindness, compassion, equanimity, patience, wisdom, non-injury, and generosity

These concepts rest on: (1) a general principle of consequentialism (the gravity and impact of one's actions judged by their consequences), moderated by (2) a teleology (a larger purpose or particularity of ends towards which each species strives even in the apparent absence of agency – hence the mountain having its own silent *telos*), and (3) a deontology (*dharma* for *dharma*'s sake), intended as a check against excessive altruism, unmitigated utilitarianism, and ritualized narcissism or a 'grand narrative' teleology.

Notes

1 *Anguttara Nikaya*, vol. III, p. 262.
2 Lewis Lancaster, 'Buddhism and Ecology: Collective Cultural Perception', in M.E. Tucker and D. Williams (eds), *Buddhism and Ecology*, p. 11.
3 *Ariyapariyesana Sutra, Majjhima Nikaya*, cited in Donald K. Swearer, 'Buddhism and Ecology: Challenge and Promise', *Earth Ethics*, Fall, pp. 19–22, p. 21, 1998.
4 Ibid.
5 *Sources of Indian Tradition*, vol. I, rev. edn by A.T. Embree, New York: Columbia University Press, pp. 144–5, 1988.
6 *Middle Length Sayings*, cited in Padmasiri de Silva, *Environmental Philosophy and Ethics in Buddhism*, New York: St Martin's Press; London: Macmillan Press, p. 31, 1998.
7 Swearer, op. cit., p. 20.
8 P. Bilimoria, 'Of Suffering and Sentience: The Case of Animals (revised)', in H. Odera Oruka (ed.), *Philosophy, Humanity and Ecology: Philosophy of Nature and Environmental Ethics*, Kenya: African Centre for Technology Studies, pp. 329–44, 1994; for a humorous counter-argument on 'me-eat', see, Arindam Chakrabarti, 'Meat and Morality in the Mahabharata', *Studies in Humanities and Social Sciences*, III (2), pp. 259–68, 1996.
9 De Silva, op. cit., p. 31.

See also in this book

Bashō, Gandhi, Tagore

Buddha's major writings

Samyutta Nikaya, ed. L. Freer, London: Pali Text Society, 1884–1904.

Anguttara Nikaya, eds H. Morris and H. Hardy, vols I–V, London: Pali Text Society, 1885–1900.

Digha Nikaya, ed. T.W. Rhys Davids, London: Pali Text Society, 1920–1.

Gradual Sayings, vols I, II, V, trans. F.L. Woodward, London: Pali Text Society, 1932–6.

Suttanipatta, ed. T.W. Rhys Davids, London: Pali Text Society, 1948.

Dialogues of the Buddha, parts I, II, trans. T.W. Rhys Davids and C.A.F. Rhys Davids, London: Pali Text Society, 1956–7.

Further reading

Bilimoria, P., 'Duhkha & Karma: The Problem of Evil and God's Omnipotence', in *Sophia, International Journal for Philosophical Theology and Cross-cultural Philosophy of Religion*, 34 (1), pp. 92–119, 1995.
Callicott, J.B. and Ames, R. (eds), *Nature in Asian Traditions of Thought: Essays in Environmental Philosophy*, Albany, NY: State University of New York Press, 1989.
Gross, Rita, 'Toward a Buddhist Environmental Ethic', *Journal of the American Academy of Religion*, 65 (2), pp. 333–53, 1997.
Tucker, M.E. and Williams, D. (eds), *Buddhism and Ecology: The Interconnection of Dharma and Deeds*, Cambridge, MA: Harvard Center for the Study of World Religions, pp. 3–18, 1997.

PURUSHOTTAMA BILIMORIA

CHUANG TZU fourth century BCE

All the fish needs is to get lost in water. All man needs is to get lost in Tao.[1]

The two most famous and enduring works of philosophical Taoism, both composed during the classical period of Chinese thought (c.500–200 BCE), are the *Tao Te Ching* (or *Dao De Jing*) and the *Chuang Tzu* (or *Zhuang Zi*). They are works, moreover, in which subsequent generations, right down to the present, have claimed to find an enlightened attitude towards the natural world, a 'doctrine of harmony with the natural environment'.[2] Traditionally the *Tao Te Ching* was attributed to one Lao Tzu or Lao Tan, supposedly a contemporary of Confucius (sixth–fifth century BCE), and the *Chuang Tzu* to a later disciple. Modern scholars, however, favour the view that the latter work was the earlier, with the *Tao Te Ching* being a third-century BCE compilation by unknown authors who, in a manner then familiar, annexed their thoughts to the name of an ancient, and perhaps mythical, sage.[3]

CHUANG TZU

Unlike Lao Tzu, the actual existence of the reputed and eponymous author of the *Chuang Tzu* is reasonably well attested. A later chronicle asserts that Chuang Tzu (Master Chuang) was an official in a lacquer garden in present-day Honan and that he refused higher royal office on the grounds, according to his own Book, that he would prefer to live like an ordinary tortoise, free to 'drag its tail in the mud', than to live artificially like one pampered at court (17.11). Other anecdotes in the Book suggest that Chuang Tzu was an engaging and ironic individualist, with scant respect for artifice and convention, especially for the Confucian rites of burial. We find him, from his deathbed, chiding his disciples for preparing a 'sumptuous burial' for him (32.14).

Although the Book is traditionally attributed to Chuang, it is now accepted that he wrote only some of the thirty-three chapters, no more, perhaps, than the 'inner chapters' (1–7), and that some sections of other chapters were assembled by thinkers of his 'school'.[4] Like one of his translators, then, 'when I speak of Chuang Tzu, I am referring ... to the mind, or group of minds, revealed in the text called *Chuang Tzu*', rather than to a specific, historical individual.[5]

The *Chuang Tzu*, as noted, is a classic of philosophical Taoism. The qualifier is important, since the Taoism it represents should be sharply distinguished from the 'religious' or 'magical' Taoism which developed after the second century CE. The distance between the two may be gauged from reflecting that whereas Chuang taught calm acceptance of, even indifference to, death, the main obsession of 'magical' Taoists was discovery of the elixir of eternal life. Before characterizing philosophical Taoism, three points should be borne in mind. First, the division of classical Chinese philosophers into 'schools' – Taoist, Confucian, Legalist, etc. – was the work of a later taxonomist and encourages exaggeration of the differences between, and similarities within, these 'schools'. Thus Confucius, though often a critical target of Chuang's, sometimes appears in the Book as an admired sage. Second, Taoism is not to be distinguished by its concern for the Tao (Way, Path), since it was the common and primary concern of Chinese philosophers to determine the proper Way for human beings to follow. Third, while there is significant affinity between the two Taoist classics, their emphases are different. The *Tao Te Ching* pre-eminently addresses the problem of how rulers should govern in turbulent, warring times, while the *Chuang Tzu* is occupied with how the private, indeed apolitical, individual should live in these, or any other, times.

Distinctive of Taoism is the very general and abstract notion of

8

the Tao which it invokes. For Confucius, the Tao was the proper Way for human beings, but for Lao Tzu and Chuang Tzu the central notion is that of a 'Great Tao', the 'complete, universal, all-inclusive' Way of the universe itself, a Way with which human lives should harmoniously accord (22.6). This 'Great Way' cannot be precisely articulated. In keeping with the famous opening lines of the *Tao Te Ching* – 'The Way that can be told is not the constant Way' – Chuang says that 'The Great Way is not named ... If the Way is made clear, it is not the Way' (2.2). Nevertheless, some things can be said about it, and lessons for human conduct drawn.

The Tao is the Way of nature as a whole, so that 'the true man', who is 'lost in the Tao', is one who lives 'naturally'. The ills of social and individual life, for Chuang, stem from the fact that, uniquely among living creatures, men are able to, and for the most part *do*, live unnaturally. This means, above all, that most people think and act on the basis of artificial distinctions – between good and bad, beautiful and ugly, men and animals, and so on – whose dependence on partial, pragmatic perspectives they fail to recognize, treating them instead as rigid and as mirrors of reality itself. 'Those who discriminate fail to see' (2.2), for the Tao itself, as the *source* of all differences and distinctions in the world, is itself seamless and fluid. Moreover, as the Way of nature *as a whole*, the Way is 'spontaneous' and 'free', since it faces no obstacles which it must endeavour to overcome and by which it is limited. Likewise, therefore, 'the true man' will behave spontaneously: indeed, his life, like that of the Tao itself, will be one of 'non-action' (*wu wei*), in the sense of being non-deliberative and non-striving. Many of the most attractive stories in the *Chuang Tzu* are in praise of skilled craftsmen – like the butcher whose carving effortlessly responds to the natural grain and joints of the ox (3.1) – who dispense with rules and verbal instructions in favour of a spontaneous, wordless 'know-how'. What is true of the butcher or bell-maker is also true of the Taoist sage. Ignoring the doctrines and principles taught by philosophers – 'the dregs of the men of old' – the sage lives in intuitive appreciation of the Way, recognizing that 'he who knows does not say; he who says, does not know' (13.10).

To live 'naturally', 'lost' in the Tao, is, then, not to live like a caveman, but to act spontaneously, flexibly and intuitively, without rigid attachment to conventional rules and distinctions, linguistic, moral or other. The 'true man', as it were, 'hangs loose'. In the so-called 'primitivist' chapters (8–12) of the *Chuang Tzu*, however – though they are not written by Chuang himself – there are passages which advocate a life of extreme simplicity. These, and corresponding passages in the *Tao Te Ching*, have encouraged the idea that, in Lin Yutang's words, the person 'with a hidden desire to go about with bare feet goes to Taoism'.[6] But this 'primitivism' is contradicted by other passages, and is anyway of too extreme a kind to be relevant to modern environmental discussion. The real relevance of the *Chuang Tzu* resides in its rejection of attitudes which, arguably, have played pernicious roles in the comportment of human beings towards their natural environments.

To begin with, Chuang is critical of anthropomorphic perspectives on which the lives and the good of animals and other living beings are judged by criteria applicable only to humans. Several anecdotes highlight the difference between our and an animal's own conception of its good: for example, the one about the marsh pheasant which, though it must struggle to get food and drink, does not therefore desire to live in a well-provided cage (3.3). Second, Chuang's rejection of rigid distinctions incorporates a criticism of those between man and other creatures, between 'great' and 'small', which encourage an anthropocentric elevation of humans above the rest of nature. Looked at 'in the light of the Tao, nothing is best, nothing is worst ... seen in terms of the whole, no one thing stands out as "better"' (17.4). Finally, the sage's life of *wu wei* is incompatible with precisely those types of desire and ambition – for profit, esteem, control – which have led men to exploit nature. Such men are 'driven', 'penned in by things ... Pitiful, are they not?' (24.4). For Chuang, the person who has put aside such pernicious attitudes, and in whom, therefore, 'the Tao acts without impediment', will 'harm no other being' (17.3) – not because he now adheres to some moral principle, but because he now lacks any motivation to cause harm.

The authors of the *Chuang Tzu* and the *Tao Te Ching* have been as influential as any of those included in this volume. For more than two millennia, Taoism – despite both the imperial sanctioning of Confucian precepts and, later, Mao Tse Tung's brutal animosity – has powerfully shaped aspects of Chinese life, not least an intimacy with nature attested to in Taoist landscape painting and poetry. No less important was its impact upon religious development in China and Japan, where its true descendant has not been 'magical' Taoism, but that intriguing blend of Taoist and Buddhist thought known as Chan – or, in Japan, Zen – Buddhism. The *haiku* verses of Bashō, discussed elsewhere in this volume, owe as much to Chuang Tzu as to the Buddha. Over the last century, philosophical Taoism has attracted many Western thinkers, not least on account of its view of human beings' relation to nature – one which, in many Western eyes, favourably contrasts with that of their own societies. Martin Heidegger, arguably the twentieth century's most penetrating critic of technology, once began a translation of the *Tao Te Ching*, and the Taoist influence on his thinking was larger than his occasional acknowledgements suggest. During the last few decades, many environmental ethicists have enthusiastically invoked Taoist ideas.[7] It would, however, be misleading to speak of Chuang's 'environmental ethic'. Certainly he would have been without sympathy for talk of, say, the 'rights' of animals or our 'obligations' to nature. In his view, the need for talk like that – talk of morality, justice, righteousness, benevolence – is a sure sign that men have 'lost the Way' and, therefore, are no longer 'lost *in* the Way' (22.1). Those who naturally 'let things be' do not stand in need of moral principles.

Notes

1 6.11. References to the *Chuang Tzu* in the text are to the chapters and sections into which translators standardly divide the text. My citations are from various of the translations listed under 'Major writings', but primarily from Watson's.
2 Fritjof Capra, *The Tao of Physics*, London: Fontana, p. 340, 1983.
3 See A.C. Graham, *Disputers of the Tao*, pp. 215ff.
4 Ibid., pp. 170ff.
5 Burton Watson, *The Complete Works of Chuang Tzu*, p. 3.
6 *My Country and its People*, London: Heinemann, pp. 109–10, 1936.
7 See, e.g., Capra, op. cit.

See also in this book

Bashō, Heidegger

Chuang Tzu's major writings

Among the many translations of the *Chuang Tzu*, the following are easily available, each with its distinctive merits:

Merton,Thomas, *The Way of Chuang Tzu*, London: Unwin, 1965.
Watson, Burton, *The Complete Works of Chuang Tzu*, New York: Columbia University Press, 1968.
Palmer, Martin, *et al.*, *The Book of Chuang Tzu*, Harmondsworth: Penguin, 1996.

Further reading

Chau, S. and Song, F., 'Ancient Wisdom and Sustainable Development from a Chinese Perspective', in J.R. and J.G. Engel (eds), *Ethics of Environment and Development*, London: Belhaven, 1990.

Cooper, D., 'Is Daoism "green"?', *Asian Philosophy*, 4 (2), 1994.

Graham, A.C., *Disputers of the Tao: Philosophical Argument in Ancient China*, La Salle, IL: Open Court, 1989.

Hansen, C., 'Laozi' and 'Zhuangzi', in R. Arrington (ed.), *Blackwell Companion to the Philosophers*, Oxford: Blackwell, 1998.

Tao Te Ching, trans. D.C. Lau, Harmondsworth: Penguin, 1985.

DAVID E. COOPER

ARISTOTLE 384–322 BCE

In all natural things there is something wonderful.

Parts of Animals 645[a] [1]

The Greek philosopher and scientist, Aristotle, was born in Macedon, where his father was physician to the king. The son was himself to enter royal service as tutor of a future and more famous king, Alexander the Great – a grateful pupil, if we credit the story that he instructed his far-flung subjects to provide Aristotle with specimens for his biological research. Much of his life, from 367 to 347 and again from 335 to 322, was spent in Athens, first as pupil and teacher at Plato's Academy and later at the school he himself founded, the Lyceum. Both sojourns were ended by outbursts of anti-Macedonian sentiment in Athens. After the first he lived in Lesbos, where his most important scientific work was done; after the second he moved to Chalcis, where he died a few months later.

Aristotle's life was one of unremitting study, the voluminous writings bequeathed to us forming only some 20 per cent, perhaps, of his original output. He wrote and lectured on an extraordinary range of subjects – including biology, astronomy, logic, metaphysics, ethics, poetics, and politics – as well as compiling massive records of, *inter alia*, the Pythian and Olympic Games. To say only this, however, underestimates his unique achievement: for, in the case of many subjects, he did not so much contribute to them as *invent* them. In some areas, moreover, such as logic and zoology, the taxonomies and general principles proposed by Aristotle remained almost unchallenged for more than 2,000 years. It is no exaggeration, therefore, to hold that 'an account of Aristotle's intellectual afterlife would be little less than a history of European thought'.[2] To equal his achievement would require someone totally to redraw the map of intellectual enquiry.

Aristotle was not, of course, an environmental scientist or philosopher in the contemporary sense. The 'eco-crises' which have stimulated recent environmental concern were happily unknown in ancient Greece. Indeed, the very concept of 'the environment' was not one available to Aristotle. Nor did he address such issues as our moral obligations to non-human life. It is clear however – from my opening citation, for example – that Aristotle experienced and urged a profound regard for the living world and, as we shall see, several elements in his thinking prove attractive to contemporary environmental thought.

Some of those elements were integral to his general conception of the natural world, one which remains alive in a way that some of his pioneering studies in zoology and biology no longer do. (One should recall, though, that Aristotle was responsible for the modern notion of species, since it was he who proposed classifying animal kinds by reproductive criteria, rather than on the basis of less explanatory similarities.)

Aristotle divided the domain of enquiry into the theoretical, practical and productive sciences. The first are concerned with obtaining truth for its own sake, something of the first importance given that 'all men by nature desire to know' (*Metaphysics* 980^a); the other two – ethics and poetics, for example – with how people should behave and produce things. The theoretical sciences, Aristotle divides into 'theology', 'physics' and mathematics. The former terms are misleading: 'theology', for Aristotle includes logic and metaphysics, while 'physics' is the study of the natural world in general. The distinction between 'physics' and the other theoretical sciences is that it deals with things subject to movement and change, where 'change' includes coming into and passing out of existence.

'Physics', therefore, directly addresses one of the two main questions Aristotle raises in his *Metaphysics*: What are the most basic entities in reality, the 'primary substances', upon which everything else depends?, and What explains regular processes and changes, such as the growth and decay of organisms? The questions are related for Aristotle, since not only is it, in the final analysis, substances which 'become' and change, but we have knowledge of a substance only 'when we have found its primary causes' (*Physics* 184^a).

Aristotle rejected two views of substance or basic reality prevalent in his times: the doctrine that substance was some stuff, 'matter', out of which things are composed, and Plato's theory that what is truly real are immaterial Forms or Ideas of which ordinary things are both products and pale copies. For Aristotle, we should not confuse what something is with what it is made of, while Plato's view, by placing the Forms outside the ordinary world, is therefore 'useless' for explaining how, in that world, there are 'comings-to-be' of things (*Metaphysics* 1033^b). Aristotle's own proposal is that a primary substance is a *unity* of form and matter. It is only through having a certain form that a region of matter constitutes a man, say, and it is this form which provides the essence – the 'what-it-is-to-be-a- ... ' – of something.

Aristotle connects his concept of form with the question of the causes of change and movement, since a being's form is also 'an end', a final 'cause as that for [the sake of] which' it begins and develops (*Physics* 199^a). Achievement of its fully developed form, that is, is part of the explanation of why a plant, say, grows from a seed (which contains the form 'potentially'). Indeed, it is the main part of the explanation, the factor on which the biologist must focus to understand the process of development. Aristotle's notion of 'final causes', his teleology, has been much misunderstood. He did not mean, absurdly, that all living beings intentionally strive to attain their forms; nor, despite a notorious passage in *Politics* where he writes that 'since nature makes nothing ... in vain, ... she has made all animals for the sake of man' (1256^b), is it his considered view that nature is a divine, purposeful intelligence. (That one-off passage is inconsistent with: (1) Aristotle's usual view of 'ends' as internal to, or 'immanent' in, natural beings; (2) his general attitude of admiration for nature in its own right; and (3) his theology, in which God, absorbed in self-contemplation, is unconcerned with creaturely life.) Aristotle's notion is, rather, a 'functionalist' one. Unless we know what something is 'for', what its normal developed state should be – the tree to grow fruit, the duck to live aquatically – we cannot fully understand the changes which we observe it undergoing – the growth of roots or of webbed feet, say.

Although only humans can intentionally aim at their *telos* or end, all living beings, according to Aristotle, have 'soul'. But 'soul', with its connotation of an immaterial homunculus 'inside' the body and surviving it, is a poor translation of the Greek term *psuche* (lit. 'breath'). The 'soul', he writes, is 'the substance *qua* form of a natural body' (*On the Soul* 412^a), that which, so to speak, 'holds together' the body in a cohering whole, the 'principle' of its organization. That Aristotle intends something very different from the Christian concept is clear from his view that nutrition and

14

sensation are faculties of the soul. Only at the level of human life is the faculty for rational thought present, and even there it is only the operations of an obscurely described 'active reason' which could intelligibly occur in the absence of body.

Aristotle's picture of the natural world, then, is a graduated one of beings ranging from 'mere things' through plants, the lower and higher animals, to human beings, each with a specific form which both constitutes its essence and plays the crucial role in explaining its behaviour. Despite its variety and complexity, therefore, the natural order is precisely that, an *order* – an interconnected and intelligible whole whose 'excellence', as Aristotle puts it, resembles that of an 'orderly' army, whose individual members, its soldiers, perform their appropriate functions (*Metaphysics* 1075ª).

For many Arab thinkers from the tenth to the thirteenth century and, from the thirteenth century on, for many Christian ones too, Aristotle was 'The Philosopher', 'the master of those who know' (as Dante put it). It would be rash to assume, however, that their 'Aristotelian' conception of the natural world was the Greek's own, for it incorporated Stoic and theological elements foreign to Aristotle's thinking. In particular, the doctrine of 'final causes' was given an 'anthropocentric' twist to make it accord with the conviction that the divinely ordained purpose of each being was to serve the ends of man – a twist encouraged, admittedly, by the *Politics* passage. Later champions of science, such as Francis Bacon and Galileo, who took themselves to be refuting Aristotle, were more often than not only refuting a bowdlerized 'Aristotelianism'. The same is true of nineteenth-century Darwinian critics, although Darwin himself fully recognized Aristotle's contribution, declaring that even the greatest of recent naturalists, such as Linnaeus, were 'mere schoolboys to old Aristotle'.[3] (Incidentally, Aristotle was well aware of the 'Darwinism' of his day. Although he rejected Empedocles' theory of, in effect, random mutation and natural selection, he did so on grounds that many contemporary naturalists take very seriously.)

Aristotle was not, to repeat, an 'environmentalist', but it is not difficult to see why some environmentalists, especially perhaps 'deep ecologists', are attracted to him. To begin with, they share his perception of the natural world as an integrated, continuous whole, without sharp 'breaks', especially between human and non-human life. Second, they can applaud his concept, once properly under-stood, of the 'end', the 'for-the-sake-of-which', each creature develops: for this favourably contrasts with a purely 'mechanistic' picture of nature which is, in their view, both impoverished and dangerous. Third, they can welcome Aristotle's view that a being's 'end' is 'a good' for it. A tree or a duck whose 'end' is actualized 'flourishes' *qua* the being it is, and to prevent anything from flourishing, Aristotle strongly implies, is to do a wrong.

Finally, some environmental ethicists have looked for inspiration to Aristotle's general approach to ethics, which focuses not on questions about rights and duties, but on ones about the 'excellences' appropriate to a flourishing human life, one which realizes the human *telos*. Such a life is one of 'virtue', and some of the practical virtues Aristotle prescribes, like self-restraint in material pursuits, are ones which surely have application to the treatment of our environments. More important, perhaps, are the implications of that 'intellectual virtue', 'theoretical wisdom', which, for Aristotle, is 'the highest form of activity' (*Nicomachean Ethics* 1177ª). The 'contemplation' of 'the loftiest things' in which that wisdom consists is not, for Aristotle, a passionless investigation into truth, but something imbued with a sense of wonder at, and admiration for, the cosmos and its ingredients. The passage from which my opening citation comes – one which 'expresses some of the best in Aristotelian man'[4] – continues with an endorsement of Heraclitus' rebuke to some visitors disappointed to find the sage doing something as mundane as warming himself by a stove. 'There are gods here too', Heraclitus told them. The lesson Aristotle draws is that even the humblest creatures should be contemplated 'without aversion', for in them too 'there is something natural and beautiful' (*Parts of Animals* 645ª).

Notes

1   References are to the standard pagination of Aristotle's works, which is retained in all good translations.
2   Jonathan Barnes, *Aristotle*, p. 86.
3   Quoted by W.D. Ross, *Aristotle*, p. 112.
4   Barnes, op. cit., p. 87.

See also in this book

Bacon, Darwin

Aristotle's major writings

The Complete Works of Aristotle, 2 vols, ed. J. Barnes, various translators, Princeton, NJ: Princeton University Press, 1984.

A *New Aristotle Reader*, ed. J.L. Ackrill, Oxford: Clarendon Press, 1987.
Contains generous excerpts from all the works cited in the entry.

Further reading

Barnes, J., *Aristotle*, Oxford: Oxford University Press, 1982.
The Cambridge Companion to Aristotle, ed. J. Barnes, Cambridge: Cambridge University Press, 1995.
Ross, W.D., *Aristotle*, 5th edn, London: Methuen, 1949.

DAVID E. COOPER

VIRGIL 70–19 BCE

> Then there are all the famous cities, laboriously built, all
> the towns piled up by human hand on sheer rocks, with
> rivers gliding beneath their ancient walls. Shall I mention
> the Adriatic and the Tuscan seas? Or the great lakes? –
> you, Como the greatest lake, and you, lake Garda, whose
> rising waves imitate the roar of the sea? Or shall I mention
> the harbours, and the dykes imposed on the Lucrine lake,
> and the sea crashing out its indignation? ... This same
> land brings forth from her veins streams of silver and
> mines of bronze, and pours out floods of gold. This was
> the land that produced a fierce breed of men, the Marsians
> and the Sabine race, the Ligurians used to hardship and
> the Volscians with their javelins, this the land that
> produced the Decii, the Marii, the great Camilli, and the
> Scipios toughened in war, and you, greatest Caesar, who
> now victorious on the furthest shores of Asia drive off the
> unwarlike Indians from the hills of Rome.
>
> 'Praise of Italy', *Georgics*, 2.155–76

The Roman poet Virgil is said to have been born on 15 October 70
BCE in Andes, a village near Mantua; the late-antique writer
Macrobius said that he was 'born in the Veneto of country parents
and brought up amongst the woods and shrubs'.[1] He was educated
in Cremona and Milan before going to Rome; he then for a while
became part of an Epicurean community in Naples, a sect which
advocated philosophical retreat from urban society and politics. In
the late 40s he was writing his first major work, the *Eclogues*
(probably published in 39–38 BCE), a book of ten pastoral poems one
of whose subjects is the confiscations of land in 42 BCE for the
settlement of veterans by Octavian (the future Augustus) after the
civil war against the murderers of Julius Caesar. According to the

ancient biographical tradition, Virgil's father's farm was one of those confiscated.

Virgil now came into the circle of the literary patron and intimate of Octavian, Maecenas, to whom he dedicated his four-book *Georgics*, in form a didactic poem on farming, probably published in 29 BCE, the year of Octavian's triple triumph two years after the decisive battle of Actium in which Octavian had defeated Antony and Cleopatra and so brought to an end two decades of civil war. During the last ten years of his life Virgil was working on the *Aeneid*, an epic in the Homeric manner on the wanderings and wars of the Trojan hero Aeneas, archetypal city-founder and ancestor of Augustus (the name taken by Octavian in 27 BCE, when he consolidated his rule in Rome). On returning from a journey to Greece Virgil died of a fever at Brindisi on 20 September 19 BCE, with the *Aeneid* lacking its final touches. His dying wish that the poem be burned was overruled by Augustus.

Virgil was immediately canonized as the national poet of Rome, and his works, above all the *Aeneid*, became the central classics of the later Western tradition. Indeed the history of the classical tradition could largely be written in terms of a history of the reception of Virgil. In all three of his major works he reveals a deep interest in the individual's relationship to his wider environment and to the natural world. Virgil's complex and sympathetic sensibility has left its mark on the ways in which succeeding generations of Europeans and North Americans have conceptualized and visualized the place of their culture and society in the world. It is difficult to generalize about the nature of this influence, since Virgil was a poet, not a systematic thinker. One of the marks of his greatness as a poet is his openness to the whole range of ancient traditions and attitudes, popular and philosophical, concerning the natural world and man's place in it. Furthermore, attitudes to man and his environment are to an extent determined by the different genres in which Virgil worked.

The three major works form a seemingly inevitable sequence, sometimes viewed in antiquity as reflecting the history of human civilization, from a pastoral to an agricultural to an urban way of life. The *Eclogues* stage a simple form of human society: individual herdsmen bonded to each other by friendship, ideally enjoying a close and unproblematic relationship with the animals they tend and the landscape they inhabit. The *Georgics* deal with the expertise and technology required to farm the land; the relationship of man to nature — animal, mineral, and vegetable — is now as much one of an imperialist and militaristic domination, as of a more collaborative coexistence. The *Aeneid*'s ultimate subject is the foundation of the great city of Rome and of a people whose military machine will conquer the world. But it is also a poem about Italy, and the agricultural societies and landscapes of Italy (like many famous Romans born outside the capital, Virgil felt a loyalty both to Rome and to his home town, Mantua); echoes of the earlier *Eclogues* and *Georgics* are not out of place in a poem written for an urban ruling élite many of whom had working country estates and a real interest in agriculture,[2] and the walls of whose houses were painted with romantic landscapes.[3]

For much of the last two thousand years the pastoral idea has been a mainly Virgilian tradition. The inaccessibility to Greekless centuries of Virgil's own model, Theocritus' *Bucolics*, obscured the origins of pastoral as a semi-realist, earthy genre. Virgil is often credited with the invention in his *Eclogues* of 'Arcadia', a dream landscape which men attempt to enter either through art or direct manipulation of their physical surroundings.[4] While the world of the *Eclogues* is a more stylized and artificial creation than its Theocritean predecessor, Arcadia is but one of the Virgilian pastoral landscapes, and arguably marginal. The modern notion of an idyllic Arcadia is the product of a Renaissance elaboration of Virgilian hints, above all in the Italian poet Jacopo Sannazaro's *Arcadia*.

Central to Virgil's pastoral vision is a sense of life in harmony with nature, but under threat from disruptions both external, in the shape of civil war and land confiscations, and internal, above all in the form of erotic passion. Perfection may wear either a private face, in the shepherd happy with his girlfriend and whose love songs are echoed back by a sympathetic nature (*Eclogue* 1), or a public face, in the apotheosed hero at whose ascension all nature rejoices (*Eclogue* 5), alluding to the deification of Julius Caesar. The idea that the natural world flourishes or fails in sympathy with the justice or injustice of the city and its rulers is an ancient one, and deeply embedded in the political imagery of all Virgil's works. Virgil is also chiefly responsible for the widespread currency in the Western tradition of the Hesiodic idea of the Golden Age, both as a primitive Eden, but also as a paradise to be regained through the intervention of a salvific ruler (*Eclogue* 4; *Aeneid* 6.791–4).[5] In the *Eclogues* philosophical (Epicurean and Stoic) notions of a life lived according to nature intersect with a popular moralizing tradition that opposes the simple and contented life of the countryside to the discontented luxury of the city; this complex of ideas also plays an important role in the *Georgics'* advocacy of a virtuous life on the farm, programmatically in the epilogue of the second book, and, in more nuanced forms, in the various pictures of life in primitive Italy in the *Aeneid*.

The last two books of the *Georgics* deal with animals, viewed from two very different perspectives. On the one hand animals are to be exploited without mercy for their utility to mankind. The old horse is to be put away without pity. On the other hand there is a sustained and often sentimental anthropomorphism in the description of animal behaviour and feeling, which reaches a climax with the instructions on bee-keeping in book four, where the hive is at times a miniature replica of an idealized Roman society. The view that the bees are a uniquely advanced species goes back at least to Aristotle, but in general pagan antiquity, in contrast to the speciesism prevalent in Christian cultures, accommodated a wide range of views favourable, at least in theory, to the claims to respect of the animal kingdom.

History and an acute sense of time enter the landscape in the *Georgics* and the *Aeneid*. The environment bears the traces of the lives of past generations, evoking a patriotic and antiquarian nostalgia, as in the description in the opening quotation of the rivers flowing at the foot of the ancient walls of hill-towns perched on their rocky prominences. The landscape may bring a remote past before the eyes of the present day, but it may also be changed beyond recognition. In *Aeneid* 8 Aeneas visits the site of Rome, hundreds of years before the birth of Romulus and Remus, and is guided by the virtuous Arcadian king who then lived at the place round a settlement of primitive huts set amid scrub and cattle. But the narrator Virgil constantly reminds his contemporary reader of the marble and gilt buildings that now dazzle the eye. There is a typical Roman pride in the staggering growth of their civilization from humble beginnings, but also a nostalgia for a simpler and more virtuous past when luxury held no temptations, and also a half-conscious anxiety that the process might be reversed, and Rome return to the semi-pastoral landscape over which Edward Gibbon would have gazed as he sat on the Capitol while the barefooted friars sang Vespers, and came to the idea of writing *The Decline and Fall of the Roman Empire*.

Virgil's sense of the past merges with a post-classical nostalgia for antiquity in seventeenth- and eighteenth-century landscape painters such as Poussin and Claude Lorrain, both profoundly influenced by Virgil. Claude's paintings were particularly popular in England, where their look and classical associations were imitated in the art of landscape gardening. Some landscape gardens, such as that at Stourhead, were designed to an explicitly Virgilian programme.[6] A Virgilian vision of the world was also transmitted through eighteenth-century descriptive poetry modelled on the *Georgics*, particularly in James Thomson's very popular *The Seasons*.[7]

Virgil's view of man's relationship to his environment is multifaceted. The Roman empire is now the historical realization of a Stoically coloured cosmic sympathy between man and the natural world, now the violent and morally questionable subjugation of indignant peoples and landscapes. Man cuts down the forest to bring the blessings of agriculture, but trees are also sacred living objects that should not be violated. The urban landscape is proof of man's cultural and political progress, but the city is also the scene of the luxurious corruption of a virtuous primitivism. Human science and technology are objects of wonder and admiration, but there is a place for mystery and awe in approaching the secrets of the natural world. One could accuse Virgil of inconsistency, or one could see in him a supremely sensitive commentator on the complexities and dilemmas of an advanced urban civilization. For all the great differences between Rome of the late first century BCE and a post-Christian and high-technology global society of the twenty-first century, some of these complexities are still familiar.

Notes

1 *Saturnalia*, 5.2.1.
2 On the upper-class Roman's relationship to the natural world see G.B. Miles, *Virgil's Georgics: A New Interpretation*, chap. 1.
3 On Roman landscape painting and its cultural contexts see E.W. Leach, *The Rhetoric of Space. Literary and Artistic Representations of Landscape in Republican and Augustan Rome*, Princeton: Princeton University Press, 1988.
4 B. Snell, 'Arcadia: The Discovery of a Spiritual Landscape', in *The Discovery of the Mind*, Oxford: Blackwell, chap. 13, 1953.
5 P.A. Johnston, *Vergil's Agricultural Golden Age: A Study of the Georgics*, Leiden: Brill, 1980; H. Levin, *The Myth of the Golden Age in the Renaissance*, London: Faber & Faber, 1970.
6 M.J.H. Liversidge, 'Virgil in Art', in C. Martindale (ed.), *The Cambridge Companion to Virgil*, Cambridge: Cambridge University Press, pp. 99–101, 1997.
7 L.P. Wilkinson, *The Georgics of Virgil. A Critical Survey*, Cambridge: Cambridge University Press, pp. 299–305, 1969.

Virgil's major writings

Georgics, trans. L.P. Wilkinson, Harmondsworth: Penguin, 1982.
Eclogues, trans. G. Lee, Harmondsworth: Penguin, 1984.
Aeneid, trans. D. West, Harmondsworth: Penguin, 1991.

Further reading

Jenkyns, R., *Virgil's Experience. Nature and History; Times, Names, and Places*, Oxford: Oxford University Press, 1998.
Miles, G.B., *Virgil's Georgics: A New Interpretation*, Berkeley and Los Angeles, CA, and London: University of California Press, 1980.
Putnam, M.C.J., *Virgil's Pastoral Art: Studies in the Eclogues*, Princeton, NJ: Princeton University Press, 1970.

PHILIP R. HARDIE

SAINT FRANCIS OF ASSISI 1181/2–1226

> When [St Francis] considered the primordial source of all things, he was filled with even more abundant piety, calling creatures no matter how small, by the name of brother and sister because he knew they had the same source as himself.[1]

At first sight, the life of Saint Francis of Assisi presents us with a paradox. On the one hand, Francis is one of the most popular and venerated saints within Christendom. His love and care for creation has become legendary. When Pope John Paul II in 1980 declared Francis Patron Saint of Ecology he was doing nothing less than acknowledging the universal appeal of his powerful creation-friendly example. Yet, on the other hand, the Christian tradition which canonized him, and which now venerates, lauds and champions him, is the same tradition which – not without justification – has itself been charged with a distinct lack of care for creation, even to the point of being directly responsible for current environmental crises. Understanding this paradox may provide the key both to the life of St Francis and its contemporary eco-relevance.

Although soon swallowed up in legend, basic details of Francis' life are still recoverable. He was born in 1181 or 1182 in Assisi, the son of a wealthy cloth merchant, Peter Bernardone. As a young man, Francis obtained a reputation as a profligate and a squanderer. In 1204, he was ill for a prolonged period which put an end to his military career. A series of encounters and experiences then drastically changed his life. At the end of 1204 or early 1205, Francis apparently received his first visionary experience. During that same year, he was brought face to face with poverty and suffering through chance encounters with paupers. But it was his meeting with a leper, the most despised and feared of all medieval outcasts, which apparently changed his life.

Much to his father's chagrin, he renounced his early military and commercial ambitions, sold his possessions, and embraced a life of poverty. Charged with having brought humiliation on his father's house, he was brought before the episcopal tribunal in 1206, but Bishop Guido II of Assisi befriended him. At San Damiano in about 1206, Francis experienced his famous vision in which a voice called upon him to rebuild the Church. From 1206 to 1208, he restored the chapels of San Pietro and Santa Maria degli Angeli at the Portiuncular while living as a hermit. Around 1209/10 Francis compiled his Rule and sought papal approval. Eager to secure reform of the Church, Pope Innocent III granted Francis an audience and subsequently authorized Francis and his followers as an itinerant preaching order within the Catholic Church. 'The friars' zeal for the proclamation of the Gospel, their highly acclaimed ministry of preaching, their rejection of material possession in imitation of Jesus Christ and their itinerant lifestyle recommended them to Innocent III.'[2] The community grew and expanded over the following ten years and became an instrument of papal reform of the Church culminating in the decrees of the Fourth Lateran Council in 1215.

From the start, we can see that Francis' work was a licensed reform experiment within the Catholic Church. Although Francis was impeccably loyal to the Church, and especially to the papacy which endorsed him, his unusual status granted him free rein to preach the Gospel in all its radical simplicity as he saw it. It is said that Francis' life was decisively transformed when he attended Mass at the Portiuncular in February 1208 and heard 'the Gospel passage in which the apostles were commissioned to preach'.[3] From the standpoint of ecological theology, there are four aspects of his ministry which deserve particular attention.

The first concerns *simplicity*. As we have seen, Francis caused scandal by his rejection of his father's wealth and by dressing in a threadbare tunic and sandals. This was not affectation. It was an attempt to imitate Jesus in his identification with the poor and outcast. In doing so, Francis lived the notion deeply rooted in the Gospels that material wealth is a handicap to spiritual progress.

Unlike most other Christians of his day – including it must be said bishops and priests – who saw no difficulty in the accumulation of riches, Francis saw simplicity of life as a moral requirement of the Gospel. Accordingly, his Rule forbade his friars from eating luxurious food,[4] wearing expensive garments or accumulating money. Simplicity required living as the poorest of the poor and sharing all things in common.

The second concerns *kinship*. Francis took literally the claim that the Gospel should be preached to 'all creation'. As the above lines from his biographer, St Bonaventure, show, Francis celebrated the kinship of all creatures created by the same God and whose Gospel of love extended to the smallest thing, both animate and inanimate, within creation. Fellow creatures are our 'brothers' and 'sisters'. Although such a notion of kinship or cosmic fellowship is implicit in the Gospels, and arguably required by a doctrine of God the Creator, Francis' high regard for creation was – in terms of conventional theology – highly eccentric. Medieval theology saw sharp distinctions between humans and animals and was deeply dualistic in its thinking, making contrasts (as most of the tradition has done) between things earthly and things spiritual. Francis' sense of friendship and kinship with other creatures, while wholly orthodox, was nevertheless deeply counter-cultural.

The third concerns *generosity*. Francis did not just perceive an ontological bond between all creatures by virtue of their common Creator, he sought to manifest that unity through acts of moral generosity. 'He overflowed with the spirit of charity', writes early biographer Thomas of Celano, 'pitying not only men who were suffering need, but even the dumb brutes, reptiles, birds, and other creatures without sensation.' [5] The key to understanding Francis at this point is to be found in his profound sense that humans were called to imitate Christ, hence they were to reflect a Christlike generosity even and especially to the least of all. Innumerable stories of Francis testify to his filial relations with other creatures. He loved even the worm not solely because it reminded him of the saying that 'I am a worm and no man', but primarily because – as Celano put it – 'he glowed with exceeding love ... wherefore he used to pick them up in the way and put them in a safe place, that they might not be crushed by the feet of the passers-by'.[6]

In order to appreciate the radicality of this approach, one has only to contrast it with the thought of Francis' near contemporary, St Thomas Aquinas. For St Thomas, there was an absolute distinction between animals and humans, and humans could have 'no fellowship' with animals because they were non-rational. Although both were canonized saints and celebrated figures within the Catholic Church, the difference between them is almost total. While Francis accepted that humans had dominion over animals, he interpreted this power Christologically, that is, in terms of service. As Paul Santmire notes, the saint displayed 'a concrete Christocentric devotion [to others] of radical proportions ... He became the Christ-like servant of nature'.[7]

The fourth concerns *celebration*. Again in contrast to wholly instrumentalist views of creation as simply here for our use, Francis saw the world of creation as a place of celebration. He took seriously those verses in the Psalms which speak of creatures praising their Creator and saw in all things, even inanimate ones, a response to the love of God. His famous 'Canticle to Brother Sun' is a tremendous theophany of creation in praise of its Creator. Normally viewed as unconscious matter, he sees the sun, moon, wind, water, and fire as part of the divine cosmic consciousness. As one commentator observes, 'for Francis, what we refer to as "dumb nature" is far from dumb; it is eloquent in singing and testifying to the beauty of its creator'.[8]

The theological significance of Francis' life may be understood as a prefiguring of that state of peaceableness within creation which will finally be accomplished at the end of time. Such eschatological consciousness was prevalent in Francis' time and, as several writers suggest, the saint's anticipation of the immanent consummation of the Kingdom of God led him to live those laws of the coming kingdom – poverty, humility, selfless love, obedience – in this world. As Roger Sorrell explains, 'there is no doubt that Francis shared his hagiographers' conceptions [that] ... creatures' responses to him demonstrated their respect for God's servant and the beginning of the restoration of harmony between God, humanity, and the rest of creation'.[9] The accounts of Celano and Bonaventure lend strong support to this view.

For example, Celano believed that when Francis was submitted to Brother Fire and was not injured, 'he had returned [the fire] to primitive innocence [*ad innocentiam primam*], for whom, when he wished it, cruel things were made gentle'.[10] Bonaventure similarly reports, 'so it was that by God's divine power the brute beasts felt drawn towards him and inanimate creation obeyed his will. It seemed as if he had returned to the state of primeval innocence, he was so good, so holy.'[11] If such an eschatological motivation is accepted, Francis' writing and ministry, far from being romantic rhetoric or eccentric practice, is a manifestation in time and space of God's eternal purpose.

Perhaps inevitably, Francis' example has been eclipsed by the centuries of Christian thought and practice which followed. The sharply contrasting approach of St Thomas – in many ways the founding father of modern Roman Catholicism – has been vastly more influential and has ushered in centuries of neglect of, and even callousness towards, the non-human world. Francis is remembered and honoured, and even lip service is paid to his example, and yet he has had little effect on the development of scholastic theology. It must be said that still many Christians, even and especially Franciscans, play down the eco- and animal-friendly dimensions to his ministry.

But there are some signs that increasing dissatisfaction with the instrumentalist and utilitarian attitudes to creation embodied in historical theology are encouraging churchpeople and theologians to re-examine the tradition and rediscover genuine but neglected creation-friendly elements within it – and not least of all, Francis himself. 'St Francis is before us as an example of unalterable meekness and sincere love with regard to irrational beings who make up part of creation', maintained Pope John Paul II in his sermon at Assisi on 12 March 1982. 'We too are called to a similar attitude', he continued. 'Created in the image of God, we must make him present among creatures "as intelligent and noble masters and guardian of nature and not as heedless exploiters and destroyers".'[12]

Notes

1 St Bonaventure, in *The Life of St Francis*, ed. Ewert Cousins, New York: Paulist Press, pp. 254–5, 1978.
2 Michael Robson, *St Francis of Assisi: The Legend and the Life*, p. 90.
3 Ibid., p. xxxi.
4 Francis' vegetarianism is disputed but it is clear that his community followed an ascetical, frugal diet which very rarely made use of flesh foods.
5 Thomas of Celano, *Vita Prima* 59, in H. Paul Santmire, *The Travail of Nature*, p. 108.
6 Thomas of Celano, *Vita Prima* 59, in Roger Sorrell, *St Francis of Assisi and Nature*, p. 46.
7 Santmire, op. cit., p. 109.
8 David Kinsley, 'Christianity as Ecologically Responsible', in *This Sacred Earth: Religion, Nature, Environment*, ed. Roger Gottlieb, London: Routledge, p. 123, 1996.
9 Sorrell, op. cit., p. 54.
10 Thomas of Celano, *Vita Secunda* 166, in Sorrell, ibid., p. 52.

11 St Bonaventure, *Legenda Minor* 3: 6, in Sorrell, ibid.
12 Pope John Paul II, Message on 'Reconciliation', *L'Osservatore Romano*, 29 March 1982, pp. 8–9. The final two lines are a quote from the previous encyclical, *Redemptor Hominis*, 15.

St Francis' major writings

The Writings of St Francis, in *The Omnibus of Sources*, ed. Marion Habig, O.F.M., Chicago, IL: The Franciscan Herald Press, 1973.
The Life of St Francis, ed. Ewert Cousins, *The Classics of Western Spirituality*, New York: Paulist Press, 1978.

The following are early biographies, all of which appear in *Omnibus*:

St Bonaventure, *The Little Flowers of Saint Francis* (*I Fioretti di San Francesco*).
—— *Major Life of Francis* (*Legenda Maior*).
—— *Minor Life of Francis* (*Legenda Minor*).
Thomas of Celano, *Treatise on the Miracles of Blessed Francis* (*Tractatus de Miraculis Beati Francisci*).
—— *First Life of Francis* (*Vita Prima*).
—— *Second Life of Francis* (*Vita Secunda*).

Further reading

Armstrong, Edward, *St Francis: Nature Mystic, The Derivation and Signifiance of the Nature Stories in the Franciscan Legend*, Berkeley, CA: University of California Press, 1973.
Cunningham, Lawrence, *Saint Francis of Assisi*, New York: Harper & Row, 1976.
—— *Brother Francis: An Anthology of Writings by and about Saint Francis of Assisi*, New York: Harper & Row, 1972.
Linzey, Andrew and Cohn-Sherbok, Dan, *After Noah: Animals and the Liberation of Theology*, London: Mowbray, 1997.
Robson, Michael, *St Francis of Assisi: The Legend and the Life*, London: Geoffrey Chapman, 1997.
Santmire, H. Paul, *The Travail of Nature: The Ambiguous Ecological Promise of Christian Theology*, Minneapolis, MN: Fortress Press, 1985.
Sorrell, Roger, *St Francis of Assisi and Nature: Tradition and Innovation in Western Christian Attitudes toward the Environment*, Oxford: Oxford University Press, 1988.

ANDREW LINZEY AND ARA BARSAM

WANG YANG-MING 1472–1528

Man is the mind of the universe: at bottom Heaven and Earth and all things are my body. Is there any suffering or bitterness of the masses that is not disease and pain in my own body? Those who are not aware of disease and pain in their body are people without the sense of right and wrong. The sense of right and wrong is knowledge possessed by men without deliberation and ability possessed by them without their having acquired it by learning. It is what we call innate knowledge (liang-chih) ... [1]

Wang Yang-ming (or Wang Shou-jen), the most influential Confucian thinker in the Ming dynasty (1368–1644) in China, was a critical inheritor of the two main tendencies of Neo-Confucianism: that is, the philosophies of Ch'eng I-ch'uan (1033–1107) and Chu Hsi (1130–1200) on the one hand and those of Ch'eng Ming-tao (1032–85) and Lu Shiang-shan (1138–92) on the other. Thus he was thought to be a kind of synthesizer who had perfected the philosophy of Neo-Confucianism. Wang was, in his youth, much influenced by Zen Buddhism and Taoism, though he rejected them later on the ground that they represent a sort of quietism which escapes from social relationships. Like a Zen master, he was suddenly enlightened at the age of 37, after long years of concentration in thinking in very harsh situations. Once when he was young Wang had concentrated his thoughts on the things outside of his mind, because his forerunner Chu Hsi interpreted the important thesis of 'Great Learning' as meaning that 'the investigation of things' will lead to 'the extension of the knowledge', thus rectifying the will. So, Wang concentrated his thought by watching the bamboos in his garden for seven days and became sick. Later he changed his course to concentrate on the inside of the mind. And he attained the idea of 'good knowledge' that combines knowledge with action. If one makes a division between one's mind and things outside, and separates the former from the latter, and if one is concerned only with things outside of one's mind, then that concern will not be combined very easily with one's mind, that is, with one's will to act. He was a keen learner, and on his own wedding day he was so involved in discussion with a Taoist that he forgot to attend the ceremony. He was the ablest general in his age and won a high reputation as general, having put down many rebellions. Even for him it was not a very easy task to defeat the selfish desire in his mind that beclouded the good knowledge as the Heavenly Principle. He said, 'it is easy to defeat the rebels in the mountains, but it is difficult to defeat the rebels in the mind'. In 1527 he was asked to subjugate a rebellion while he was suffering from a serious disease. After he defeated the rebellion and came back to his home he died, aged 57; on his deathbed he must have felt that he had defeated his innate enemy, his human desire. His last words were: 'My mind is full of light; I have nothing to say any more'. The most basic requirement of the moral philosophy of Wang Yang-ming is unity of knowledge and action. For all moral purposes the only thing which needed to be done was to bring forth 'the good knowledge' ('intuitive knowledge', or 'good conscience') of the mind. If one knows that he ought to do something and does not do it, this knowledge, for Wang, means that he does not in fact know. This reminds us of the contemporary theory of what R.M. Hare calls 'prescriptivism' in contrast with 'descriptivism'.[2] The decisive moral question in environmental ethics is not only what is the matter with the environment, but what are we to do. This is one of the reasons why Wang Yang-ming's moral philosophy is most promising when applied to environmental ethics. 'Knowing is the beginning of action, and doing is the completion of knowledge. When one knows how to attain the desired end, though one speaks only of knowing, the doing is already included; likewise, though he may speak only of action, the knowing is also implied.'[3]

Originating from some versions of Confucianism is the popular expression 'Heaven knows'. Heaven was said to watch our good acts as well as our evil acts, even if no one on earth knows. So, if someone has escaped punishment in doing some evil acts, heaven will punish him some day, because heaven was believed to be something that is completely impartial. ('Heaven's vengeance is slow but sure.') People, therefore, are recommended to 'self-care in solitude'. And also when one falls into a difficult situation, not due to their own failure, one could find consolation in such beliefs as: 'Heaven gives one severe trials, before heaven gives him a mission' or 'Sincerity can move heaven'. Wang said once that his philosophy of 'good knowledge' was born from 'a hundred deaths, a thousand difficulties'. Such ways of thinking have certainly helped people under the influence of Confucian culture in enforcing their impartialist morality.

Every one knows the Confucian golden rule of 'Do not do to others what you do not want them to do to you' (*The Analects*, 12:2, 5:11). This is the spirit of 'jen' ('benevolence' or 'love', usually translated as 'humanity', but jen reaches far beyond humanity to all things). 'Jen' is sometimes juxtaposed with five other virtues (filial piety, loyalty, orderly love among spouses, among brothers, trust between friends). But it is often thought not only as representative, but as fundamental, in the sense that it forms the basis by which other virtues are justified. If this interpretation is possible, what Confucius wanted to point out was a logical thesis about morality which is, in a sense, shared by recent Western moral philosophers like Hare, Peter Singer and others, who argue that 'universalizability' (but not 'universality' in the sense of 'generality') is the fundamental requirement of a moral judgement. If we interpret 'jen' as something like the utilitarians' 'impartial benevolence', then what is the difference between the two positions?

The difference is this: the utilitarian motto in the old version of the theory is 'the greatest happiness of the greatest number'. The utilitarians expanded our moral concern beyond our species to include the wellbeing of animals (Bentham and Singer), and further expanded our moral concern to future generations. Confronting environmental crisis on a global scale, people have realized that if the environment is endangered, there is no longer any happiness for any being. So an expanded utilitarianism will need to be supported by some kind of eco-holistic view. If we, as moral agents, go one step further, expanding moral subjects further than sentient beings, and include in our moral consideration the natural environment that is relative to human activities, then we will come very close to the position of Wang Yang-ming.

So far we have seen that the two logical requirements of moral judgements (i.e. prescriptivity and universalizability) which are made explicit by Western moral philosophers are already implicit in Wang Yang-ming's philosophy. However, this may seem rough and biased in the eyes of people who are experts in Chinese philosophy. If we further add an eco-holistic view to these two logical requirements, what will happen is as follows.

Wang Yang-ming had critically inherited another thesis of Ch'eng Ming-tao: 'jen is the love of all things in the universe as one body'. This thesis was related to, inherited from, the Buddhist thesis that 'Heaven and Earth have the same roots as myself and all things are one body with me' and the Taoist thesis of Chuang Tzu that 'Heaven and Earth live alongside me and all things are one body with me'. One must understand, Wang said, that 'jen' is the unity of all things. According to Wang, each and every one of us possesses the original mind, which is one with the universe. 'The man of jen regards Heaven and Earth and all things as one body. If a single thing is deprived of its place, it means that my jen is not yet demonstrated to the fullest extent' (89). Thus, jen is not only the basis of human virtue, but is the original principle according to which heaven and earth make everything live. Jen is 'the principle of unceasing production and reproduction. Although it is prevalent and extensive and there is no place where it does not exist, nevertheless there is an order in its operation and growth. That is why it is unceasing in production and reproduction' (93). 'Our nature is the substance of the mind and Heaven is the source of our nature. To exert one's mind to the utmost is the same as fully developing one's nature. Only those who are absolutely sincere can fully develop their nature and know the transformation and nourishing process of Heaven and Earth' (6).

It is in this way, that is, according to the jen, which is innate original knowledge and the principle of the universe at the same time, that a ruler is expected to rule society and the whole country. This is called 'jen-politics' or moral politics. If one somehow unifies oneself with the society that is one body, and if one knows that people are suffering, then this will become enough incentive for one to save people from suffering, because knowledge and action are united.

This element of the social philosophy of Wang Yang-ming, because it is holistic, can be extended to a view of nature. What is most important for environmental ethics is that jen is not only a matter of human concerns.

> When one hears the cry of birds and animals, one will have compassion, because the jen is one with the birds and animals. If one says that animals have senses, then one will have compassion when one sees the grasses and trees faded and broken, because the jen is one with the grasses and animals. If you say that grasses and trees are animated beings, then one will regret when one sees tile-stones collapse; this is because the jen is one with tile-stones.
>
> And yet the Grand Master [Confucius] was extremely busy and anxious, as though he were searching for a lost son on the highway, and never sat down long enough to warm his mat. Was he only trying to get people to know him and believe him? It was rather because his jen, which regarded Heaven and Earth and all things as one body, was so compassionate, keen, and sincere that he could not stop doing so even if he wanted to ... Alas! Aside from those who truly form one body with Heaven and Earth and the myriad things, who can understand the Grand Master's intention?
>
> (182)

31

If the community is a closed one, people tend to see it as the whole and would be able to sacrifice themselves for it. If people could lift their eyes a bit higher and expand their concern to include nature as a whole, they could be prepared to devote their labours to enriching the natural environment. When people were settled and earthbound they knew their survival depended on a sound natural environment. Thus social ethics in the East was severely restricted and shaped by the limits the natural environment imposed upon the society. Yang-ming's eco-holistic tendency was for many years one of the strongest ideological backgrounds of pre-modern Japan, where the natural environment was marvellously enriched and sustainable. Kumazawa Banzan (1619–91), a samurai scholar who belonged to the Yang-ming school, is well known for his ecological policies and achievements.

Yang-ming's social ethics and the vitality of his thoughts inspired samurai revolutionaries when Japan opened its door to the West and caused the Meiji Restoration (1868). They fought for what they thought was the whole, that is, for the country to keep independence, not for the interests of their own class, and after the revolution was achieved they eliminated their own class. On the other hand, the opinion leaders' ideological model in the period of Japanese modernization, after the long period of being a closed country, was mainly influenced by British utilitarianism. What both the Yang-ming school and utilitarianism could somehow share was their social ethics. Yet the very gap lies in their views on nature. While utilitarian concern focuses on the interests of sentient beings, Yang-ming's concern was with all beings interrelated under heaven. But such eco-holistic views were abolished and instead the Western dualistic tendency was, under the pressure from Western powers, imported and the dominion of nature had prevailed. But an industrial and economic giant means also an environment-degrading monster. The cost of the modernization has not yet been generally noticed. Only a few philosophers have started to take another look at the traditional Confucian views on nature, among them a prominent environmental philosopher, J. Baird Callicott, who classified Confucianism as a form of deep ecology.

Notes

1   Quoted on p. 179. Page references in the text are to *Instructions for Practical Living*.
2   See R.M. Hare, *The Language of Morals*, Oxford: Oxford University Press, 1952.

3 W. Liu, *A Short History of Confucian Philosophy*, p. 171.

See also in this book

Callicott, Chuang Tzu, Singer

Yang-ming's main writings

Chan, W. (trans.), *Instructions for Practical Living and Other Neo-Confucian Writings by Wang Yang-ming*, New York: Columbia University Press, 1963.
Ching, J. (trans.), *The Philosophical Letters of Wang Yang-ming*, Charleston, SC: University of South Carolina Press, 1972.
Chan, W. (ed. and trans.), *A Source Book in Chinese Philosophy*, Princeton, NJ: Princeton University Press, 1973.

Further reading

Callicott, J.B., *Earth's Insights: A Multicultural Survey of Ecological Ethics from the Mediterranean Basin to the Australian Outback*, Berkeley, CA: University of California Press, 1994.
Chang, C., *Wang Yang-ming: Idealist Philosopher of Sixteenth-Century China*, New York: St John's University Press, 1962.
Liu, W., *A Short History of Confucian Philosophy*, New York: Delta, 1964.
Tu, W., *Confucian Thought: Selfhood as Creative Transformation*, Albany, NY: SUNY Press, 1985.

T. YAMAUCHI

MICHEL DE MONTAIGNE 1533–92

> When I am playing with my cat, who can know whether she is not amusing herself with me, rather than I with her?[1]

Montaigne is the most congenial of intellectual companions. He was an exceptionally well-educated member of the local gentry, who spent most of his life on his estate near Bordeaux. A trained lawyer, he served two terms as mayor of Bordeaux and was a minor player on the national political stage, at a period when France was ravaged by religious civil war of unparalleled savagery. In 1571, he vowed to retreat to the 'peace and security' of his library and his own reflections. This retreat was much interrupted by his political responsibilities and by a tour of Italy in 1580–1, but 1571 marked the beginning of the years of study, rumination and writing that resulted in the *Essays*.

Two books of essays were published in 1580, and successive editions made alterations and additions, the most important of which was a third book, appended in 1588. Montaigne was working on the essays up to his death, and a posthumous edition in 1594 contains very extensive insertions into the body of the text. Well before then, Montaigne had discovered the subject of his writing, and that subject was himself. This was a revolution in the history of European thought. No authors prior to Montaigne had made themselves the matter of their book, except to present a partial view of themselves, as examples of God's grace or as witnesses to historical events. Montaigne's *Essays* are loosely structured, but extraordinarily intelligent and critical, ponderings on his own responses to the total diversity of his own experience, his reading, his social interactions, his habits, his environment, his mental cogitations, his sensations and his bodily proclivities. The book found avid readers throughout western Europe. John Florio published it in English in 1613, Francis Bacon's *Essays* could not have been conceived without it. Descartes, Pascal, and all the major thinkers of the seventeenth century start from questions raised for them by Montaigne, even though, under the influence of the Scientific Revolution, they came to a very different world-view. The *Confessions* of Jean-Jacques Rousseau, Europe's next great essay in autobiography, is clearly of Montaigne's progeny, but so is the modern preoccupation with the self, and every essay that was ever written. As an environmental thinker (a concept he could not have recognized, though he did think about the natural environment), Montaigne may best be considered under three heads: his reaction to attitudes typical of his social class; his reflections on the 'wild'; and the place of animals in his sceptical account of human pretensions to a superior place in the natural order.

There was no appropriate language already in existence for Montaigne's novel investigation of his own psychology, and Montaigne invented one out of metaphors. One of his favourite metaphors is that of the hunt, a pastime closely linked to the social status of a gentleman. When his essay on Cruelty (Book II, no. 11) gets round to his own sense of what is cruel, he confronts us straightaway with the reality of 'the hare squealing when my hounds get their teeth into it'. As always, Montaigne mistrusts any simple analysis of his behaviour. It may be mere squeamishness, too weak to underpin a moral position. He is fully conscious that his distaste for the spectacle of the hunted beast at bay is likely to be mocked by his peers. Yet, his self-awareness forces him to stand aside from his social group and triggers more general speculation about man's natural propensity to cruelty, about the respect and affection religion and our very humanity enjoin that we give to our fellow creatures, beasts and plants, and about the lessons animals have for our presumption: 'I willingly lay aside that imaginary rule over other creatures that we have been assigned'.

Montaigne's capacity to imagine the other finds its most startling expression in his essay on Cannibals (Book I, no. 31), which also turns on questions of cruelty and presumption. This is one of two essays on the New World in which Montaigne deplores the depredations of the European conquerors with a fierce bitterness rarely articulated before him. His subject is the human inhabitants of South America, not the natural environment, but his basic distinction between the wild and the cultivated, the natural and the artificial, has implications for both. For Montaigne, the 'savagery' of Brazilian cannibals is akin to the vigour and virtue of uncontaminated nature. It is European culture that has corrupted nature by artifice. So, the refined cruelty Europeans inflict on their colonial subjects, and on each other in the name of religion, is a barbarity far in excess of the 'barbarity' we ascribe to cannibals. While condemning cannibalism in absolute terms, Montaigne contextualizes the practice in a description derived from explorers' accounts of a very simple, 'natural' society living in equilibrium with the environment, using its resources without cultivating or altering it. The Brazilians desire nothing beyond what their environment liberally provides, so they have no concept of conquest, of property, of trade or of social division. Their cannibalism derives from a competitive sense of honour, as does their polygamy. Montaigne focuses on these two practices so repugnant to his own culture to demonstrate that they are not alien to nature, that they are conceivable within a different environment and make sense within a different social structure. His capacity to imagine the other is also a tool for attacking Europeans' presumption that they are morally and culturally superior: 'there is nothing barbarous or savage about the Brazilian tribes, except that all call "barbarous" anything they are not used to'. Moreover, this view from the other side can be very disturbing. When his 'savages' visited France, they 'naively' marvelled at the disparity between rich and poor and wondered why the destitute did not 'seize the others by the throat or set fire to their houses'.

Montaigne's most sustained discussion of man's general relationship with other inhabitants of his environment is to be found in the very long Apology for [or, Defence of] Raymond Sebond (Book II, no. 12). It serves as an introduction to a translation Montaigne had been asked to make of Sebond's *Natural Theology*, written in the early fifteenth century. Sebond had argued that truth can be read in the Book of Nature, but only if those who observe nature and interpret it do so in the light of Christian revelation. His subject matter here forces Montaigne to investigate traditional theological attitudes to the natural world and to explore the cosmological, psychological and biological science of his day. The strategy of his essay requires him to be sceptical, for he has chosen to undermine those who object that Sebond's arguments are weak by demonstrating how fallible all human reasoning is. Montaigne accumulates evidence on two counts: first, to show that opinions held about the workings of nature are incoherent and self-contradictory; second, to show what a feeble creature man is, despite his much lauded faculty of reason on which is founded his presumption to rule the rest of creation. He pursues this second theme through a copious inventory of examples where animals put man to shame, in their ability to communicate, their creative skills, their ingenuity, their powers of deduction, their memory, their moral virtues of fidelity and courage, and many more. Fact and fable are all grist to his mill. Montaigne revels in the literature from which he takes his examples, and that shows him to be a man of his time. In the middle of the sixteenth century lavish books were printed reproducing all that was known about animals, with detailed, realistic illustrations. They adhered to a basic grouping of species, but it is the profusion of animal life, rather than its taxonomy, that still entrances the browsing reader of these works, where the fabulous is interleaved with familiar creatures from the Old World and exotic beasts from the New. Their text is exhaustive about anatomy, habitat, feeding, breeding, and so on, but it dwells just as much on references to the animal in historical texts, poems, fables, proverbs, sayings and emblems. Animals are literary and cultural objects, as much as objects of scientific observation. The same could be said about the study of the natural environment outside books. The sixteenth century was the great period of the curiosity cabinet, filled with a heterogeneous collection of animal and mineral objects designed to excite wonder at the uninhibited variety of natural forms rather than to initiate scientific research.

Montaigne's apparently undisciplined gathering of the more amazing feats of animals recounted in books, interspersed with the occasional personal observation, has analogies with the encyclopedism and collecting mania of his contemporaries (and exactly the same appeal as exotic 'wildlife series' on television). There are, however, features of Montaigne's discourse that betray a rather different preoccupation. In emphasizing the role of animals in human culture and in subjecting their remains to the wondering gaze of the possessor of a curiosity cabinet, encyclopedias and collections tended to promote an anthropomorphic view of the animal world just as effectively as the moral and Christological lenses through which their medieval predecessors had read the Book of Nature. Montaigne's purpose is not to show how man can know nature and therefore feel easy with it, but to discomfort man, to show that his claim to be superior to animals is undermined by counter-examples at every turn. He claimed in his essay on Cruelty that humility with respect to the rest of creation is both a proper human attitude and a Christian one, but it does go against the grain of a certain theological attitude that puts man and his immortal soul on a level above all other living things and also identifies the 'bestial' with the degenerate and morally corrupt. Moreover, there runs through Montaigne's catalogue of animal behaviour that same sense of the other that allowed him to grope towards an anthropological understanding of the alien cannibals. He does not rest in a state of wonder, but conceives imaginatively a world where other forms of language operate, other values hold, in which animals have an incomplete sense of what makes humans tick, but no less complete than our insight into them, a world where cats amuse themselves with humans no less than we amuse ourselves with cats.

Montaigne's kind and generous attitude to non-human creatures was the product of a pre-scientific mentality. For him, the natural environment was an array of mobile forms, a playground for his agile mind. When Descartes, in his *Discours de la méthode* of 1637, defined the natural world as the scientific object of man's investigating reason, he constructed it as a machine. Animals for Descartes were functioning mechanisms without thought, language or sensation. Montaigne's respect for animals survived the Scientific Revolution in the fables of La Fontaine. In La Fontaine, too, animals have their mode of communication and teach mankind a lesson.

Note

1   Book II, Essay 12 of *The Complete Essays.* All references to Montaigne in the text are to Book and Essay numbers of this work.

See also in this book

Bacon, Rousseau

Montaigne's major writings

The Complete Essays, trans. M.A. Screech, with good introduction, London: Allen Lane, Penguin Press, 1991.

Further reading

Burke, P., *Montaigne*, Oxford: Oxford University Press, 1981.
Daston, L. and Park, K., *Wonders and the Order of Nature, 1150–1750*, New York: Zone Books, 1998.
Sayce, R.A., *The Essays of Montaigne: A Critical Exploration*, London: Weidenfeld & Nicolson, 1972.
Screech, M.A., *Montaigne and Melancholy: The Wisdom of the Essays*, London: Duckworth, 1983.

ANN MOSS

FRANCIS BACON 1561–1626

> Human knowledge and human power meet in one; for where the cause is not known the effect cannot be produced. Nature to be commanded must be obeyed; and that which in contemplation is as the cause is in operation as the rule.[1]

Bacon was a politician, jurist, royal councillor, natural scientist and essay writer, who spent his entire life within the highest political, courtly and intellectual circles around Queen Elizabeth and King James I. His maternal uncle, Lord Burghley, was the most powerful statesman of his age. After Cambridge University and a period in France, he became a lawyer and Member of Parliament. Despite his earlier friendship with the charismatic Earl of Essex, he was active in the prosecution of Essex for treason – a deed which has inspired some, probably unfairly, to accuse Bacon of the worst sort of betrayal. Bacon was also involved in the prosecution of Sir Walter Raleigh, once an associate of Essex.

The first edition of Bacon's *Essays and Counsels* (1591) dates from his early career, and was eventually expanded into a third edition containing fifty-eight essays in 1625. Upon the succession of King James I in 1603, Bacon moved upward in the court hierarchy even more rapidly, eventually becoming, in 1618, Lord Chancellor and Baron Verulam. Bacon's writings during these politically active years reflect his many interests – in English Law, the Church of England and the 'Advancement of Learning', which offered a sweeping survey of the current state of knowledge in every field. In 1621, Bacon was created Viscount St Albans and finally published the *Novum Organum*, the first part of his vast systematic natural philosophy. But he had also made some serious enemies, who had him removed from office and convicted for taking bribes. Released from prison by King James, Bacon retired to his country home, where he could devote his undivided attention to carrying out many of his principal works. He died in April 1626 from pneumonia contracted while testing the preservative effects of snow on a chicken.

Perhaps no other person in the history of modern ideas has provoked such incompatible and one-sided assessments, towards which adherents maintain an almost sectarian zeal. One reason for this is that there is little agreement on what his actual intentions were or the scientific status of his achievements. The seventeenth century praised and imitated him, the eighteenth century glorified him as the precursor of Enlightenment, while the nineteenth century devoted effort to debunking him and making him the villain of the Jacobean period. The 'Secretary of Nature and All Learning' came to be despised as a charlatan, an enemy of real science, and quite recently was even described as a Satanist. In Mathews' summary of twentieth-century opinion, Bacon is dubbed an atheist and hailed as a religious thinker; acclaimed for his prophetic insights in natural history, his understanding of logic, his theory of forms, and his powerful imagination; while at the same time he is decried for his ignorance of natural history and logic, his absurd notion of forms, and his entire lack of imagination.[2]

One must resist the temptation to describe Bacon's life and works, let alone his attitude towards environmental issues, in cartoon terms. It is far too facile to condemn him for his comments about the merits of human domination and exploitation of nature. Bacon likened himself to a honey bee; the proper philosophical conduct is to work together in the accumulation of material in order to then transform it into something sweet and nutritious. The traditional thinkers, especially the medieval scholastics, were like spiders who spun intricate webs entirely from the inside and then imposed their structures on the world. The empirics, especially alchemists, astrologers and other pseudo-scientific dabblers, were like ants who merely collect curiosities and arcane lore, unable to articulate a coherent intellectual framework. Bacon described three defective methods in the pursuit of knowledge: the 'disputatious' erudition of scholastics; the 'delicate' learning which preserved the errors of revered authorities; and the 'fantastic' learning of the occultists and Hermeticists who catalogue dubious instances of isolated marvels.

Bacon attempted to address all these issues, and more, in the various parts of the *Instauratio Magna*, 'the Great Setting-Forth'. The first part appeared as a revision and expansion of *The Advancement of Learning*, while the second part, the *New Organon*, recasts Aristotle's *Organon* (the Logical Texts) in new terms and contains Bacon's most detailed though incomplete exposition of his criticisms of the false path of natural philosophy and his outline for a cooperative programme in the various natural sciences.

At the core of Bacon's notion of scientific knowledge are the doctrines of induction, hidden forms, and maker's knowledge. In his doctrine of hidden forms, Bacon resuscitated an old idea, that it is the form which gives a thing its true nature. Baconian forms are the simple constituents of matter and, though there are only a small number of them, they can be combined in an infinite number of arrangements, like the letters of an alphabet which can be combined to generate an infinite number of words. The aim of his whole project is, in his words, 'the inquisition of forms' which leads to works, the fruits of correct experimental procedures; he defines natural philosophy as 'the inquiry of causes and the production of effects'. The canon of basic physical properties is the discovery of those true forms which are 'nothing more than those laws and determinations of absolute actuality which govern and constitute any simple nature'. In Perez-Ramos' adroit words, the scientist as a human knower is first and foremost a maker or doer, and his warrant for claims to knowledge depends on his credentials as a maker: 'Bacon's idea of science ... establishes that to know something (a natural phenomenon) amounts to being able to (re)produce that very phenomenon on any material substratum susceptible of manifesting it'.[3]

In order to fully comprehend the grandeur (or grandiosity) of Bacon's entire project one must realize that the doctrines of hidden forms and maker's knowledge are aspects of a scheme which concerns 'the advancement of ideas about moving and persuading things and human beings'. The overall scheme is embraced under Bacon's notion of rhetoric which combines psychological, economic and material dynamics. The very idea that one can persuade *things* seems, to modern readers, to be utterly strange, unless one bears in mind that for Bacon there are hidden spiritual forms which compose the nature of all things, including human beings. 'In the new learning, experiment is more than a method of discovery; it is an ordeal, a test of a subject's true nature. Ultimately, all experiments work upon the matter and spirit of the created world, including the minds and passions of human beings.'[4]

In Bacon's theistic picture, the Creator moves the created world in a cryptic manner; the surface language of the perceptible properties of animals, vegetables and minerals conceals a secret code, which the scientist must decipher in order to interpret the latent or deeper language. This notion helps to explain Bacon's repeated references to ciphers and codes, encryption and decryption. The New Learning endorses secrecy, an adept's privileged knowledge, and this is most appropriately expressed through aphorisms and riddles: 'God's encryption of the world is an enigma, and its maker is hidden to all but those who can discover the signs of God's wisdom by suffering the scourging of their vanities in the sweet ordeal of Solomonic inquiry'.[5] This aphoristic and riddling format is featured most prominently in the *Essays*. But if one grasps the twofold power of Bacon's rhetoric, then one can appreciate that the recommendations addressed to civil servants and power-brokers, for instance, in regard to domination of natural forces and the production of works beneficial to humankind, are designed both to persuade them in terms of their own self-interests and (more secretly) to obey the commands of nature hidden at the deepest levels: 'Each essay stands by itself as a separate counsel fitted to move those peculiarly susceptible to its appeal ... Together they are a paradigm of enlightenment. They are perhaps the classic example of the art behind the light ... "which gradually, by imperceptible degrees, would illuminate the world".'[6]

This artful light is the philosophical force behind the statement that 'Nature to be commanded must be obeyed; and that which in contemplation is as the cause is in operation as the rule.' One can only command natural material and forces in order to shape them into works insofar as one has already understood the deeper inalterable structure of hidden forms; and further, that the 'object' of theoretical insight has its own causal dynamics which must be strictly followed in practical terms in order to (re)make the object for one's own purposes. To violate this fundamental principle, to attempt to alter nature's hidden laws and work against its intrinsic dynamics, is a dangerous enterprise. In the *Wisdom of the Ancients,* Bacon meditated on the fable of Daedalus to draw the lesson that careless fooling with mechanical techniques can have malicious and even deadly consequences. But one should be wary of an overly optimistic reinterpretation of his attitude towards scientific progress, for 'Bacon's works are various mirrors of one another, some darker than the rest'[7] – the darkness often concealed beneath the advocacy of a philanthropic practice.

Despite Bacon's awareness of the dangers of even carefully controlled experiments, he was willing to risk these for the material improvement of human life. If some of his pronouncements on these issues are obscure and ambiguous, his vision of a scientific utopia in *New Atlantis* is both unambiguous and frightening. European travellers in the South Seas are blown off course and arrive at the island of Bensalem. They are provided with the benefits of this strange welfare state – food, shelter, and medical care – shown the island's indigenous customs and rituals, and given a guided tour of Solomon's House, the realization of Bacon's scientific research institute. The guide shows them through many rooms where various 'research and design' programmes are being carried out, such as the transformation of birds, beasts and plants into new, barren or super-fertile, kinds, the manufacture of more violent weapons and munitions, and 'houses of deceits of the senses'. The catalogue of twenty-four 'improvements' in scientific knowledge presages some of the most dreadful nightmares of human reason: genetic modification of living things, drug trials on animals, nuclear armaments, powerful machines for the pursuit of luxury or idleness, the ideological apparatus for the control of human behaviour, and more. Equal weight is given to the trivial and the profound, the beneficent and malevolent, reflecting Bacon's ideal of experimental inquiry focused on 'nature under constraint and vexed; that is to say, when by art and the hand of man she is forced out of her natural state, and squeezed and molded'. This utopian vision is the culmination of Bacon's twofold strategy: a physical science capable of dealing with powerful natural motions and a rhetorical 'science' capable of dealing with human emotions.

Bacon's influence on natural science and politics has been pervasive and paradoxical. An epic poem at the front of Thomas Sprat's *History of the Royal Society* (1663) treats Bacon as nearly godlike in the breadth of his vision, one whose instructions could be substituted for divine commandments. At the height of the French Enlightenment, D'Alembert thought that Bacon's grand plan was like the light after the dark, that he was 'the greatest, the most universal, and the most eloquent of philosophers'; his Great Setting-Forth was the model for the *Encyclopédie, ou Dictionnaire Raisonée*. But in Britain about the same time, Bacon's grand scheme was the object of vicious satire and ridicule in Jonathan Swift's *Gulliver's Travels*, where Solomon's House is turned into an asylum for crackpot inventors. Under the impact of the early nineteenth century's revaluation of the history of scientific ideas, Bacon was accorded a more dubious honour, having failed to realize the importance of mathematics in an understanding of physical laws. Lord Macaulay's once famous eulogy of Bacon's philosophy praised this beneficent promoter of human advancement but damned him for moral turpitude in his betrayal of close friends and pandering after honours and riches. In the twentieth century, the Frankfurt School theorists Horkheimer and Adorno denounced Bacon as the initiator of the worst forms of human domination and oppression under the aegis of an instrumental rationality in the service of a capitalist state. Perhaps one can attain a measured balance between these two extremes by considering the arguments of the eminent biologist Loren Eiseley, who arrives at an ambivalent assessment of Bacon, an Elizabethan scientist-magician who promised so much for the good of the human species, but was willing to destroy or distort at least as much to achieve this end.

Notes

1   *The New Organon*, Bk. I, p. 1.
2   N. Mathews, *Francis Bacon: A History of a Character Assassination*, chap. 1.
3   Perez-Ramos, in M. Peltonen (ed.), *Cambridge Companion to Bacon*, p. 115.
4   J.C. Briggs, *Francis Bacon and the Rhetoric of Nature*, p. 3.
5   Ibid., p. 9.
6   R.K. Faulkner, *Francis Bacon and the Project of Progress*, p. 29.
7   Briggs, op. cit., p. 12.

See also in this book

Aristotle

Bacon's major writings

The Works, 7 vols, and *The Letters*, 7 vols, ed. James Spedding, R.L. Ellis and D.D. Heath, London: Longman, 1857–74.
The Advancement of Learning and *New Atlantis*, Oxford: Oxford University Press, 1915.
The New Organon and Related Writings, Indianapolis, IN: Bobbs-Merrill, 1960.
Essays and Counsels; The Advancement of Learning, London: J.M. Dent/ Everyman, 1968.

Oxford University Press is in the process of publishing a new, complete edition of Bacon's works.

Further reading

Adorno, T. and Horkheimer, M., 1944, *Dialectic of Enlightenment*, trans. J. Cumming, London: Verso, 1979.
Briggs, J.C., *Francis Bacon and the Rhetoric of Nature*, Cambridge, MA: Harvard University Press, 1989.
Eiseley, Loren, *Francis Bacon and the Modern Dilemma*, Lincoln, NE: University of Nebraska Press, 1962.
Faulkner, R.K., *Francis Bacon and the Project of Progress*, New York and London: Rowman & Littlefield, 1993.
Mathews, N., *Francis Bacon: A History of a Character Assassination*, New Haven, CT: Yale University Press, 1996.
Peltonen, M. (ed.), *Cambridge Companion to Bacon*, Cambridge: Cambridge University Press, 1996.
Perez-Ramos, A., *Francis Bacon's Idea of Science*, Oxford: Oxford University Press, 1988.

PAUL S. MACDONALD

BENEDICT SPINOZA 1632–77

> The highest good is ... the knowledge of the union that
> the mind has with the whole of Nature.[1]

Spinoza was born in Amsterdam in 1632, the son of Jewish emigrants from Portugal; his father was a merchant and a respected member of the Jewish community's board of elders. Spinoza was brought up in the orthodox fashion, studying Hebrew, Holy Scripture and the Talmud. It is possible to trace the various cultural influences in his life through his choice of first name, from Bento (Portuguese) to Baruch (Hebrew) to Benedict (Latin), each of which means 'blessed'. Sometime about 1656, Spinoza was excommunicated from the Jewish church, on the grounds of heretical beliefs. The bookseller and freethinker Franz van den Enden played a pivotal role in his life and thought; he brought Descartes' works to Spinoza's attention, taught him Greek and Latin, and his mystical views about God or Nature as an infinite substance probably had a decisive influence on the young philosopher's unusual perspective, substance monism, as early as the treatise *On the Emendation of the Intellect*. Spinoza moved to Rijnsburg near Leiden in 1660, where his close friends persuaded him to set down his careful, though uncompleted, exegesis of Descartes' metaphysics, *Descartes' Principles of Philosophy*, published in 1663. Spinoza lived a solitary, almost reclusive life, grinding lens and working slowly on the text of the *Theological-Political Treatise*, published anonymously in 1670. The Council of the Reformed Dutch Church condemned the book as 'a treatise of idolatry and superstition', while one professor at Utrecht wrote that it was 'the most pestilential book'.[2] In 1672, the glorious Dutch Republic came to a disastrous close with an invasion of the French and German armies. The Republic's leader, Jan de Witt, was murdered by an angry mob, and the Dutch Estate Holders brought back into power the young Prince William III. Spinoza was much distressed at the death of de Witt and the unfinished *Political Treatise* demonstrates his unyielding advocacy for the rational foundations of a legitimate state; today it shows his readers the immediate and direct manner in which a philosopher can be engaged in important social and political issues. But it is his final work, the *Ethics*, left incomplete at his premature death in 1677, which has had the most decisive influence in the history of modern philosophy.

In many respects, Spinoza's systematic philosophy in the *Ethics* is the most beautiful, perfectly ordered picture of the universe and humans' place in it. Every aspect of every dimension of human experience is consistently explained in terms of the greater whole. For Spinoza, to explain something is to know its cause, that which not only brings that being into existence but also makes that being just what it is and not something else. A cause also necessarily produces the effect that it does. Understanding, therefore, consists in showing how some feature of the universe necessarily has the role it has as some kind of essential property of substance which is the cause of all things *and* the cause of itself. In contrast with Descartes' dualism, Spinoza propounds substance monism: there is only one substance which has two principal attributes, thought and extension. In this fashion he rejects the dualisms of God and created world, and of mind and body. There are an infinite number of particulars in the world, each of which can be considered a dependent part of that one substance. There is one substance, God or Nature, with two infinite attributes.

These attributes should be thought of as different ways of 'seeing' one and the same reality. We think of extended substance as divided into separate bodies which occupy a limited area in space and time, but extension *itself* cannot be thought of as other than limitless in time and space. The way in which we think of thought will depend upon the level of knowledge which our particular finite mind has reached. The infinite and eternal mode of extension is motion-and-rest; the finite mode, which constitutes individual bodies, or the medium-scale things in our environment, are configurations of simplest particles. The configurations which compose individual physical objects are elements in a *hierarchy* of such organized systems in which there is an ascent from the simplest particles to the whole world; there is one complete cosmic substance in which all other entities are components. All individual things then are configurations of particles in a charged energy state which possess a drive (*conatus*) to maintain themselves in being. The hierarchy of beings then is a plenary order of *power*: the higher an individual is on the scale, the less it is acted on by external forces and the more its changes come from within itself. Moreover, there is an equation between being more or less active as a causal agent upon others and being more or less real. In ascending order of power, these are: the inorganic, the organic, the animal, and the human. The human body is more real than merely animal bodies because it maintains itself in being more effectively than others, does so more under its own control, and interacts with its environment with greater foresight.

The ordered arrangement of beings corresponds with the hierarchy of levels of knowledge. The highest level is intuitive knowledge which approaches the 'infinite idea of God'. 'The more each of us is able to achieve in this kind of knowledge, the more he is conscious of himself and god, that is, the more perfect and blessed he is.' Since God is the same as Nature as a whole, and since Nature is defined as perfect, every being is oriented towards its own perfection or completeness of essence. From this vantage point arises the individual's striving to unite with the source of that which causes the experience of joy or bliss. 'The mind has had eternally the same perfections which now come to it and that is accompanied by the idea of god as an eternal cause. If joy then consists in the passage to greater perfection, blessedness must surely consist in the fact that the mind is endowed with perfection itself.'[3] One can readily appreciate how the German Romantic poet Novalis later referred to Spinoza as 'the god-intoxicated man' and Goethe dubbed him 'the most Christian one'.

The key to Spinoza's moral theory and thus to his attitude towards environmental concerns can be found in his theory of ideas. Corresponding to each level of knowledge or class of idea, there is an ideatum or 'object' of that same idea; degrees of rationality and degrees of reality must be linked at every stage. Thus, insofar as we purify our understanding in order to consider ideas of the highest order of rationality, we come close to the condition of godhood; in this way we cease to be merely parts of nature. Our status as 'natural' beings under the aspect of extension wholly depends on the class of idea (confused, adequate or intuitive) which constitutes our minds, and vice versa. Spinoza has an unusual and seemingly paradoxical claim about the union of mind and body in the human being: the complex idea which the human body has of itself is its mind. This union under two aspects which constitutes a person is only a special case of a general, uniform principle.

There is thus an equal novelty in his notion of psycho-physical causation: changes in one do not *produce* or generate changes in the other, rather every bodily change *is* a mental change and vice versa, since there is only one Nature conceived under two different attributes. Spinoza was well aware of consequent paradox in identifying mental with physical changes. The particular finite mode of extension which is my body exchanges energy with its own proximate environment; and every such 'interaction' is reflected in an idea.[4] Since Spinoza construed the moral dimension as coterminous with the perfectibility of things as parts, exchanges which diminish living beings' energy states are poisons, and thus evil, and exchanges which augment their energy states are healthy, and thus good.

Human beings maintain their identity by preserving a constant adjustment of their parts. This self-maintenance is not the result of some decision by the person, but occurs as a natural process. Other things are susceptible to fewer changes because their structure is less complex and have less 'reality' than human beings. They can manage only a lesser field in their environment and hence the cohesion of their parts is liable to disruption by a more narrow range of external causes. Human beings have a high degree of complexity which, under the attribute of thought, is captured by saying that they have *mind* and that they are *self-conscious*. Thus, a human mind consists of ideas which reflect the effects of external causes insofar as they modify the balance of motion-and-rest which constitutes the human body. Such an alteration arises out of the body's interaction with other things and may be either an increase or a decrease in energy, its 'life-force'. There is thus a wide range of internal energy states within which a human cohesion of parts may remain united, without the individual being destroyed. These changes in state can be described both in physical terms as an increase or decrease in the organism's life-force; and they can be described in mental terms as pleasure and pain.[5] Thus, every increase in the 'life-force' is experienced as pleasure and every decrease is experienced as pain; by 'pleasure' Spinoza means 'the passion by which the mind passes to a higher state of perfection, and by pain the passion by which it passes to a lower state of perfection'.[6] Any increase in the power or perfection of the human body must be an increase in the power or perfection of the mind and vice versa. The moral principle here is that all things which contribute to one's perfection are good and all things which detract from it are evil.

The degree of power or perfection of any finite thing depends on the degree to which it is causally active in relation to things other than itself. The one infinitely powerful and perfect being is God or Nature, who is in every respect active and not passive. A human being has greater power and perfection insofar as the succession of ideas which constitutes its mind are linked together as cause and effect; a human is active insofar as the succession of ideas in its mind is a logical one. A human being has less power or perfection as a thinking being insofar as its present ideas are not explicable as the logical consequences of previous ideas in its mind. In God, there would be an infinite sequence of ideas each one of which would be logically entailed by its predecessors. But human minds, for the most part, consist of more or less *random sequences of ideas*, in the sense that the causes are *external to the sequence*. The sequence of ideas is not self-contained and hence cannot be completely intelligible – there are always gaps. The power and perfection of an individual mind is increased in proportion as it becomes *less passive* and *more active* in the production of its ideas. The equivalent for the individual human body of this increased cognitive activity is the internal stability of the organism, which enables it to carry on living without any violent perturbations produced by external causes. Thus, the mind is relatively free and active in its thinking when the body is in a relatively *constant state vis-à-vis* its own proximate environment.[7]

Human beings, unlike animals, can be *aware* of the tendency towards self-maintenance which constitutes their real 'nature'. The reflection in a *conscious idea* of this *conatus*, the drive to maintain oneself in being, is called *desire*. Spinoza defines desire as *appetite* together with conscious awareness of its occurrence and the 'object' towards which it is directed. Now pleasure and pain are not to be found in the 'objects' which desire and aversion afford, nor can they be discovered by any form of abstract reasoning. They represent a change in the psycho-physical state of the whole person; they are the mental reflection of a rise or fall in the power or activity of the organism. Which specific things will promote or depress the life-force of any organism depends on the constantly changing 'nature' of the individual organism. It may be difficult to understand how *conatus* pertains to inanimate things – how could a *stone*, for example, be said to have a *drive* to maintain itself in being? The problem for the common-sense view is that we think of a stone as an individual *thing* or substance – but of course, for Spinoza, this is incorrect. A stone, a plant, or an animal are each no more than temporary configurations of finite modifications of the infinite attributes of *one thing*, God or Nature; they are all *parts* of the cosmos which work together towards the maintenance of the whole. So, plants consume soil and water, animals consume plants and animals, and so forth, each thereby participating through exchange of energy in the greater whole, from whence all ultimately derive their life-force. This grand conception of the world-whole is perhaps more familiar to contemporary readers from Lovelock's Gaia Hypothesis.

The Norwegian 'eco-philosopher' Arne Naess proposed that Spinoza was the most important philosophical source for inspiration regarding concern for environmental issues. He claims that nature conceived by ecologists is not the passive, inert, value-neutral nature of mechanistic science, but more like Spinoza's Nature – all-inclusive, creative, infinitely diverse, and alive in the broad sense of panpsychism. Further, Spinoza's reflections on morality are 'important for striking a balance between a submissive, amoral attitude towards all kinds of life struggle, and a shallow moralistic and antagonistic attitude'. Future societies will achieve an equilibrium with their environment by following a 'third way' between the two extremes. In Spinoza's world-picture, every thing is connected with every other thing. Nothing is really causally inactive, there is nothing wholly without an essence which it expresses through a cause. And finally, every thing strives to preserve and develop its specific essence or nature, and since every thing is a part of God's perfection, this striving is an active shaping of its environment.[8] 'The highest good is ... knowledge of the union which the mind has with the whole of nature.'

Notes

1 *On the Emendation of the Intellect*, p. 5.
2 Wim Klever, in D. Garrett (ed.), *The Cambridge Companion to Spinoza*, p. 40.
3 *Ethics*, V P32, 33.
4 Stuart Hampshire, *Spinoza*, pp. 72–3.
5 Ibid., pp. 98–9.
6 *Ethics*, III P11Schol.
7 Hampshire, op. cit., p. 100.
8 Arne Naess, *Freedom, Emotion and Self-Subsistence*, pp. 19–20.

See also in this book

Goethe, Lovelock, Naess

Spinoza's major writings

The Ethics and Selected Letters, ed. and trans. Samuel Shirley, Indianapolis, IN: Hackett, 1982.
On the Emendation of the Intellect, Short Treatise, Letters, Principles of Philosophy and *The Ethics* are contained in: *The Collected Works*, vol. I, ed. and trans. Edwin Curley, Princeton, NJ: Princeton University Press, 1985. *The Theological-Political Treatise* and *The Political Treatise* are planned for vol. II.

Further reading

Delahunty, R.J., *Spinoza*, London: Routledge & Kegan Paul, 1985.
Donagan, A., *Spinoza*, Chicago, IL: University of Chicago Press, 1988.
Garrett, D. (ed.), *The Cambridge Companion to Spinoza*, Cambridge: Cambridge University Press, 1996.
Hampshire, Stuart, 1951, *Spinoza*, rev. edn., Harmondsworth: Penguin Books, 1988.
Nadler, Steven, *Spinoza: A Life*, Cambridge: Cambridge University Press, 1999.
Naess, Arne, *Freedom, Emotion and Self-Subsistence*, Oslo: Universitets Vorlaget, 1975.

PAUL S. MACDONALD

BASHŌ 1644–94

The chestnut by the eaves
In magnificent bloom

Passes unnoticed
By men of this world.[1]

Regarded by many as amongst Japan's finest literature, Bashō's work
is an important development and summation of medieval Japanese
cultural attitudes to the natural world, and emphasizes a heightened
sense of unity with nature, much stressed in later artistic expressions
of Zen Buddhism. Held in high regard in Japan, his authority as a
literary figure is matched by his growing influence on religious and
artistic responses to the environment, especially in contemporary
Western circles.

There is a dearth of material concerning Bashō's early life. It is
generally believed that Matsuo Kinsaku, who would later take the
name Bashō, was born in 1644 to a samurai family in service to the
lord of Ueno, south-east of Kyoto, then the capital of Japan. As a
boy he was a page to Tōdō Yoshitada, the eldest son of the ruling
feudal lord of the area – thus his duties as companion and page
brought him into contact with the literature of the ruling classes.
Both boys shared an interest in poetry, and as their friendship grew
they influenced and encouraged each other, particularly in the
writing of haiku. It was an old literary tradition amongst the more
affluent classes of medieval Japan, to engage in the team
construction of *renku* (or *renga*), linked poems of thirty-five, fifty
or one hundred lines. By the end of the fifteenth century it had also
become popular to generate the first stanzas alone as haiku
(originally *hokku*), and haiku competitions were a fashionable
leisured activity. The earliest known recorded verse by Bashō dates
from his time with Yoshitada in the early 1660s.

In 1666 Yoshitada died suddenly and Bashō's grief prompted him
to leave the service of the ruling family, and thereby renounce his
samurai status. It is generally believed that during the period of
wandering that followed he studied in Kyoto, and it is certain that he
continued to write and gain a name for himself as a poet. Between
1667 and 1671 his verses were included in four anthologies, and by
1672 he was able to publish his own record of a haiku competition,
The Seashell Game (*Kai Ōi*), which also marks the beginning of
Bashō's recorded critical prose.

In 1672, at the age of twenty-eight, he left Kyoto to journey to Edo. He joined in the writing of *renku* with several local poets there, and it is assumed that around 1675 he became a professional writer, his work appearing with greater frequency in haiku anthologies of the time. He was presented with a hut or cottage surrounded by banana trees by the local people of Edo, Banana Tree Hermitage (*Bashō An*), from which Bashō took his name. However, he was not entirely satisfied with a static lifestyle and set out on what became his first major journey in 1684 – he often referred to himself as homeless and certainly had few possessions. Whilst travelling he continued to compose *renku* and haiku and wrote his first travel journal, *The Records of a Weather-Exposed Skeleton* (*Nozarashi*). During the remaining years of his life he undertook several such journeys, writing journals alongside his poetry, including *A Visit to the Kashima Shrine* (*Kashima Kikō*), *The Records of a Travel-Worn Satchel* (*Oi no Kobumi*) and *A Visit to Sarashina Village* (*Sarashina Kikō*). Two poetry anthologies of 1686, *Frog Contest* (*Kawazu Awase*) and *A Spring Day* (*Haru no Hi*), include his famous frog haiku, which is often used as the paradigmatic example of Bashō's poetic style:

The old pond;
A frog jumps in—
The sound of the water.[2]

The Narrow Road to the Deep North (*Oku no Hosomichi*) is an account of a two-and-a-half-year trek, begun in 1689, taking in the villages and country north and west of Edo. It is regarded as Bashō's greatest literary achievement, combining concise, crisp prose and poetry in a breathtakingly unified piece. He wrote one other major journal, *The Saga Diary* (*Saga Nikki*), before focusing solely on poetry and the encouragement of younger writers composing in his style – anthologies of this final period include *The Monkey's Cloak* (*Sarumino*) and *A Sack of Charcoal* (*Sumidawara*). In 1694 Bashō died on a pilgrimage to Osaka. Around one thousand haiku are attributed to Bashō. He established the haiku as a serious and deep poetic form that captures a purity and unity in the immediacy of experiences of the natural world.

The haiku translator and historian, R.H. Blyth, once commented that 'Nature *is* Japanese Literature'.[3] Although this is an exaggeration, there is much evidence to support the spirit of this claim. Japan has marked seasons, and seasonal poetry dates back to the earliest recorded anthologies, nature and natural cycles remaining key subjects for writers throughout the development of the *renku* and *hokku* and up to the present day. Many words and phrases acquired connotations of seasons and seasonal activities, bringing to mind more than the picture inspired by the literal meanings of the word. For example, 'blossom' (*hana*) in a poem means ornamental cherry tree blossom and the associated image of its fluttering in a warm spring breeze. This 'logopoeia' allowed poets to condense a great deal of imagery into a simple phrase. Sōgi (1421–1502) mastered this technique in his *renku* two hundred years before Bashō, and it subsequently became a mainstay of Japanese verse. Through use of these linked and complex images the seasons and nature remained central to later poetry as the haiku came into its own, so much so that all haiku – in their traditional form at least – must refer to a season to be complete. It is the concentration of associated 'images', across all the senses, that paradoxically makes haiku so pure. The 'plop' of the frog jumping into the old mill pond, together with the stillness, the ripples of the water and the flash of colour evoked by 'the sound of the water', brings us to an imagining of the moment that lengthier descriptions fail to evoke. So in haiku we find a supreme aesthetic expression of the experience of a thinking being in a relationship with the natural world. And Bashō was undoubtedly one of the great haiku masters. Yet his genius lies in more than the cleverness of his style.

Whilst there seem to have been no major developments in philosophy during the Tokugawa period (1600–1867) of Japanese history, the closing of the borders to foreign influences around the time of Bashō accentuated the purely Japanese aspects of art and literature produced by the 'home-grown' talent of the time. Writers and artists looked back to their own cultural forefathers, such as Sōgi, in whose writing Bashō obviously found great inspiration. But Bashō also explicitly drew on the Japanese ' ... *conflation* of the religious and literary dimensions of human experience',[4] a deliberate refusal to separate out different aspects of an experience where a separation could be made. Consequently it is disingenuous to treat Bashō's poetry as 'pure' literature alone, separated from its religious and philosophical roots in Shintō, Buddhism and earlier Chinese thought. In this spirit of co-existence Buddhism and Shintō stood side by side in Bashō's Japan, despite differences in practice and origin, as they do today, often in the same shrine. Bashō's poetry is in tune with this tradition of 'conflation' and often incorporates into more sophisticated responses to nature an explicit animism derived from Shintō:

> Making the uguisu [warbler bird] its spirit
> The lovely willow-tree
> Sleeps there.[5]

However, to see Bashō as presenting a simple, romantic view of the living world would be to ignore a much deeper Buddhist component. Bashō almost certainly learnt to meditate under a Zen master and direct references to Buddhism are scattered throughout his verse:

> The anniversary of the Death of the Buddha;
> From wrinkled praying hands,
> The sound of the rosaries.[6]

Even when it does not point directly to Buddhism, Bashō's verse often exemplifies key components of Buddhism – awareness of impermanence, the non-existence of the self, emptiness, the suffering of all living beings, and the compassion we should feel for these beings (and that we would feel if we paid attention to our true nature):

> Singing, singing,
> All the long day,
> But not long enough for the skylark.[7]

And in reference to his dead mother's hair:

> Should I take it in my hand,
> It would melt in my hot tears,
> Like autumn frost.[8]

Above all, it is the sense of sympathetic compassion with all life that pervades Bashō's writing:

> The ancient poet
> Who pitied monkeys for their cries,
> What would he say, if he saw
> This child crying in the autumn wind?[9]

He attempts the impossible for us, to lead us to an unmediated glimpse of the real world of natural things in which blossom is glorious and beautiful and then fades and dies, to return the next year, each flower new and unique. It is only through reaching a state of enlightenment that such an insight could be perfected, for only then would we be aware of the connectedness of all living beings and true Buddha-mind. And yet, like all the great Zen masters, Bashō strives always to point the way, showing us moments of experience of frogs and flowers and muddy roads. Bashō reminds us, from a Zen perspective, that the silence of an autumn moon reflected in a lake will have much more to tell us about ourselves and reality than the chatter of our thoughts and theories.

By combining his mastery of the haiku form, his Buddhist insights and his personal commitment to a life largely liberated from material concerns, Bashō produced an account of nature, in its broadest terms, that has been deeply influential in Japan's culture and literature. After Bashō, haiku writing flourished with renewed vigour, his travels having spread his teaching and style throughout the country. There were many students and imitators, but none of Bashō's contemporaries and immediate successors reached his standard of equanimous simplicity. Of later poets only Buson (1716–83) and Issa (1763–1827) can compete for his impact on modern poetry in Japan. Certainly Issa, in particular, made more of the ineliminability from pure aesthetic experience of compassion for living beings, but it was Bashō who crystallized the use of modern haiku for such a Zen purpose, superseding all earlier models for haiku writing whilst refreshing a well-established tradition.

Bashō's verse now appears in almost all inspirational Zen collections.[10] He is also quoted and used by Buddhist writers attempting to forge a connection between Zen and deep ecology, but in the end, it is the uncluttered purity of his prose and poetry that keeps his concern for the natural world alive and continually attracts new readers.

Notes

1 *The Narrow Road to the Deep North and Other Travel Sketches*, p. 108.
2 R. Aitken, *A Zen Wave*. This haiku has attracted more discussion and analysis than any other; in Japanese it is a perfect balance between sound, form and content and, as with all haiku, presents problems of translation. See Hiroaki Sato, *One Hundred Frogs*.
3 R.H. Blyth, *The Genius of Haiku: Readings from R.H. Blyth on Poetry, Life and Zen*, London: The British Haiku Society, p. 72, 1994.
4 W.R. LaFleur, *The Karma of Words*, p. 149.
5 R.H. Blyth, *A History of Haiku*, p. 111.

6 Ibid., p. 119.
7 Ibid., p. 127.
8 Ibid.
9 *The Narrow Road to the Deep North and Other Travel Sketches*, p. 52.
10 It would be impossible to survey them here, but see, for a good balanced
   example, K. Tanahashi and Tensho D. Schneider (eds), *Essential Zen*,
   New York: HarperCollins, 1994.

See also in this book

Buddha

Bashō's major writings

There are several collections of Bashō's work, some more scholarly than
others. For a survey of around 250 haiku with commentaries, see:

Uedo, M., *Bashō and His Interpreters*, Stanford, CA: Stanford University
Press, 1992.

A good collection of many of the travel poems and prose is:

Bashō, *The Narrow Road to the Deep North and Other Travel Sketches*, trans.
Nobuyuki Yuasa, Harmondsworth: Penguin, 1966.

Further reading

Aitken, R., *A Zen Wave: Bashō's Haiku and Zen*, New York: Weatherill,
   1978.
Blyth, R.H., *A History of Haiku*, 2 vols, Tokyo: Hokuseido Press, 1963.
LaFleur, W.R., *The Karma of Words: Buddhism and the Literary Arts in
   Medieval Japan*, Berkeley, CA: University of California Press, 1983.
Sato, Hiroaki, *One Hundred Frogs: From Renga to Haiku to English*,
   New York: Weatherill, 1983.

DAVID J. MOSSLEY

JEAN-JACQUES ROUSSEAU 1712–78

> Man's proper study is that of his relation to his
> environment ... this is the business of his whole life.[1]

Born in Geneva, Rousseau was raised by his aunt and eccentric
watchmaker father, who instilled in him an abiding love of literature,
especially classical. After an unstable childhood and several years as
a vagabond, Rousseau moved in 1743 to Paris, where he met Diderot

and other *philosophes* involved in the great *Encyclopédie, ou Dictionnaire Raisonée,* for which he contributed an article on music. In 1749 Rousseau experienced an overwhelming inspiration from which he later claimed all his philosophical speculations were derived. He won a prestigious prize with his *Discourse on the Arts and Sciences* in 1750, and wrote two operas. In 1754, on a return visit to Geneva, he reconverted to Calvinism and regained his citizen status, of which he was always proud. During the following eight years, living mainly in the country, he published most of his principal works, including *Émile,* and *The Social Contract.* These works were condemned in Paris and Geneva, and Rousseau moved to England, on the instigation of David Hume, with whom he soon quarrelled. Returning to France in 1767, he became mentally disturbed and was always in fear of being arrested. He finally settled in Paris in 1770, where he finished work on *The Confessions,* only to have his former friend and confidante Madame d'Epinay issue a police ban against him. His final, unfinished work, before his death in 1778, was the more serene and meditative *Reveries of a Solitary Walker.*

In the *Discourse on the Arts and Sciences,* Rousseau answered the question 'Has the rebirth of the arts and sciences contributed to the purification of morals?' with an emphatic negative. In direct opposition to the view espoused by the *philosophes,* he asserted that the progress of the arts and sciences in every society has been accompanied by the corruption and diminution of morality. In this essay he broached the concept of a natural human being, characterized by simplicity, lack of vanity and basic virtue, a natural state eroded by the acquisition of politeness, superfluous ornaments and dependence on artifice, including the machinery of warfare. He drew numerous examples from ancient history to show that the arts and sciences have not inspired humans with courage or patriotism, but instead deflected their energies into unnecessary inventions, the flattery of paintings and sculptures, and the display of erudition. Even our most valued sciences have developed out of idleness and trivial pursuits: astronomy from superstition, geometry from avarice for property, and physics from excessive curiosity. Rousseau's vigorous condemnation of modern morality is drawn from a conjectural history of humanity. He argues that the human species has declined from the innocence of its original condition and the most praised civilizations are decadent under the weight of their own cultural progress.

Despite its confident tone, this first *Discourse* suffers from incoherence, lack of originality, and indecisiveness about a remedy for the parlous situation. In this essay, he is not clear whether the general decline of culture is the cause or the effect of the erosion of morality.

In the *Discourse on the Origin of Inequality*, Rousseau carries forward his central theme of the denaturation of human beings, their progressive removal from the sources of their natural being. The second *Discourse* is an ingenious, tightly argued essay which ran counter to the then-accepted view that humans in their original state were motivated solely by self-interest and aggression towards their fellows, and remained fractious until they were coerced into accepting governance under the rule of law. Rousseau distinguishes between natural inequality, which results from discrepant physical and mental abilities, and moral or political inequality, which depends on social conventions and is authorized by mutual consent. The subject of this essay then is 'the moment at which … nature became subject to law, and to explain by what sequence of miracles the strong came to submit to serve the weak, and the people to purchase imaginary repose at the expense of real felicity.'[2] Previous political theorists, such as Hobbes, made the mistake of imputing to their hypothetical natural humans ideas which were only acquired by socialized humans. Rousseau constructs a conjectural history in order to make sense of the origins of moral and political notions such as natural right and justice. He resists the temptation to retroject notions which the civilizing process has conferred upon humans and considers instead an entirely natural human, a creature whose basic needs of hunger, thirst and sex are satisfied in the most immediate manner.

Rousseau follows Descartes in considering the animal in its bodily dimension to be an intricate machine, driven by its senses to seek what would nourish it and to guard against or avoid what would damage it. But where non-human animals carry out their actions for need-satisfaction by the internal operations of instinct, humans have a freedom to choose; they are at liberty to acquiesce or forbear to carry out what their natural desires impel them towards. 'In the power of willing or rather choosing, and in the feeling of this power, nothing is to be found but acts which are purely spiritual and wholly inexplicable by the laws of mechanism.'[3] This account of the freedom prefigures the dualism of spirit and body in the Savoyard Priest's discourse in *Émile*. Rousseau thus expressly sides with the philosophical view that only the bodily aspect of humans can be explicated in mechanistic terms. But the fact that non-human animals are sentient creatures means that *they ought to partake of natural rights*; humans are subject to an obligation even towards the brute. 'This is less because they are rational than because they are sentient beings; and this quality, being common both to men and beasts, ought to entitle the latter at least to the privilege of not being wantonly ill-treated by the former.'[4] Rousseau clearly expresses here one of the first conceptions of the intrinsic moral standing of non-human animals.

The first step beyond this entirely natural human condition was made by the first person who declared a piece of ground to be his own; civil society is founded on the notion of private property. But the satisfaction of natural humans' basic needs might not be immediate due to variations in circumstances, climate, soil and so forth which provoked the additional needs to build shelter, storage and implements. Reflection on the best way to achieve these ends would have inspired a sense of prudence which required that only in some cases would pursuit of private interest be to one's best advantage, whereas in other cases cooperation with one's fellows' pursuit of their interests would best serve one's deferred needs. Freed from the demand to be incessantly in pursuit of one's own needs, socialized humans had the opportunity to sing and dance, 'the true offspring of love and leisure', as Rousseau charmingly phrases it. It was from the desire for public esteem that the first moves towards inequality were made – on the one hand, vanity and contempt, and on the other, shame and envy. Moral sentiments are judgements conferred upon persons and actions which are deemed to endorse or contravene a suitable estimate of a person's or an action's worth.

Rousseau extols a conjectured golden age, 'the real youth of the world', whose best exemplar is the noble savage who maintains 'a just mean between the indolence of the primitive state and the petulant activity of our *amour-propre*'.[5] The next stage was the specialized labour of metal-working and agriculture, but variable distribution of natural resources ensured that those who had more property and power accumulated greater riches. It was in the interests of those with more property and power to retain the services of the poor, and for the poor to offer their labour, even their liberty, in exchange for protection. Since the rich enjoy greater physical goods and the talented enjoy greater public esteem, it becomes a new interest for those less well blessed to appear to be what they really are not. Flattery, trickery and deceit become valued skills. But since even the rich and powerful might have to contend with dangers and even rebellion from everyone else, they devised an ingenious plan: 'to make allies of his adversaries, to inspire them with different maxims, and to give them other institutions as favorable to himself as the law of nature was unfavorable'. Thus the first version of the social contract is tendered, in which the supreme power which governs everyone is invested in the rule of law. 'All ran headlong to their chains in hopes of securing their liberty'; the contract 'bound new fetters on the poor and gave new powers to the rich; which irretrievably destroyed natural liberty ... and for the advantage of a few ambitious individuals, subjected all mankind to perpetual labor, slavery and wretchedness.'[6]

This ringing denunciation of the misfortunes which result from the progressive denaturation of human beings is taken up again in *The Social Contract*: 'man is [or was] born free, and everywhere he is in chains'. *The Social Contract* portrays an association by contract which draws citizens together instead of driving them apart and protects egalitarian ideals of public engagement which enhance liberty. Rousseau argues that our proper passage from the original, natural condition to civil society must not suppress true liberty, but instead realize our freedom by transforming appetite and desire into obedience to laws which we prescribe for ourselves. His radical vision centred around the notion that this association by contract ensured that the various parties were able to fulfil ambitions which they could not have managed without the contract. By renouncing freedom from 'each other's control, ... citizens acquire moral personalities and cooperative interests unimaginable to solitary savages'.[7]

Rousseau's most complete, mature exposition of two themes little discussed in *The Social Contract* – humans' natural condition and the process of denaturation – is in *Émile*. This is divided into five books which roughly correspond with the five ages of man – infancy, childhood, puberty, adolescence and adulthood. The central theme of this convoluted work is that the proper education of children must take account of the maturation of their cognitive and affective abilities, leading their natural desires towards goals which will be of value to them as adults, and not impose adult expectations on each stage of growth. Rousseau's own experiences as a private tutor taught him that the only way to compel a child to obey one's commands was to prescribe nothing, forbid nothing, exhort nothing, and avoid boring him with useless book-work.[8]

Rousseau profoundly disagreed with John Locke's *Treatise on Education* and its numerous adherents who, he claimed, distorted the child's natural inclinations and inculcated ambitions for useless pursuits, vain conceits and superfluous social niceties. Rousseau's astonishing advice was to employ two other inborn motives for learning which do not corrupt the pupil's natural goodness. In childhood, this basic drive is for food, and after puberty it is for sex. In Alan Bloom's excellent analysis of these themes, the child seeks out desirable foods, whereas the adolescent and young adult seeks out other ideals because he does not yet know what he really longs for. 'The task is to enrich his desires before they are satisfied ... The goal is to sublimate his desires prior to his capacity to distinguish sex from love, so that when he learns about the distinction it no longer interests him.'[9] The tutor's task in the life-long education of Émile is to prepare him for his encounter with Sophie, the embodiment not merely of his sexual desires but also his longing for an ideal in this world.

Every child before the onset of education lives in the golden age of his world, a natural creature whose source of action is a surfeit of self-love. But the immediate environment does not always satisfy the child's desires, nor can the child count on the ability to manipulate persons and things to achieve its ends; however, nature has also endowed humans with imagination and this cognitive power compensates for what nature in general does not supply for the child's own existence. It is through imagination that the maturing child comes to understand that others have desires and feelings and that through compassion the child can extend its world. The adult needs other persons' compassion, their fellow-feeling for his own desires and their realization; this mutual compact with other adults is founded on an even balance between the self-serving primitive mode of human being and dependence on the esteem of others in the socialized mode. 'Man's proper study is that of his relation to his environment. So long as he only knows that environment through his physical nature, he should study himself in relation to things; this is the business of his childhood; when he begins to be aware of his moral nature, he should study himself in relation to his fellow-men; this is the business of his whole life.'[10]

Rousseau is often assimilated into the broad current of the Enlightenment project, but although he concurred with the *philosophes* in their attempt to eliminate religious prejudices, he was their sharpest critic in rejecting the élitist notion that human reason should hold sway over our passions. He rejected the Baconian and Cartesian advancement of humans' dominance over the natural order and their exploitation of the precious gifts of God's creation. Rousseau argued passionately for the natural goodness of the ordinary person and championed the idea of collective self-expression and popular self-rule. His epistolary novel *Julie, or the New Heloise*, with its evocation of ideal love and an earthly paradise, was highly influential and much imitated. *Émile* became the most important treatise on education since Plato's *Republic* and the *Reveries of a Solitary Walker* became the vade mecum of the Romantic Naturalist movement. Through his entire life and writings runs one of his deepest concerns – the implacable commitment to prevent an individual's dominance or submission, which would chain him to worldly things and negate his natural liberty.

Notes

1 *Émile*, pp. 209–10.
2 *The Discourses*, pp. 44–5.
3 Ibid., p. 54.
4 Ibid., p. 42.
5 Ibid., p. 82.
6 Ibid., pp. 88, 89.
7 Robert Wokler, *Rousseau*, p. 61.
8 Ibid., p. 94.
9 Alan Bloom, *Love and Friendship*, p. 61.
10 *Émile*, pp. 209–10.

See also in this book

Bacon, Goethe

Rousseau's major writings

Émile, or On Education, 1911, trans. Barbara Foxley, London: Everyman, 1992.
The Social Contract and the Discourses, 1913, trans. G.D.H. Cole, London: J.M. Dent, 1973.
The Confessions, trans. J.M. Cohen, Harmondsworth: Penguin, 1954.
Julie, or the New Heloise, abr. and trans. J.H. McDowell, Pittsburgh, PA: State University Press, 1968.
The Collected Writings, ed. R.D. Masters and C. Kelly, 5 vols, Hanover, NH: University Press of New England, 1990–2000.

Further reading

Bloom, Alan, *Love and Friendship*, New York: Simon & Schuster, 1994.
Cranston, Maurice, *Jean-Jacques*, vol. I, *The Noble Savage*, vol. II, *The Solitary Self*, vol. III, Chicago, IL: University of Chicago Press, 1983, 1991, 1997.

Green, F.C., *Jean-Jacques Rousseau: A Critical Study of His Life and Writings*, Cambridge: Cambridge University Press, 1955.

Grimsley, Ronald, *Jean-Jacques Rousseau: A Study in Self-Awareness*, Cardiff: University of Wales Press, 1969.

Wokler, Robert, *Rousseau*, Past Masters Series, Oxford: Oxford University Press, 1995.

PAUL S. MACDONALD

JOHANN WOLFGANG VON GOETHE 1749–1832

> The alarming increase in machines torments and frightens
> me, they are rolling down upon us like a thunderstorm,
> slowly, slowly, but they are on their way, they will come
> upon us.[1]

The Germany into which Goethe was born on 28 August 1749 was a pre-industrial collection of statelets. By his death on 22 March 1832 this pre-eminent genius, a poet, dramatist, novelist, artist, critic, lawyer, civil servant, statesman and scientist, had lived through a period which took Germany to the very threshold of its delayed industrial revolution. After a childhood in Frankfurt, Goethe studied law at Leipzig and Strasbourg. During convalescence from serious illness he dabbled in alchemy, the influences of whose underlying philosophy are still evident in Goethe's later approaches to both science and literature. In August 1771 he began to practise as a lawyer, but the tumultuous success in 1774 of his drama *Götz von Berlichingen* and, especially, of his epistolary novel *The Sorrows of Young Werther*, catapulted him to European-wide fame as a writer. In 1776 Goethe was called to the court of the Duchy of Sachsen-Weimar, marking the start of a life-long career in Weimar as a civil servant and minister under the patronage of the Duke Carl August. In 1782 he was elevated to the aristocracy and in the same decade began to develop his interest in the natural sciences, in the course of time covering fields including geology (he was for a time Minister of Mines), botany, optics, zoology, anatomy, morphology and meteorology. His exploration of botany and geology in particular developed during a sojourn in Italy between 1786 and 1788.[2] During the 1790s Goethe not only worked on a number of long-lasting literary projects which were to become world classics (especially *Faust*, the second part not finished until 1831), but he also began a lengthy endeavour to discredit Isaac Newton's theory of optics in favour of his own chromatics. Goethe finally published his *Theory of Colours* in 1810, by which point his literary reputation was reaching new heights with the publication of the first part of *Faust* in 1806 and of the novel *Elective Affinities* in 1809. By the 1820s Goethe's fame and acknowledged importance were such that his friend Eckermann made detailed notes of his dinner conversation over several years; this along with other sources, and the huge number of words that Goethe wrote, have provided a profoundly rich source for an assessment of his views.

It is primarily Goethe's view of nature that makes him attractive to those interested in environmental thought. Having abandoned Christianity early in life in favour of a Hellenic neo-paganism (though not in any organized or evangelical manner), Goethe allowed his holistic view of nature to inform every aspect of his work. Though he himself was wary of the term pantheism, which is conventionally attributed to him,[3] there is no doubting his holistic understanding, a spiritual dimension to his approach to 'God-Nature', and above all and everywhere apparent, his passionate veneration of the natural world. Goethe rejected a view of nature which concentrated solely on the totality, however. A perception of nature as an external, complete, static given is as limiting, indeed false, as an excessively analytical, taxonomic approach which concentrates on the detailed elements in isolation. In Goethe's view the question of the whole and of the parts is inseparable; one cannot be viewed without the other, and both must be seen as part of a process, in constant change, growth, death, rebirth. He was convinced that for this reason there was an intimate relationship between 'the demands of science' and 'the impulses of art and imitation'.[4] Accordingly, in an uncanny foreshadowing of Heisenberg's uncertainty principle, and in contradiction to the secure objectivity of the eighteenth- and nineteenth-century scientific method, Goethe insisted that there can be no separation between subject and object, between observer and the observed. The interweaving of humankind and nature precludes any such division; the very act of observation affects the observed, while the observed is capable of profoundly altering the observer. The fundamental processes of nature, the polarities of bonding and separation, of breathing in and out, as he understood it, are reflected in the human spirit; Goethe's holism would admit no other. The corollary is that there must be an ethical dimension to the relationship between nature and humankind; nature demands respect, even veneration. Nature, as observed by the scientist, is imbued with values. Herein, surely, lies much of the attraction of Goethe to the modern Green movement. Writing in an age before human beings had the capacity to shape nature in a thoroughgoing post-industrial fashion (although humankind had been leaving its mark on the planet for thousands of years), Goethe nevertheless recognized that inner nature and external nature are indistinguishable, and thus came near to the concept of inner or constructed nature which was only fully developed by Horkheimer and Adorno in the 1940s and 1950s. A further dimension of the attraction Goethe's view of nature holds for modern Greens is his insistence that nature can only be properly comprehended by means of *Ahnung*, or intuition. This does not mean a rejection of science; but it does mean a rejection of the conventional scientific method; and indeed an understanding of Goethe as a scientist is fundamental to an understanding of his thought in ecological terms.

The lasting achievement of Goethe's scientific work is also his earliest in the field of natural sciences: the discovery of the intermaxillary bone in human beings. Until Goethe's discovery, the absence of a bone in the human jaw which in animals houses the canine teeth was taken as evidence of the essential distinction between the two. The suture which remains as the indication that human beings also retain such an anatomical structure bears Goethe's name still. But it is in the theological and social, not to say scientific, importance of the recognition of a relationship between human beings and animals that the importance of this discovery lies. It points to an essential cornerstone in ecological thinking; that human beings, while in Goethe's view the crowning achievement of nature and clearly distinct from animals, are a part of nature like any other.

Although with the exception of this anatomical discovery none of Goethe's scientific revelations are of acknowledged lasting significance, his writings on science nevertheless remain the subject of lively debate. Distinguished physicists including Walter Heitler, Werner Heisenberg and Max Planck have written on Goethe. The reason for this enduring interest lies in his idiosyncratic scientific methodology.

This is nowhere more clearly or fully expressed than in the substantial *Theory of Colours,* the work which he regarded as his most important.[5] On the basis of a chance observation through a prism, Goethe became convinced that Newton's spectral theory of light was wrong, in contrast to his own understanding of light as a unity of white which achieved colour by varying admixtures of shade. To his lasting chagrin, Goethe was unable to convince his contemporaries of the correctness of this thesis, partly since he was of course utterly in the wrong. It has been argued that Newton and Goethe were in fact talking about two different things; Newton about the composition of light and Goethe about the human perception of it.[6] And it is on subjective perception that Goethe's scientific method relied. The attack on Newton was anything other than objective; indeed, a 'Polemical Section' of the work is devoted in part to denigrating Newton's character in the most scurrilous fashion. In fact, the basis for Goethe's deep disquiet was Newton's analytical methodology, which allegedly embodied the nature-dominating techniques of the scientific method. Spectral analysis using optical instruments was a dispassionate dissection, objectification and subjugation of nature. For Goethe, an account of an experiment was not a formula setting out aim, method, equipment and results, but a story in itself, which included his own feelings, the origins of the experiment, the effect on his senses; in short, a contextual narrative, the whole deriving from subjective evidence. Experiment must also be experience, easily repeatable for the reader with the most rudimentary equipment. Only such 'zarte Empirie' (delicate empiricism) could do justice to the wholeness of 'God-Nature'. Accurate detail and linear causality were of less importance to Goethe than broad-ranging context, the network of interconnections. To be absolutely clear: Goethean science is a rejection not of science, but of a science which is contemptuous of nature. The extent of Goethe's influence can be gauged by the fact that there are today scientists working in ecology and other fields who pursue their research in an explicitly Goethean fashion.[7]

Goethe is conventionally celebrated for his literary achievements, where proto-ecological elements have also been discovered.[8] Merely on the level of content, Werther's despair at the cutting down of ancient nut-trees is an emotion with which many modern Green activists could sympathize, while the fear expressed in *Wilhelm Meister* at the ubiquity of machines (quoted at the outset) also has contemporary resonances. Goethe's refusal to distinguish between art and science often led him to give literary expression to scientific results. His poem *Metamorphosis of Plants* encapsulates the results of his essay of the same name, but the poem *Metamorphosis of Animals* is even more directly relevant for our topic. The apparent foreshadowing of Darwin is so startling as to make it worth quoting (my translation):

Thus the form determines the animal's way of living
And the way of living powerfully affects all forms
In turn. The ordered formation is thus clearly shown
Which, through the operation of outside elements, tends to
change.[9]

As the biochemist Friedrich Cramer argues, this does sound very
much like Darwinism.[10] At the very least there is a clear recognition
here of the way in which creatures adapt to their environment, and,
perhaps, of the interplay between organism and environment
without which any ecological view is unthinkable. But on a more
fundamental level too, Goethe's assumptions concerning nature
inform his literary work. In particular, his masterpiece *Faust* has
been interpreted as an attempt using alchemical metaphors to show
the way in which the economy depends on the exploitation of
nature.[11] Similarly, Jost Hermand argues that the long-standing
misreading of the text as a paradigm of technical progress and
individual ambition requires correction; in fact, it is a celebration of
the natural virtues of harmony, holism and mutuality. Faust's
destructive drives arise, Hermand argues, because he has lost 'all
sense of human solidarity or empathy with nature'.[12] Both the
Sorrows of Young Werther and *Elective Affinities*, as well as a
number of poems, have also been refracted through an ecological
prism. 'The Magician's Apprentice', for example, a poem known to
every German-speaking school-child, is routinely used to demonstrate the dangers of meddling with powerful forces one does not
properly understand.

Goethe's influence on the history of ecological thought is
manifest: Darwin, without whose work there could be no science
of ecology, cites him in the *Origin of Species*. Ernst Haeckel's late-
nineteenth-century fusion of science and mysticism in the form of
monism, which invested a holistic nature with spiritual qualities, is
explicitly derived from Goethe. Rudolf Steiner, the founder of
anthroposophy and an originator of organic farming, was deeply
indebted to Goethe, as are contemporary Green campaigners of the
stature of Fritjof Capra.[13] Was Goethe himself an early Green
campaigner? Clearly not; despite the opening quotation, the steam
engine is only mentioned explicitly a handful of times in the vast
number of words he wrote, though it was invented in 1776. And
there is a distinct thread of anthropocentrism, to be expected in his
era, running through all his work. It would be dangerous and
misleading, then, to instrumentalize Goethe in the light of

contemporary concerns (though each age has appropriated him for its own purposes). But it is beyond dispute that, among his many accomplishments, Goethe remains a lasting source of inspiration to the ecological imagination.

Notes

1   *Wilhelm Meister's Travelling Years*, 1829; quotation translated from the original German by Colin Riordan.
2   He was especially impressed by Rousseau's work on botany.
3   He feared its use might lead to a simplistic categorization of his views. See letter to C.F. Zelter, 31 October 1831, WA, IV, 49.
4   See WA, II, 6: 9 (my translation).
5   Indeed, Goethe felt that it was his scientific work which would be his lasting monument.
6   See H.A. Glaser (ed.), *Goethe und die Natur*, Frankfurt am Main: Peter Lang, p. 29, 1986.
7   For examples, see especially Part II of D. Seamon and A. Zajonc (eds), *Goethe's Way of Science*, entitled 'Doing Goethean Science'.
8   Non-specialists frequently and mistakenly associate Goethe with Romanticism. In fact he was an overarching figure whose relations to the Romantics were ambivalent; there were fundamental differences in philosophy.
9   WA, I, 3: 90 (my translation).
10 Friedrich Cramer, ' "Denn nur also beschränkt war je das vollkommene möglich" ... Gedanken eines Biochemikers zu Goethes Gedicht "Metamorphose der Tiere" ', in Glaser, op. cit., pp. 119–32.
11 See Hans-Christoph Binswanger, 'Die moderne Wirtschaft als alchemistischer Prozeß – eine ökonomische Deutung von Goethes "Faust" ', in Glaser, op. cit., pp. 155–76.
12 Jost Hermand, *Grüne Utopien in Deutschland. Zur Geschichte des ökologischen Bewußtseins*, Frankfurt am Main: Fischer, p. 58, 1991 (my translation). See also Jost Hermand, 'Freiheit in der Bindung. Goethes grüne Weltfrömmigkeit', in Jost Hermand, *Im Wettlauf mit der Zeit. Anstöße zu einer Okologiebewußten Ästhetik*, Berlin: Sigma Bohn, 1991. Gerhard Kaiser makes a very similar argument in his *Mutter Natur und die Dampfmaschine. Ein literarischer Mythos im Rückbezug auf Antike und Christentum*, Freiburg im Breisgau: Rombach Verlag, 1991.
13 See especially Fritjof Capra, *Wendezeit: Bausteine für ein neues Weltbild*, Munich: Droemer Knaur, 1999.

See also in this book

Darwin, Rousseau

Goethe's major writings

Sorrows of Young Werther, 1774, Harmondsworth: Penguin Books, 1989.
Faust: Parts One and Two,1806 and 1831, London: Nick Hern Books, 1995.

Elective Affinities, 1809, Oxford: Oxford Paperbacks, 1999.
Theory of Colours, 1810, Cambridge, MA: MIT Press, 1970.
Conversations of German Refugees / Wilhelm Meister's Journeyman Years or The Renunciants, 1829, in *Goethe: The Collected Works in 12 Volumes*, vol. 10, Princeton: Princeton University Press, 1996.
Scientific Studies, in *Goethe: The Collected Works in 12 Volumes*, vol. 12, Princeton: Princeton University Press, 1995.

Further reading

Bortoft, H., *The Wholeness of Nature: Goethe's Science of Conscious Participation in Nature*, Hudson, NY: Lindisfarne Press, 1996.
Hoffmann, N., 'The Unity of Science and Art: Goethean Phenomenology as a New Ecological Discipline', in D. Seamon and A. Zajonc (eds), pp. 129–77.
Seamon, D. and Zajonc, A. (eds), *Goethe's Way of Science. A Phenomenology of Nature*, Albany, NY: SUNY Press, 1998.
Whyte, L.L., 'Goethe's Single Vision of Nature and Man', *German Life and Letters*, 2, pp. 287–97, 1949.
Williams, J.R., *The Life of Goethe. A Critical Biography*, Oxford: Blackwell, 1998.

COLIN RIORDAN

THOMAS ROBERT MALTHUS 1766–1834

> I think that I may fairly make two postulata.
> First, That food is necessary to the existence of man.
> Secondly, That the passion between the sexes is necessary and will remain nearly in its present state.
> Assuming then my postulata as granted, I say, that the power of population is indefinitely greater than the power in the earth to produce subsistence for man.
> Population, when unchecked, increases in a geometrical ratio. Subsistence increases only in an arithmetical ratio.
> A slight acqaintance with numbers will shew the immensity of the first power in comparison of the second.

The above frequently used quotation is from *An Essay on the Principle of Population as it affects the Future Improvement of Society, with Remarks on the Speculations of Mr. Godwin, M. Condorcet and other Writers* by the English economist, mathematician and clergyman, Thomas Robert Malthus. His principle that population growth will constantly tend to outrun subsistence unless there are severe limits on reproduction is regularly cited to this day. But more cited than read, he is the most misinterpreted scholar in population studies. Published anonymously in 1798 as a long mainly theoretical pamphlet when he was a country vicar aged thirty-two, it was succeeded by five attributed and much more documented editions between 1803 and 1826, becoming a massive and very different work retaining the initial thesis but entitled *An Essay on the Principle of Population, or A View of its Past and Present Effects on Human Happiness, with An Inquiry into Our Prospects Respecting the Future Removal or Mitigation of the Evils which It Occasions.*

Even by the time of the second edition in 1803, Malthus knew much more about the English population from the returns of the 1801 census as well as from parish registers, and he knew more about the population of Europe from visits to Ireland and several other European countries, leading him to change the way that he discussed the relationships between production and reproduction and to become rather less pessimistic about people breeding themselves into poverty. He came to realize that people were doing something about their lot. But as the work grew in size and scholarship it became less easy to read, so supporters and critics have tended to concentrate on the *First Essay*, although Malthus later regarded it as an unsatisfactory statement of his views. Nevertheless, it has had an immense influence on subsequent thought during the nineteenth and twentieth centuries, so much so that few publications on the relationships between population, economics, environment and development have failed to mention Malthus or Malthusianism. Moreover, the use of statistical data to support his wide-ranging theories in the more mature work of the later editions effectively established Malthus as the father of modern population studies, although some later scholars have felt that the data were not effective in empirically validating those theories.

The second of eight children of a cultured country gentleman, Malthus was destined to be a clergyman. Educated first privately and then at Jesus College, Cambridge, where he excelled, he took holy orders in 1788 and was elected a fellow of Jesus College in 1793. It was a post that enabled him to travel and to nurture his theories and which he retained until after his marriage to Harriet Eckersall in 1804. In the following year he became professor of history and political economy at the East India Company's college at Haileybury, Hertfordshire, the first time that political economy had been used for an academic office in Britain. Although he lived there quietly for the rest of his life, he publicized his views widely and became renowned in his lifetime, his later years being marked by numerous distinctions, including a fellowship of the Royal Society in 1819, a royal associateship of the Royal Society of Literature in 1824, and election to the French Académie des Sciences Morales et Politiques and the Royal Academy of Berlin in 1833. A good-natured, kindly and tolerant man to friends and foes alike, he is an unlikely figure to become one of the most controversial social scientists of all time.

His *First Essay on the Principle of Population* was written partly as a reaction to the unbridled optimism of his father, Daniel Malthus, who was an admirer of the eighteenth-century Enlightenment and a friend of contemporary philosophers of humanitarian optimism, including David Hume, Jean-Jacques Rousseau and Adam Smith. It was also written to prove the imperfectibility of mankind and to demonstrate the error of two recent utopian visions that extolled a future of reason, science, abundance, equality, peace and prosperity: *An Enquiry Concerning Political Justice* (1793), by the radical English political writer and novelist William Godwin, and *Esquisse d'un Tableau Historique des Progrès de l'Esprit Humain* (1795), by the French statesman and philosopher the Marquis de Condorcet.

Malthus was much more pessimistic, seeing poverty and misery as mankind's inescapable future. The contemporary reality in Britain at the end of the eighteenth century was a largely pre-industrial society in economic and social revolution, with acceleration of the enclosures, imports of corn replacing exports, the decline of rural industries and some rural depopulation, along with early industrial and urban growth, all at the same time as an increasing population. Among the traditional remedies for social troubles were hanging and the Poor Laws, which entitled the poor to increasing public assistance. Malthus did not envisage an egalitarian society and regarded poverty as inevitable, stressing that the poor had no claim by right to be given subsistence and that in helping them population would grow and suffering would be extended. His views have been oversimplified by critics, but in his opposition to the Poor Laws, he influenced social policy by claiming that they encouraged large families and limited the mobility of labour. Workhouses should be established for the most unfortunate, but they should not be 'comfortable asylums'. Although Malthus had humanitarian motives, his views on society were rigid, and it is not surprising that he was regarded by utopians like Godwin as a hard-hearted conservative and a prophet of misery and gloom.

Living at a time when the need for subsistence was the most pressing, Malthus was very concerned with the constant tendency among human beings to increase their species more than the amount of food available to them. Using the youthful populations of the American colonies and United States as an example, he asserted that population could grow geometrically by doubling every twenty-five years, whereas food production can at best only increase arithmetically during this period. In his view, the disparity between the two rates of growth would act as a brake on unlimited population growth, but instead of the natural 'positive checks' of high mortality caused by poverty, disease, wars, famines and 'excesses of all sorts', he wanted to substitute a voluntary mechanism of 'preventive checks'. However, he regarded 'self-restraint' as the only acceptable check, rejecting others such as adultery, prostitution, sexual deviation, birth control and abortion as 'vicious customs with respect to women'. Voluntary limitation of births was rare at the time, and he believed in celibacy and chastity until a person was able to accept the responsibilities of marriage, but after delayed marriage he thought that a family of six was normal. In later editions he recognized that social or preventive checks were more important than he had earlier realized, although he did not believe that they would reduce reproductivity sufficiently that positive checks would not operate more or less continuously.

Malthus continued to publish a variety of tracts and pamphlets on economics, but his only other major published work was *Principles of Political Economy Considered with a View to Their Practical Application* (1820). It was much less influential than *Essay on the Principle of Population*, but demonstrated his desire to formulate rigorous economic propositions. Among these were advocacy of public works and private luxury investment to counter economic distress, and criticism of saving as a virtue, which 'pushed to excess, would destroy the motive to production', stating that to maximize wealth a nation had to balance 'the power to produce and the will to consume'. By encouraging low wages and discouraging charity, it put a brake on economic optimism.

Malthus has been called 'a prophet of the past'. Like other classical economists, he did not foresee the impressive power of technology to influence both production and population dynamics. In these circumstances it is surprising that his name has persisted in the public eye for so long, though less surprising that his works have inspired such deeply conflicting intellectual, religious and public reactions throughout the eighteenth and nineteenth centuries.

Other British classical economists, including David Ricardo, a close friend, and John Stuart Mill, acknowledged Malthusian population theory as important to their economic theories of wages and of the stationary state. Much later they were joined by John Maynard Keynes. Malthus also had a catalytic effect on Charles Darwin's studies of evolution, by the idea that a surplus of population would be compensated by excess mortality of the least fit, and by making him realize that the struggle for existence mainly occurs within species. More ironically, the nineteenth-century British radical Francis Place and fellow 'birth controllers' adopted Malthusian theory to publicize birth control methods as a check on population growth, and called the doctrine Neo-Malthusianism despite Malthus's adamant opposition to contraception. Many Western Protestant countries, where individualism was strong and fertility high, later adopted anti-natalist policies usually for social rather than demographic aims, with the result that the name of Malthus has been more associated with family planning than delayed marriage, which became largely redundant as a population policy.

The basic message of Malthus – that production will be outrun by reproduction – saw a considerable resurgence in the second half of the twentieth century especially in English-speaking countries, including those of south Asia. Increasing public concern about acceleration of world population growth, widespread poverty in many countries of the so-called Third World, excessive use of finite resources, and human impact on global environmental change resurrected the concept of 'limits to growth', epitomized for example in reports for the Club of Rome on the predicament of mankind. Although Malthus was no environmentalist, this was regarded as a Malthusian concern, and helped to encourage the wide diffusion of family planning around the world, assisted by rising female status and a growing desire for smaller families. The terms Malthusian and Malthusianism have been so popularized and garbled that their meanings now often owe little to the original works of Malthus; they are sometimes used to indicate little more than either a pessimistic and gloomy view of the relationships between population and resources or even just the advocacy of family planning to solve economic problems.

Anti-Malthusianism has always abounded, academically as well as socio-politically, but expressed most strongly by socialists, Marxists and the Catholic Church, who have all emphasized the advantages of population growth. Socialist writers of the nineteenth century unanimously attacked the morals of Malthusian theory, regarding him as cruelly unfeeling and the incarnation of Manchester individualism. Equally, Malthus was an anathema to Karl Marx and Friedrich Engels, as he justified the persistent impoverishment of the poor. Marx attacked him vehemently because he believed that overpopulation did not result from an overall lack of the means of subsistence but from their maldistribution in society, a view which has gained credibility as the gap between the 'haves' and 'have-nots' has grown. During the first half of the twentieth century, Anti-Malthusian populationist policies were adopted enthusiastically by the communist countries of USSR and China as well as by Italy and Germany under fascist, militaristic dictatorships, but all were later abandoned and now all experience low fertility. And despite the prolonged antagonism of the Vatican to Neo-Malthusianism, some of the Catholic countries of southern Europe have the lowest fertility rates in the world. For or against, Malthus lives on as a major influence on thinking in many aspects of social science.

See also in this book

Darwin, Ehrlich, Marx, Wilson

Malthus's major writings

Wrigley, E.A. and Souden, D. (eds), *The Works of Thomas Robert Malthus*, 8 vols, London: William Pickering, 1986.

Further reading

Coleman, D. and Schofield, R. (eds), *The State of Population Theory: Forward from Malthus*, Oxford: Basil Blackwell, 1986.
Dupâquier, J., Fauve-Chamoux, A. and Grebenik, E. (eds), *Malthus Past and Present*, London and New York: Academic Press, 1983.
Hollander, S., *The Economics of Thomas Robert Malthus*, Toronto: University of Toronto Press, 1997.
Petersen, W., *Malthus: Founder of Modern Demography*, New Brunswick, NJ: Transaction Publishers, 1998.
Teitelbaum, M.S. and Winter, J.M. (eds), *Population and Resources in Western Intellectual Traditions*, Population and Development Review, A Supplement to vol. 14, New York: The Population Council, 1988.

JOHN I. CLARKE

WILLIAM WORDSWORTH 1770–1850

Nature never did betray / The heart that loved her.[1]

The name William Wordsworth is almost synonymous with 'nature poet' (and with the landscape of the English Lake District); paradoxically, Wordsworth is also the 'poet of the self' (of the inner landscape). Indeed, when Wordsworth writes, 'Nature never did betray / The heart that loved her', we see him draw together his sense of external nature both as a ministering agent, one ministering 'to' the self, and as a patient recipient of the responses of the 'heart', receiving 'from' the *inner* landscape of the 'self' the promise of both their futures.[2] Here is not the science but the experience of ecology.

Wordsworth's external and internal 'natures', while literally as old as the hills (and the Lakes of his native District), were startlingly new and paradoxical ones too. His reinvention of ancient nature worship or pantheism, for example, was both a challenge to and easily reconcilable with Christian humanism, Enlightenment individualism, the heady power and energy of the industrial age, and rural Toryism.

Wordsworth was born in Cockermouth in West Cumberland, just outside the English Lake District. He grew up in the Lake District in Hawkshead near Esthwaite Lake; attended St John's College, Cambridge (1787–91); spent time in France during the early part of the French Revolution; came back to England and endured an emotional crisis of some five years' duration, precipitated by severed personal relationships, confused national loyalties, and a growing disillusionment with the progress of the French Revolution; lived in Racedown, Dorsetshire with his sister Dorothy, whose own mind and writing reveal a startlingly original though usually neglected contribution to environmental thought; and then moved with Dorothy to Alfoxden, Somersetshire (1797), to be near their new friend S.T. Coleridge. There, according to one traditional account, Wordsworth recovered, in his growing sense of a personal relationship to the natural rhythms and agency of the pastoral Somersetshire landscape, his sense of purpose.

Donald Worster writes, 'The Romantic approach to nature was fundamentally ecological; that is, it was concerned with relation, interdependence and holism.'[3] For Wordsworth, these three concepts are as much psychological as ecological, a key correspondence in Wordsworth's most significant contribution to environmental thought: his steps to an ecology of mind and feeling. Indeed, in Somersetshire, his sense of the organic wholeness of *nature* appears to have grown out of his sense of a need for *personal* wholeness (whole = hale = health). However, as some critics suggest, Wordsworth's recovery was rather an escape from awkward political and personal responsibilities than an affirmation of an intrinsic wholeness in nature itself.[4] Nonetheless, rather than undermining the centrality of Wordsworth to modern environmental philosophy, such controversy has served to keep him at its centre.

In 1798, Wordsworth and Coleridge published *Lyrical Ballads*. To speak boldly, this book instituted a Copernican-like shift in poetry and in how we think about the relationship of our inner nature to (our?) outer nature. Copernicus replaced the geocentric (and human-centred) model of the solar system with a heliocentric model. While no such absolute shift is made in *Lyrical Ballads*, Wordsworth and Coleridge seek in their early poetry to replace the anthropocentric model of experience with what today we would call a biocentric one: indeed, in this new view, 'experience' is a general biological category not just a human one. In 'Lines Written in Early Spring', from *Lyrical Ballads*, Wordsworth writes,

> The budding twigs spread out their fan
> To catch the breezy air;
> And I must think, do all I can,
> That there is pleasure there.   4
> If I these thoughts may not prevent,
> If such be of my creed the plan,
> Have I not reason to lament
> What man has made of man?   8

Here, in a key biocentric image, 'the twigs' experience pleasure! This is, of course, a far cry from the mechanistic view of René Descartes (1596–1650), who believed that animal cries are merely the organic equivalent of the squeaking gears of machines. However, even for Wordsworth, separating himself from Descartes' belief in the essential separation of matter and spirit (of 'pleasure' from 'twigs') is no easy task. When Wordsworth writes of the 'twigs' that he '*must* think' (emphasis added [line 3]) '[t]hat there is pleasure there' (line 4), such a conclusion, he tells us in the same poem, is only after he does all he can ('do all I can' [line 3]) to prevent such an irrational thought. In dramatizing his own struggle to accept the biocentric view of experience, in using the words '*must* think', Wordsworth implies that his thoughts are somehow beyond his control. In philosophical terms, he dramatizes his discovery that his thoughts are not, as in the Cartesian tradition, self-evident or immediately knowable. For Wordsworth, the mind is not fully present to itself but is always only to be understood as an encounter with the living agency of nature, an agency that Wordsworth later in *Lyrical Ballads* calls 'One impulse from a vernal wood'. As Charles S. Peirce (1839–1914) asserts, 'that every thought is an external sign, proves that man is an external sign'.[5] Wordsworth's locating the agency of his own thoughts in part outside himself (that is, within his environment) represents a displacement of consciousness from the presumed internal locus of the rational Cartesian mind.[6] Here is a key sense in which *Lyrical Ballads* represents a Copernican-like displacement.

Later, in Book IX of the long philosophical poem *The Excursion* (1814), Wordsworth presents us with a more developed image of his 'environmental mind'. The existence of such a mind entails something more than merely thinking *about* the environment. Wordsworth writes,

> Whate'er exists hath properties that spread
> Beyond itself, communicating good,
> A simple blessing, or with evil mixed;
> Spirit that knows no insulated spot,
> No chasm, no solitude; from link to link,   5
> It circulates, the Soul of all the worlds.
> This is the freedom of the universe;
> Unfolded still the more, more visible,
> The more we know; and yet is reverenced least,
> And least respected in the human Mind,   10
> Its most apparent home.

While the 'human Mind' is a key node (or 'home' [line 11]) in this great web of being, 'being' always spreads beyond itself: 'I think; therefore, *you* are (or he, she, or it is)'. For Wordsworth, as for present-day ecologists and semioticians, a thing, a person, or an idea is always, in addition to itself, something other than or supplementary to itself. Therefore, as environmental scientist Garrett Hardin wrote in 1973: 'We can never do merely one thing'.[7] Note in Wordsworth's passage above that the 'Spirit' (line 4) that circulates 'from link to link' (line 5), while not energy flowing through the links of a food chain, represents a spiritual recycling along a food chain of signification (or meaning). Here, then, Wordsworth offers us a precise psychological equivalent of a modern ecological process. Indeed, he offers us the *psychology or experience* of ecology before the *science* of ecology.

Today we understand that women, minorities and children suffer disproportionately from environmental pollution and other environmental degradations. It is thus no coincidence that the poems in *Lyrical Ballads* are not only about the tenets of an emerging 'environmental' manner of knowing or being but about female vagrants, displaced pastoralists, mad women, cold and hungry people, and even an 'Idiot Boy', in other words, the dispossessed and the voiceless. Another great central insight dramatized by *Lyrical Ballads*, then, is that environmental and social issues are inseparably linked, and thus Wordsworth has further reason to lament 'what man has made of man'.

Wordsworth settled at Grasmere, in the Lake District, with Dorothy in 1799, a move that marked a permanent return to the region, and married Mary Hutchinson (1802). Wordsworth became Poet Laureate in 1843. His places of residence and the literary landscape that he created in the Lake District became tourist attractions – the man and the place are now understood as inseparable.

The central drama of Wordsworth's poetry is the mutual creation of Being through the reciprocal relationship between an active 'self' and an active 'nature', a relationship Wordsworth calls 'interchange' in *The Prelude* (1805 [1933]; 1850):

> From Nature doth emotion come, and moods
> Of calmness equally are Nature's gift:
>
>   * * *
>
> Hence Genius, born to thrive by interchange
> Of peace and excitation, finds in her
> His best and purest friend.

Thus, nature is not just dead matter but a 'being', our 'best and purest friend', and, in other poems, a 'Power', a 'Presence', and a 'spirit': 'Touch – for there is spirit in the woods', writes Wordsworth in 'Nutting'.

Wordsworth, in 'Nutting' and elsewhere, takes what was for the ancient pantheist a spirit's individual embodiment in a particular object and transfers that individuality from the object itself to each human subject's (potential) individual response to that object. Wordsworth's great achievement, then, is to transform an outmoded pantheistic (spectator–spectacle) ontology of being into a modern (participant-observer) one: the spirit indwells in the mutually constituting *relationship* between nature and human beings – not in the trees themselves. This is a view that while finding some sympathy in Enlightenment sensibility anticipates twentieth-century phenomenology, the philosophy of experience.

In 'Nutting', following a boy's 'savage treatment' of a 'shady nook of hazels', those trees 'patiently g[i]ve up / Their quiet being'. But as the boy says, 'Ere from the mutilated bower I turned', 'I felt a sense of pain when I beheld / The silent trees'. For Wordsworth, as for the present-day phenomenologist Drew Leder, 'the universal or the "spiritual" need not be conceived of as something opposed to the flesh and blood. The body itself proclaims spirit in our lives, that is, transcendence, mystery, and interconnection.'[8] The body of the boy in 'Nutting' makes this proclamation. His 'pain' draws his attention to what was his own previously 'absent' body: when we do *not* hurt, our bodies often are in the background of our awareness. (Indeed, as Aldo Leopold later demonstrated, the hurts or 'wounds' of the body of the natural world are for many people below the threshold of their awareness.) The boy's new awareness of his own body (emerging out of the background of his self) parallels his awareness of the bodies of the trees (emerging out of the background of nature); these trees, once only a romantic 'nook' or 'bower', have become individuals. One body (the inner body of the boy) 'calls out' the other (the outer bodies of the trees), and vice versa. Importantly, the boy's inner body, his viscera, the gut from whence his pain comes, is as much a mystery to him and as outside his own control as the life forces of the 'body' of the external natural world, the 'shady nook of hazels'.[9] Wordsworth's great achievement here is that two awarenesses (the two bodies, the two mysteries) become one for the boy. Thus the boy himself concludes the poem as if nature had a 'body' sensitive to touch: 'Then, dearest maiden, move along these shades / In gentleness of heart; with gentle hands / Touch'. Wordsworth helps us to em*body* the earth in our experience of it.

In the context of intellectual history, Wordsworth dramatizes in 'Nutting' and elsewhere what French phenomenologist Maurice Merleau-Ponty (1908–61) codifies more than a century later. As David Abram tells us, 'Merleau-Ponty sensed (1) that there was a unity to the visible–invisible world that had not yet been described in philosophy, that there was a unique ontological structure, a topology of Being that was waiting to be realized, and (2) that whatever this unrealized Being is, we are in its depths, and of it, like a fish in the sea, and that therefore it must be disclosed from *inside*'[10] – which in fact it was for the boy of 'Nutting'. From this perspective, to be put off by Wordsworth's 'egotistical sublime', as his poetic and personal (supposedly self-centred) orientation to outer nature was called in his time, is to fail to understand a great insight of Wordsworth's: environment cannot be conceived of as distinct from a unique individual and the uniqueness (and the unity and diversity) of that environment is only revealed through a parallel revelation of the uniqueness (and the unity and diversity) of the individual.

In today's parlance, Wordsworth's central interest in nature as agent and in the mutually constitutive or reciprocal relationship between the 'self' and 'nature' both anticipates and parallels the new interest on the part of nineteenth-century natural scientists in the analysis of reciprocal relations between 'organism' and 'environment'. For example, in the theory of ecological succession, plants are understood both to be fitted to particular environments and ultimately to change those environments so as to make them, ironically, more hospitable to the plants' competitors of the next generation. Living things are agents of environmental change – not merely passive objects. In ecological succession, too, all's well that ends well, as the process culminates in the creation of a climax or mature natural community: a place (a forest or field) of stability, protection, and quality – measured in part in the diversity of mutually interdependent plant and animal species and homes (habitats).[11] Wordsworth, who invented the psychology of ecology before the invention of ecological science, writes similarly, in 'Tintern Abbey' (from *Lyrical Ballads*) and in *The Prelude*, about the growth (or personal succession) of his mind from the 'glad animal movements' of youth to the 'Abundant recompense' of a more thoughtful maturity in which 'the still, sad music of humanity' becomes Wordsworth's image for his renewed love of human nature when that nature is seen as a part of (not apart from) natural life.

Again, in terms that represent both a psychology *and* biology of 'becoming' (succession) over 'being', Wordsworth writes, 'Praise to the end! Thanks to the means which Nature deigned to employ' in his mind's development. Continuing in the tradition of a Wordsworthian or Romantic ecology, American poet Theodore Roethke (1908–63) picks up on Wordsworth's 'Praise to the end!' (exclamation point and all) for the title of a volume of poetry (1951) and for the eponymous poem in that volume, one that develops the implications of Wordsworth's ecology of mind (of his concern with the growth or succession of a poet's mind from stage to stage, in part through its reciprocity with an active nature) within the Darwinian tradition, 'the end!' being for Roethke and Darwin the goal-directedness (though not necessarily goal intention) of evolution. For the Wordsworthian Roethke, then, we not only succeed to the climax of our own mature self from youth (as for Wordsworth) but we see our individual lives as a climactic recapitulation of the youth of the world – which is our own. As Wordsworth writes about this active agent 'nature' and its goal-directedness, 'nature' is 'the nurse, / The guide, the guardian of my heart, and soul' ('Tintern Abbey'). Also, Wordsworth's continuous personal re-creation and its emphasis on 'becoming' parallels and then presages Lamarck's and Darwin's models of continuous creation at the species level.

Wordsworth not only envisions a *personal* climax or maturity in his steps to an ecology of mind. He also envisions a *community* climax, a mature community of minds. In his pastoral poem 'Michael' and in his *A Guide Through the District of the Lakes* (1835, 5th edn), he describes the natural and cultural histories of the vale and the people of Grasmere. Wordsworth desired to preserve, as 'a sort of national property', the mature, interdependent natural and human communities of the 'Lake District', what he calls 'a perfect equality, a community of shepherds and agriculturalists'. This community, the product of a long succession of generations (as Wordsworth also details in the *Guide*), reflects in its *social* organization the old growth or climax values of stability, protection, and quality, aspects of its *ecological* organization. In a kind of paradigm for tensions in political ecology today, these ecological or conservation-oriented values may also be seen as conservative and élitist. Tim Fulford, for example, recently asks: '[C]an we derive a political lesson about the importance of ecological consciousness from a Wordsworth whose rural Toryism is included in the account?'[12] Fulford refers in part to Wordsworth's desire – expressed in his *Guide* and in two letters to the Editor of the *Morning Post* (1844–5) – to protect (and preserve) the Lake District from vacationing industrial workers and the certain commercialization that would follow. This desire may be seen either as a significant anticipation of the British National Trust and Park System, as Jonathan Bate argues,[13] or as a selfish élitism, a charge often levelled at upper-(middle-)class environmentalists today. More broadly, Wordsworth's politics of nature in the nineteenth century raise the important question for the twenty-first century of the extent to which political 'conservativism' and environmental 'conservation' (etymologically rooted as they are) are or should be ideologically aligned.

Wordsworth has recently been 'upgrade[d]' from nature poet to 'proto-ecologist'.[14] However, in the *Guide*, Wordsworth appears more the ecologist than the proto-ecologist. Wordsworth speaks of 'plants' that are fashioned 'by those that have preceded them', and of a 'tree' that is 'compelled to conform itself to some law imposed upon it by its neighbors' – startling anticipations of the theoretical dimensions of ecological succession in that living things themselves – not just their environments – set their own limits for and act as their own agents of change. In conclusion, Wordsworth is the poet of the ecology of mind because he understood something of real ecology too.

Notes

1 'Tintern Abbey', lines 122–3.
2 See Michael Polanyi's 'From-to' structure as described in Drew Leder, *The Absent Body*, Chicago, IL: University of Chicago Press, pp. 15–17, 1990.
3 *Nature's Economy: A History of Ecological Ideas*, 2nd edn, Cambridge: Cambridge University Press, p. 58, 1994.
4 See, for example, Jerome McGann, 'The Anachronism of George Crabbe', in *The Beauty of Inflections*, Oxford: Oxford University Press, pp. 310–11, 1985.
5 'Some Consequences of Four Incapacities', in *The Essential Peirce*, Indianapolis, IN: Indiana University Press, p. 54, 1992.
6 Elizabeth Fay argues, in ways that again help place Wordsworth at the centre of contemporary debates on the role of ecology (and nature generally) in the politics of the person and the state, that Wordsworth's poems perform the act of making their performance *appear* natural or purely descriptive. Much interest exists today about what is or isn't 'natural' about persons, governments and everything in-between. See *Becoming Wordsworthian*, Amherst, MA: University of Massachusetts Press, 1995.
7 *Exploring New Ethics for Survival*, London: Pelican, p. 38, 1973.
8 Leder, op. cit., p. 68.
9 Here I apply to Wordsworth another of Drew Leder's phenomenological insights.
10 'Merleau-Ponty and the Voice of the Earth', in *Minding Nature*, Guilford: The Guilford Press, pp. 98–9, 1996.
11 Eugene Odum, in 'The Strategy of Ecosystem Development', *Science*, pp. 164, 262–70, 1969, first characterized old growth ecosystems according to these three terms.
12 'Wordsworth's "Yew-Trees": Politics, Ecology, and Imagination', *Romanticism*, 1 (2), p. 273.
13 *Romantic Ecology*, London: Routledge, pp. 10, 47ff, 1991.
14 Peter Coates, *Nature: Western Attitudes Since Ancient Times*, Berkeley, CA: University of California Press, p. 134, 1998.

See also in this book

Clare, Darwin, Leopold

Wordsworth's major writings

Descriptive Sketches, 1793; ed. Eric Birdsall and Paul M. Zall, Ithaca, NY: Cornell University Press, 1983.

Lyrical Ballads, 1798, with S.T. Coleridge; ed. W.J.B. Owen, Oxford: Oxford University Press, 1969.

Lyrical Ballads, with Preface, 1800.

The Excursion, 1814; reprint edn, London: Cassell, Woodstock Books, 1991.

Collected Works, 1815; see *The Complete Poetical Works of William Wordsworth*, London: Macmillan & Co., 1988, on-line edn, July, 1999, Bartleby.com.

A Guide Through the District of the Lakes, 5th edn, 1835; see W.J.B. Owen and Jane Worthington Smyser, *The Prose Works of William Wordsworth*, 3 vols, Oxford: Oxford University Press, 1974.

The Prelude; or Growth of a Poet's Mind, 1805, 1850, 1933; see, for example, Jonathan Wordsworth (ed.), *The Prelude: A Parallel Text*, London: Viking Press, 1996.

The standard scholarly edition is Ernest de Sélincourt (ed.), *The Poetical Works of William Wordsworth*, 5 vols, Oxford: Clarendon Press, 1958–65. The standard paperback is Stephen Gill and Duncan Wu (eds), *William Wordsworth*, Oxford: Oxford University Press, 1994. See also in paperback *William Wordsworth: Selected Poetry*, Nicholas Roe (ed.), London: Penguin, 1992.

Further reading

Buell, Lawrence, *The Environmental Imagination*, Cambridge, MA: Harvard University Press, 1995.

Harrison, Robert Pogue, *Forests: The Shadow of Civilization*, Chicago, IL: Chicago University Press, 1992.

Kroeber, Karl, *Ecological Literary Criticism*, New York: Columbia University Press, 1994.

Lacey, Norman, *Wordsworth's View of Nature*, Hamden, CT: Archon, 1965.

Meeker, Joseph W., *The Comedy of Survival: Studies in Literary Ecology*, New York: Scribners, 1972.

W. JOHN COLETTA

JOHN CLARE 1793–1864

[F]ields were the essence of the song[1]

John Clare, the self-styled 'Northamptonshire Peasant Poet', was a poet of the 'fields' in more ways than one: he himself laboured in the fields; he wrote of the life of field hands; he was a superb field naturalist; he lived through and lamented the loss of the old sustainable open-field system of agriculture; he celebrated the ecology of fields, considered not only as sites of agricultural production but as habitats (homes) of mutually dependent plants and animals; and he was, like Wordsworth, but in a much more explicitly ecological way, a great poet of what may be called phenomenological ecology: the study of fields of experience. That is, rather than the study of resources (plants, animals, and minerals considered with respect to their use value) distributed throughout 'space', phenomenological ecology is the study of 'lived' relationships (i.e. experience) considered with respect to a specific 'place'.

The classical definition of ecology is the study of the relationships between living things and their environments. In his poem 'Shadows of Taste', written before the science of ecology was codified and even before the word 'ecology' was coined, Clare provides us with a rhymed couplet that anticipates this definition while giving it a wider experiential dimension. Clare writes: 'Associations sweet each object breeds / And fine ideas upon fancy feeds'. This is to say that the ecological web of life (the 'associations' or 'relationships' bred between things or objects) cannot be separated from the phenomenological web of *being* (the perceptual and conceptual feeding of 'fine ideas' upon 'fancy', a 'fancy' that itself feeds upon the associations bred by natural objects in *a food chain or web of signification*). For Clare, all objects of all thought, then, are (re)charged with a significance beyond that of mere use; all objects have being; the objective is itself subjective. As Clare also writes in the same poem, 'Flowers in the wisdom of creative choice / Seem blest with feeling and a silent voice'. Such natural objects are subjects because they have 'feeling' and 'voice'. As *subjective* ecological objects, 'birds and flowers and insects' '[a]ll choose for joy in a peculiar way': in their ability to 'choose', they also have agency. Furthermore, biological subject–objects, unlike the passive regularities of objects in Newtonian physics, are 'peculiar'; that is, they are individuals.

In contradistinction to Albert Einstein's search for a 'universal field theory' of the space–time continuum, John Clare's 'ecological field theory' of the *place*–time continuum and its great web of being was local or situated (rather than universal) and embodied (rather than abstract) – peculiar rather than regular. Though merely thought parochial in Clare's time, such a 'situated' and 'embodied' perspective now plays a key role in the work of important contemporary historians and philosophers of science such as Donna Haraway, who seeks to replace 'relativism' (again an echo of

Einstein) with 'location', substituting 'local knowledges' for 'world system' and 'webbed accounts' for 'master theory'.[2]

Clare's poems (and the 'webbed accounts' and 'local knowledges' they embody) represent an explicit response to the following questions drawn from phenomenology (the 'ecology' of experience). We can perceive individual blades of grass, but can we perceive (with just our senses) a field, if by 'field' we mean not a congeries of things but a series of relationships, a living community involved in a mutually sustainable process of self-regulation? The answer is 'no': 'relationship' is not a sensory phenomenon. However, through the mediation of culture, ecological communities (such as a field) may be experienced (if not directly sensed). Not all cultures, though, provide a mode of *sustainable* (or ecological) experience. Therefore, what would the songs and stories of such a sustainable culture be like? What visual images are more sustainable than others? How would songs, stories, and imagery function to provide the feedback necessary to any self-regulatory, sustainable community, constituted by both the human and the non-human? Such are the situated (local) or embodied (ecological) questions that readers today may profitably ask of Clare's poetry and prose.

Like the socio-economic status of the local places that Clare defends in his verse, Clare's place in the field of English literature has, until recently, been marginal. Today, however, Clare is considered the 'finest poet of Britain's minor naturalists and the finest naturalist of all Britain's major poets';[3] the 'first true ecological writer in the English-speaking world'.[4]

Ecologist Paul Sears writes that ecology is a 'subversive subject'.[5] Natural history, for Clare, could be subversive not only because it could serve to describe healthy natural communities that would themselves serve as benchmarks against which to measure environmental devastation; natural history could also help reveal the inseparability of environmental and human concerns. As James McKusick writes: 'Clare is virtually unprecedented in the extent of his insight into the complex relation between ecological devastation and social injustice'.[6] Indeed, consider the following two lines from the poem 'Remembrances', lines that illustrate how Clare's 'eco*logical*' argument ('eco*logical*' because it sees interdependence between premises and terms that an earlier logic overlooked) subverts conventional distinctions by suggesting relationships among categories that in the nineteenth century would have been thought to belong to separate spheres, viz., natural history (ecology), religion, agricultural policy, and continental history and imperialism. Clare laments the devastation of a place he had known and loved, 'old round oaks narrow lane':

> ... its hollow trees like pulpits I shall never see again
> Inclosure like a buonaparte let not a thing remain.

Clare's 'hollow trees', also called den trees today, serve as homes for several species of living things. Foresters today use the number of hollow trees per acre to indicate the status of a woodland's health. Such trees are therefore also called ecological indicators. Anticipating such an indexical function for hollow trees, John Clare, in simultaneously ecological and religious terms, compares 'hollow trees' to 'pulpits', implying that such trees are sites that proclaim (or give indication of, as would a preacher from a pulpit) the status of both our spiritual and ecological health. But, Clare tells us, such trees are threatened by the politics of parliamentary enclosure, a socio-economic process of privatizing (enclosing or fencing off) the old open-fields and of industrializing the means of agricultural production. Significantly, Clare likens his local experience of parliamentary enclosure to the imperial politics of Napoleon Bonaparte. Indeed, Clare is one of the first to recognize the interdependent relationship between colonial or imperial *politics* (symbolized by Napoleon) and colonial or imperial *biologies* (symbolized by parliamentary enclosure's effect on the 'hollow trees' of Clare's 'round oaks narrow lane'). Clare also recognizes here the interdependence between *local* and *global* (or at least continental) processes. Napoleon's destruction of life on a continental scale in Europe is related to the destruction on a local scale (in and around Clare's home village of Helpstone) of both ecological *habitats* and local social *habits* (the customs in common – 'common' understood as a relationship and a place, the commons or open-fields). Plants and animals are reduced by biological imperialism to mere commodities and elevated (as in Kew Gardens or the *Jardin du Roi*) to signs of national identity.

Tim Fulford points out that Clare's poem 'The Fallen Elm' is unique in how it 'develops a discourse of political protest from a personal response to a local landscape'.[7] Even though Clare is in many ways not part of the English Romantic literary tradition to which he belongs by date of birth, in his use here and elsewhere of *personal experience* (the foundation of being and knowing for English Romantics) as a basis for *political protest* we find the origins of a Romantic style of ecological politics. For example, Clare takes the Romantic notion of the supremacy of the 'individual' (a notion criticized by some for having emerged with and being necessary to those less desirable aspects of capitalism) and uses it to make readers aware that biotic communities are individuals too. In a poem such as 'The Lament of Swordy Well', Clare has 'Swordy Well' (a once complex biotic community that has had its ecological capital nearly spent) speak for itself as an individual. In giving a voice and a face to 'Swordy Well', Clare succeeds at least aesthetically in claiming for biotic communities the moral standing that in Clare's (and even in our own) time has only been thought due to individual human beings. As James McKusick writes, 'Clare is certainly among the first to suggest that the earth itself should have the legal right to redress of environmental grievance.'[8] Not until some 150 years later, in 1972, does a law professor, Christopher D. Stone, begin to chart the legal path towards rights for natural objects.[9]

Clare also establishes 'poverty' as an environmental category or condition – not just an economic one: Robert Pogue Harrison writes that the last stanza of 'The Lament of Swordy Well' provides 'an ominous ending, for it gives the condition of poverty a broad, almost universal extension to nature as a whole'; poverty for Clare meant 'the state of defenselessness against the forces of assault and expropriation. It did not mean destitution, at least not intrinsically.'[10] Clare, therefore, makes the vulnerability of nature natural, a real possibility (in a time when extinction, for example, was still a categorical impossibility within the stability of the Natural Theological world-view). He also anticipates the philosophical basis for what today is called the voluntary simplicity movement, poverty not as destitution but as a sustainable personal alliance with the land.

John Clare was born in the village of Helpstone, which in 1793 was in Northamptonshire, England. He was largely self-educated and, until his declining physical and mental health no longer permitted it, worked for some twenty-five years as a ploughboy, a gardener, an inn keeper's helper, and a lime-burner. He also served as a reservist in the local militia. 'When he was thirteen', writes R.K.R. Thornton, '[Clare] was set on his path to be a poet by discovering Thomson's *Seasons*, which inspired him to write down his first poem.' 'His poems accumulated and, through contact with a local bookseller, he succeeded in having a book of poems published by Keats' publisher, John Taylor.'[11] This first book ran to four editions, receiving, as one critic writes, 'considerable though condescending acclaim'. However, his last three books did poorly. Indeed,

> Clare was doubly damned from the beginning – damned because he was associated with one locality at a time when the railways were breaking down regional boundaries and regional consciousness; and damned because he was a peasant at a time when the national imagination was being captured by the immensity of industrialism ... Clare was not only loyal to the countryside, he was part of it ...[12]

Today, in academic culture, we are constantly made aware of the force of Clare's 'sin' of being a 'part of' (rather than disinterestedly 'apart from') some one place. A mere regional poet is of relatively little value compared to a poet of universal appeal (but of no place). To present a paper at a regional (or, heaven forbid, a local) conference has little value for academic advancement. Indeed, in academic culture's aversion to the local we find the enshrinement (even in the 'incorrigible gap' between culture and nature posited by postmodernism) of the earlier class-based condescension that kept the 'rustic' Clare 'in his place' (and out of place in high culture) – while, ironically, enclosure and scientific agriculture served to remove that place from him. To read Clare well today, then, is to seek to reclaim an aesthetic and an ethic that reconciles the local and the global as well as culture and nature.

Clare's father was a thresher, whose reading was little but whose knowledge of the vanishing oral folk traditions of song (and work), of story (and field), was great. His mother, the daughter of the town shepherd of Castor, was another source of that oral folk tradition that had been an integral part of the long-sustained common- or open-field agricultural systems of medieval England. Clare writes, 'I am now Rhyming some of my Mother's "old stories" as she calls 'em they are Local Legends Perhaps only known in these Places'.[13] Here, significantly, in a language that is simultaneously ecological, literary, and historical, Clare speaks of the (ecological) habitats of 'Local Legends', of stories as inseparable from 'Places' as species are from their environments. Indeed, in the loss Clare was to experience of the cultural diversity of his own *folk* tradition is perhaps the origin of the emotional depth of his original response to the loss of diversity in the *biological* tradition.

As George Deacon writes, Clare was the 'earliest collector of the songs people actually sang in Southern England'.[14] Furthermore, Clare's own literary ballads show evidence of his desire not only to commemorate that oral tradition but to adapt it to what for Clare could only be an uncertain future community beyond his imagining. Another key question, therefore, that emerges from reading Clare today is this: If not for Clare in his time, is it possible for readers of Clare in our time to recover or to reinvent the lost ecological ethic and aesthetic once embodied in the folk song and ritual of Clare's rural Northamptonshire (agri)cultural tradition, a tradition that contemporary scientific 'narratives' such as Garrett Hardin's highly influential 'Tragedy of the Commons' erase or efface?[15]

Hardin argues that a 'commons', his metaphor for any ecosystem – a lake, estuary, grassland, or even ocean or atmosphere – subject to communal or unregulated use, is at risk of a tragic ecological collapse because of a virtual law of human behaviour. Consider a grassy commons used by several families of herders. Each herder will generally find it to his or her economic advantage, when the possibility arises, to add one cow to his or her herd – and thus to the commons. In the short term, the degradation of the commons will not be great, and the loss of profit that results from this general but moderate degradation – a degradation that itself resulted from the combined independent decisions of the herders – will be shared (and *experienced* therefore in 'diluted' form) by all. However, each individual herder who decided to add one cow will reap all of the *economic* gain from that cow. Of course, according to Hardin's model, in the middle or long term the ecological and economic viability of the commons will collapse. Here then is Hardin's insidious tragedy of the commons. Hardin's atomistic view, however, assumes the operation of self-interest only; it assumes that there are no community feedback mechanisms for assessing the condition of the commons and acting upon those assessments. For Hardin, the cows may feed but the herdsmen give no feedback. Clare's poems, however, are the voice (the ecological feedback mechanism) of the herdsmen – and of the other labourers whose voices parliamentary enclosure disrupted and Hardin never heard.

In the poem 'The Wild Bull', Clare begins:

Upon the common in a motely plight
Horses & cows claim equal common right
Who in their freedom learn mischiveous ways
& driveth boys who thither nesting stray ...   4
& school boys leave their path in vain to find
A nest – when quickly on the threatening wind
The noisey bull lets terror out of doors
To chase intruders from the cows lap [cowslip] moores.

Here, then, is a 'story' that makes the commons a place worth preserving. Clare describes the interdependent community of the commons as a self-regulating one – one that keeps 'intruders' from despoiling the nests of those birds whose habitat it is. Clare's strategy here, then, transforms the biological principle of self-regulation into the political one of self-sufficiency, which political principle is itself echoed by the 'claim' the 'horses and cows' make for 'equal common right' (line 2) and 'freedom' (line 3). The commons, then, is a place of freedom.

But here lies a terrible political irony, as we see in Clare's 'The Fallen Elm'. In this poem, Clare shows his sophistication as a writer of environmental polemic, when he writes about one of his favourite trees, felled as part of the new economics of enclosure. Speaking to the memory of the tree, Clare writes,

> Self interest saw thee stand in freedom's ways
> So thy old shadow must a tyrant be.

Here Clare shows his insight into the fact that all landscapes (even the trees in them) under enclosure's imperial gaze must themselves be made to seem tyrannical so as to justify their despoilation, ironically, in the name of *free* enterprise. But as Clare shows in 'The Wild Bull', the land is always already a free enterprise. As Robert Pogue Harrison writes: 'In an age that rallied around the cry of "freedom", that conceived of freedom as a liberation, ... in short, as a freedom *from* – in such an age, then, Clare located freedom elsewhere: in what already existed in its own right'.[16]

John Clare is important to the history of environmental thinking in at least two additional ways. His natural history poems dramatize what the twentieth-century ecologist Eugene Odum describes as the 'values' of 'old growth' ecological communities, their tendency to optimize protection, stability, and quality over production, change, and quantity.[17] For example, consider the following passage from Clare's 'The Robins Nest':

> ... each ancient tree
> With lichens deckt – time's hoary pedigree
> Becomes a monitor to teach and bless ...

* * *

Where old neglect lives patron and befriends
Their [the birds'] homes with safetys wildness – where nought
   lends
A hand to injure

We see in Clare's use of 'safetys wildness', 'ancient' and 'time's hoary pedigree' Odum's old growth values of 'protection' ('safetys wildness'), 'stability' ('ancient'), and 'quality' ('pedigree').

Clare's natural history poetry also dramatizes the operation of natural systems in what we might today call post-modern terms: these systems are ironic agents. For Clare, natural systems are sites of resistance to the closure of science or to any other form of institutionalized thought. In 'Shadows of Taste', Clare writes of the resistance to the taxonomic scientist on the part of insects, who 'e[v]en grow nameless mid their many names'. In 'May', from *The Shepherd's Calendar*, Clare writes about a ventriloqual bird of the grasslands, a rail, that resists a swain's (a country lad's) and a schoolboy's attempts to locate it even in the most regular terrain:

   ... in the grass the rails odd call
That featherd spirit stops the swain
To listen to his note again
& school boy still in vain retraces
The secrets of his hiding places

The ventriloqual voice of the rail is a deferral or displacement of its identity, one that puts a stop to the boys' search for the rail's nest – the origin or centre of its environment. Similarly, contemporary virus hunters have most often been foiled in their search for the origin of newly emergent viruses, and, parallel to Clare's rail's deceptive strategy, some viruses, in another act of deferral, present themselves, in Trojan-horse fashion, to our immune systems as something other than harmful agents. (Indeed, many literary themes, including Homer's Trojan horse, have always already been biological.) Clare, then, anticipates at the ecosystem level what pathologists have only relatively recently discovered: the ironic agency of the non-human biological world.

Notes

1   Clare, 'The Progress of Rhyme', in Eric Robinson, David Powell and P.M.S. Dawson (eds), *Poems of the Middle Period 1822–1837*, vol. III, line 144.

2 Donna Haraway, 'Situated Knowledges', *Simians, Cyborgs, and Women*, New York: Routledge, p. 194, 1991.
3 James Fisher, quoted in Margaret Grainger (ed.), 'Introduction', *The Natural History Prose Writings of John Clare.*
4 James McKusick, ' "A language that is ever green": The Ecological Vision of John Clare', *University of Toronto Quarterly*, 61, 2 (Winter), pp. 226–49, p. 233, 1991.
5 Paul Sears, quoted in Donald Worster, *Nature's Economy: A History of Ecological Ideas*, 2nd edn, Cambridge: Cambridge University Press, p.23, 1994.
6 McKusick, op. cit., p. 239.
7 Tim Fulford, 'Cowper, Wordsworth, Clare: The Politics of Trees', *John Clare Society Journal*, 14 (July), p. 47, 1995. A special 'Clare and Ecology' issue.
8 McKusick, op. cit., p. 241.
9 *Should Trees Have Standing: Toward Legal Rights for Natural Objects*, special rev. edn, New York: Avon Books, 1975.
10 Robert Pogue Harrison, *Forests: The Shadow of Civilization*, Chicago, IL: University of Chicago Press, pp. 216, 213, 1992. Contains half a chapter on Clare and nature.
11 R.K.R. Thornton (ed.), 'Note on the Author and Editor', *John Clare*, Everyman's Poetry, London: J. M. Dent, p. v, 1997.
12 Eric Robinson and Geoffrey Summerfield (eds), 'Introduction', *Selected Poems and Prose of John Clare*, Oxford: Oxford University Press, p. xiv, 1978.
13 Eric Robinson, David Powell and P.M.S. Dawson, 'Introduction', *Cottage Tales*, by John Clare, Manchester: The Mid Northumberland Arts Group, Carcanet Press, p. xii, 1993.
14 *John Clare and the Folk Tradition*, London: Sinclair Browne, p. 18, 1983.
15 *Science*, 162, pp. 1243–8, 1968.
16 Harrison, op. cit., p. 219.
17 'The Strategy of Ecosystem Development', *Science*, 164, pp. 262–70, p. 265, 1969.

See also in this book

Wordsworth

Clare's major writings

Clare published, in his lifetime, *Poems Descriptive of Rural Life and Scenery* (1820), *The Village Minstrel, and Other Poems* (1821), *The Shepherd's Calendar, with Village Stories and Other Poems* (1827) and *The Rural Muse* (1835). Eight of the projected nine volumes of *The Complete Poems of John Clare* are now published.

The Early Poems of John Clare 1804–1822, ed. Eric Robinson, David Powell and Margaret Grainger, 2 vols, Oxford: Clarendon Press, 1989.

John Clare: Poems of the Middle Period 1822–1837, ed. Eric Robinson, David Powell and P.M.S. Dawson, 4 vols, Oxford: Clarendon Press, 1996, 1998.
The Later Poems of John Clare 1837–1864, ed. Eric Robinson, David Powell and Margaret Grainger, 2 vols, Oxford: Clarendon Press, 1984.

Also of central interest to Clare's environmental thinking is:

The Natural History Prose Writings of John Clare, ed. Margaret Grainger, Oxford: Clarendon Press, 1983.

A readily available paperback selection of Clare's poetry is *John Clare*, ed. R.K.R. Thornton, London: J.M. Dent, 1997, as part of their Everyman's Poetry series.

Further reading

Barrell, John, *The Idea of Landscape and the Sense of Place: An Approach to the Poetry of John Clare*, Cambridge: Cambridge University Press, 1972.
Coletta, W. John, 'Ecological Aesthetics and the Natural History Poetry of John Clare', *John Clare Society Journal*, 14 (July), pp. 29–46, 1995. A special issue subtitled 'Clare and Ecology'.
—— ' "Writing Larks": John Clare's Semiosis of Nature', *The Wordsworth Circle (TWC)*, XXVIII (3), pp. 192–200, 1997. A special issue subtitled 'Romanticism and Ecology'.
—— 'Literary Biosemiotics and the Postmodern Ecology of John Clare', *Semiotica*, 127 (1/4), pp. 239–72, 1999.
Helsinger, Elizabeth, 'Clare and the Place of the Peasant Poet', *Critical Inquiry*, 13, pp. 509–31, 1987.
McKusick, James C., 'John Clare's Version of Pastoral', *The Wordsworth Circle*, XXX (2) (Spring), pp. 80–4, 1999.
Studies in Romanticism, 35 (3) (Fall), 1996. Special issue on Romanticism and Ecology.
Williams, Raymond, *The Country and the City*, Oxford: Oxford University Press, 1975.

W. JOHN COLETTA

RALPH WALDO EMERSON 1803–82

Ralph Waldo Emerson once penned in his *Journals*, 'Right is a conformity to the laws of nature so far as they are known to the human mind',[1] against which we can set as a retort John Stuart Mill, 'Conformity to nature has no connection whatever with right and wrong'.[2] Mill is emphatic about humans and their achievements: 'All praise of Civilization, or Art, or Contrivance, is so much dispraise of Nature'.[3] Emerson demurs, with characteristic poetic vigour: 'In their vaunted works of Art, The master-stroke is still her part'.[4] The two met once, the transcendentalist sage of New England and the British logician framing the techniques of empirical science, contemporaries setting the contrasts of their times.

Seen now, a century and a half later, Emerson was launching an ecological view, 'harmony' with nature (we might say, rather than 'conformity'), lost as this has largely been during the flowering of humanism, science and technology, the liberal 'modernism' of whom Mill is an early type specimen. What now seems clear is that humans are nowhere near a sustainable relationship with their planet Earth, and that a radical separation, humans over nature, 'dispraising' it, has been as much part of the problem as part of the solution.

Emerson was reared in nineteenth-century New England, a promising Harvard graduate, one-time Unitarian minister. He became an iconoclast critic of his establishment. He delivered a controversial Harvard Divinity School address and was not invited back for thirty years. He gained fame from his literary essays, espousing a spiritual relationship to nature, intuitively known, ultimately an idealism of self-reliance residing in a deeply sacred world. His life was spent in Concord, outside Boston, a quiet domestic life in then rural Massachusetts, but adjacent to the Boston centres of intellectual life. Over time the novelty of his views accommodated somewhat to society; society accommodated some-what to him. Along with Henry David Thoreau, Emerson was entered among the worthy geniuses of the traditions of which he had been so critical.

Emerson is a 'romantic', provided one correctly understands this now somewhat outmoded term. The reference is not to a suitor overly swayed by love, but to a philosophical movement, Romanti-cism, that reacted to an Enlightenment overemphasis on rational-ism, objectivity, Cartesian dualism, and hard science, mind versus matter, the new science that was bringing increased competence at exploiting the world, and at the same time decreased confidence about the place of humans in the scheme of things. Emerson was wondering already about the negative results. Provocatively we might say that Emerson is already a 'postmodernist', or at least that he is uncomfortable with the increasingly assertive urban, urbane 'modernism' secularizing Boston life and at once civilizing it and alienating it from the New England landscape.

Keep 'romance' in life, Emerson says; or, we might say: 'love life' in its rich fullness. Enjoy life as an 'epic, adventurous narrative' (one meaning of the French *roman* and of the English *romance*). The good life is not so much reasoned analysis, dominion over nature, rebuilt environment conquering nature; rather (as the feminists would now say) life requires appropriate respect, sensitivity and 'caring' whether in culture or nature. Humans need a deep sense of engagement with the landscape. 'Nature is the opposite of the soul, answering to it part for part. One is seal, and one is print. Its beauty is the beauty of his own mind. Its laws are the laws of his own mind.'[5]

Two of Emerson's works, similarly titled, introduce his thought. The first is his earlier small book entitled *Nature*, the original transcendentalist manifesto of 1836. The second is a later essay, 'Nature', published in 1844. 'Nature' begins with a poem:

The rounded world is fair to see;
Nine times folded in mystery:
Though baffled seers cannot impart
The secret of its laboring heart,
Throb thine with Nature's throbbing breast,
And all is clear from east to west.

The learned seers at Harvard University (rationalists, empiricists, scientists) are 'baffled' by the developing astronomy, geology and historical biology. They puzzled over the clockwork heavens, the rock strata, the fossil record. Asa Gray was filling his herbarium with strange plants from around the world. Science was upsetting old world-views; but an attuned heart throb understands. The sciences cannot teach us all we need to know about nature; indeed they cannot teach what we need most to know: how to value it. The wise person needs to 'transcend' this cold, mechanistic universe, known by reason and observation in its causal sequences, and to realize deeper truths.

Nature cannot be understood merely as a commodity, a *resource*; it can only be understood in *romance*. So Emerson revels in nature's 'sanctity', in the 'spell' of nature; its 'enchantments'. 'We ... make friends with matter', reconciling mind and matter. Nature 'shames us out of our nonsense'. 'Cities give not the human senses room enough.'[6] Richer aesthetic experiences are possible in forest and field – more to see, smell, touch, taste, more sense of space, time, place, proportion.

Less than a quarter of a mile away, at Walden Pond, Henry David Thoreau agreed: 'In Wildness is the preservation of the World'.[7] Socrates claimed: 'I'm a lover of learning, and trees and open country won't teach me anything, whereas men in town do'.[8] But Emerson and Thoreau objected.

In *Nature* Emerson argues that nature yields: Commodity; Beauty; Language; and Discipline. The planet's endless circulations give us sustenance, life, life-support, and prosperity. All the human useful arts but further embellish these natural cycles. As we now say, an ecology underlies every economy – a fact Bostonians were increasingly inclined to neglect. Nobler wants are served by the beauties of woods and sky. 'There is ... the necessity of being beautiful under which every landscape lies.'[9] Such beauty is reciprocal and ancillary to human character and intellectual life. In current vocabulary, Emerson has a 'virtue ethics'.

Nature's function is linguistic or sacramental. 'Every natural fact is a symbol of some spiritual fact.'[10] Rivers speak of the flux of things; rocks speak of permanence. Nature equally offers stability and dynamism – the everlasting hills, the timeless natural givens, wind, rain, sea, sky, land. Language, indeed all wisdom, roots in these earthy, proverbial symbols, as when we say that what you sow you reap, or that into each life some rain must fall. Nature disciplines, schools the will. As nature confronts us, and we figure life out, character unfolds.

'There are all degrees of natural influence', from the commodity of 'the bucket of cold water from the spring', across a spectrum to the sacramental and 'sublime moral of autumn and of noon'. 'We nestle in nature, and draw our living as parasites from her roots and grains.' 'It seems as if the day was not wholly profane, in which we have given heed to some natural object.'[11] We never have a bad day if we have enjoyed a snowfall, a field of waving grain, or wildflowers. 'He who knows the most, he who knows what sweets and virtues are in the ground, the waters, the plants, the heavens, and how to come at these enchantments, is the rich and royal man.'[12]

Nature has correlate aspects. *Natura naturata* (borrowing from medieval Scholasticism) is particular separated objects, passive, inert. These result from *natura naturans* – active energetic, the restless processes generating such objects, expressing itself in diverse and varied forms.[13] In myth, this is Mother Nature; etymologically, the root meaning of 'nature' is to give birth or spring forth. In science, this is creative natural history.

Though pre-Darwinian, Emerson is already accepting an evolutionary advance over long timespans: 'Geology has ... taught us to ... exchange our Mosaic and Ptolemaic schemes for her larger style. We knew nothing rightly, for want of perspective ... What patient periods must round themselves before the rock is formed, then before the rock is broken ... into soil, and opened the door for the remote Flora, Fauna ... How far off yet is the trilobite! ... It is a long way from granite to the oyster; farther yet to Plato.'[14]

There are two faces – 'secrets', as Emerson calls them – of nature:

1   Motion, process, the flux of things, an *élan vital*, catches the element of change and development. Nature is always moving on: a 'system in transition', breaking through to new achievements in know-how and power. 'Plants ... grope ever upward toward consciousness; the trees are imperfect men.'[15] Nature, to use current vocabulary, is 'self-organizing'.
2   Rest, changelessness or identity catches a complementary dimension. The same laws and materials are present in all its forms – from stars to men. Matter is conserved, as is energy; there is homeostasis and re-cycling. 'From the beginning to the end of the universe, she (Nature) has but one stuff.' 'The direction is forever onward, but the artist still goes back for materials, and begins again with the first elements on the most advanced stage.'[16] 'Nature is a mutable cloud, which is always and never the same.'[17] Nature's diversity and unity, its stability and spontaneity, are dialectical and complementary values.

Emerson sees wisdom in what we now call co-evolution. An animal is armed, given a niche, yet checked by its predators. An animal lives in an environment, yet has to maintain itself against that environment. So birds have feathers. Nature's order is enthusiastic and extravagant; nature seems to overdo it, but thereby succeeds. 'Exaggeration is in the course of things. Nature sends no creature, no man into the world, without adding a small excess of his proper quality ... Nature ... makes them a little wrong-headed in that direction in which they are rightest.'[18] We first think that an oak tree makes too many acorns or that the squirrel in the oak is too nervous. But the seeming waste of seeds and the squirrel's instinctive fear, usually groundless, ensures the propagation of their species. In the checks and balances of an ecosystem, this results in beauty and integrity in the biotic community, as Leopold later termed it. The 'calculated profusion'[19] adds excitement, efficiency, creativity and diversity.

There are similarities here to recent thought about the spontaneous generation of integrated order in decentralized systems, as happens in society with language and markets, or in nature with ecosystems. Such decentralized order is not low quality; to the contrary, it is richer and more diverse than centralized order.

Human life and society are, or ought to be, lived in continuity with nature. 'A man does not tie his shoe, without recognizing laws which bind the farthest regions of nature.' 'We talk of deviations from natural life, as if artificial life were not also natural.' Yet in humans there is novelty added to identity. We are not simply to 'camp out and eat roots; but let us be men and not woodchucks'.[20] Still, we should not look for the meaning of life in technological advances – hoping using electromagnetism to grow salads quickly (or, we might say, microwave ovens to cook chicken instantly). Such accomplishments will never replace living out our threescore and ten years with roots in the soil, enjoying the seasons, spring salads included. 'Nothing is gained: nature cannot be cheated: man's life is but seventy salads long.'[21]

Homo sapiens is a microcosm, an epitome or compendium of nature, in whom nature comes to completion. At times, Emerson can seem anthropocentrist: 'All the facts in natural history taken by themselves, have no value, but are barren like a single sex. But marry it to human history, and it is full of life.'[22] Natural phenomena have their glory unrealized, until humans wake up to this, and this is a principal destiny of humankind.

Emerson closes 'Nature' trying to make sense of a certain 'deceit' in the 'face of external nature', in contrast to his opening revelry in its beauty. We travel hopefully and never arrive. 'There is throughout nature something mocking, something that leads us on and on. All promise outruns performance.' There is 'friction' and 'inconvenience'. 'Must we not suppose somewhere in the universe a slight treachery and derision?'[23]

At first yes, but ultimately no. A better perspective sees a creative discontent in which nature satisfies, but never quite fully. She is ever 'inaccessible', always remaining at an unconquerable 'distance'. We never arrive at possessing nature – 'always a referred existence, an absence, never a presence and satisfaction'. Nature is 'a vast promise, and will not be rashly explained'. She is 'fathomless'. We only touch her 'outskirts'. We never reach the end of the rainbow. This may overwhelm us with 'uneasiness' and 'helplessness', but rightly understood this should give a sense of transcendence, a higher power, a spiritual universe.[24] If, in these secular years, this seems overly romantic, consider Loren Eiseley's exclamation, as a paleontologist: 'Nature itself is one vast miracle transcending the reality of night and nothingness'.[25]

Emerson concludes his brooding over nature in the philosophical idealism that underlies all his thought: 'Nature is the incarnation of a thought ... The world is mind precipitated'.[26] But it takes long insight to see this.

Notes

1 *Collected Works*, vol. 3, p. 208.
2 John Stuart Mill, 'Nature', 1874, in *Collected Works*, vol. 10, Toronto: University of Toronto Press, p. 400, 1963–77.
3 Ibid., p. 381.
4 'Nature II', p. 226.
5 'The American Scholar', p. 55.
6 'Nature' (1844), p. 382.
7 Henry David Thoreau, 'Walking', 1862, in *The Portable Thoreau*, ed. Carl Bode, New York: Penguin Books, p. 609, 1980.
8 *Phaedrus*, 230d.
9 'Nature' (1844), p. 386.
10 *Nature* (1836), p. 18.
11 'Nature' (1844), p. 382.
12 Ibid., p. 384.
13 Ibid., p. 388.
14 Ibid.
15 Ibid., pp. 389–90.
16 Ibid., p. 389.
17 'History', p. 8.
18 'Nature' (1844), p. 392.
19 Ibid., p. 393.
20 Ibid., pp. 390–1.
21 Ibid., p. 400.
22 *Nature* (1836), p. 19.
23 'Nature' (1844), pp. 396–8.
24 Ibid., pp. 398–9.
25 Loren Eiseley, *The Firmament of Time*, New York: Atheneum, p. 171, 1960.
26 'Nature' (1844), p. 400.

See also in this book

Carson, Darwin, Griffin, Lovelock, Thoreau

Emerson's major writings

The Collected Works of Ralph Waldo Emerson, Cambridge, MA: Harvard University Press, 1971.
Nature, 1836, in *Collected Works*, vol. 1, pp. 7–45.
'The American Scholar', 1837, in *Collected Works*, vol. 1, pp. 49–70.
'Nature', 1844, in *Emerson's Essays*, New York: Thomas Crowell, pp. 380–401, 1961.

Journals, Boston, MA: Houghton-Mifflin Company, 1910.
'Nature II' (a poem), in *The Complete Works of Ralph Waldo Emerson*, vol. 9, Boston, MA: Houghton-Mifflin Company, p. 226, 1918.
'History', in *Collected Works*, vol. 2, pp. 3–23.

Further reading

Richardson, Jr, Robert D., *Emerson: The Mind on Fire*, Berkeley, CA: University of California Press, 1995. Contains a chapter on 'Nature'.

HOLMES ROLSTON III

CHARLES DARWIN 1809–82

> It is interesting to contemplate an entangled bank, clothed with many plants of many kinds, with birds singing on the bushes, with various insects flitting about, and with worms crawling through the damp earth, and to reflect that these elaborately constructed forms, so different from each other, and dependent on each other in so complex a manner, have all been produced by laws acting around us.[1]

Charles Darwin was born in Shrewsbury, the son of a prosperous medical doctor. He was educated privately and then at Shrewsbury School. After a brief period unsuccessfully studying medicine at Edinburgh University, he went to Cambridge University in 1827 with a view to becoming a parson. This never came about. After graduation he joined Captain Robert FitzRoy on the *Beagle* voyage (1831–6). The voyage was the key event in Darwin's life, transforming the way he thought about the natural world. The ship was commissioned by the British Admiralty to survey the coastal waters of southern South America, especially in and around Tierra del Fuego. In the event, the ship went round the world, visiting Brazil, Argentina, Patagonia, Chile, the Galapagos Islands, Tahiti, New Zealand and the Cape of Good Hope, as well as various other ports of call. Darwin explored and collected extensively in all these areas, afterwards writing an account of his experiences that is today recognized as an important record of natural environments as well as a classic of travel literature. It should perhaps be noted that Darwin's appreciation of natural landscapes and the beauties of nature did not preclude his ambitious programme of collecting: he accumulated specimens with all the fervour of a big-game hunter, and his ability to shoot was as valuable to him at that time as any growing understanding of intellectual issues. During the voyage he learned from Charles Lyell's *Principles of Geology* (1830–3) that the environment is constantly changing. He applied this idea usefully to the origin of mountain ranges, coral reefs and other natural history questions. He returned to England brimming with fresh ideas and perspectives.

On his return Darwin soon became convinced of the truth of evolution in living beings. While the Galapagos finches were highly significant in this conviction, it was only when they had been properly identified after the voyage that he began to understand the relationships between them. This was the most exciting, intellectually fertile period of his life. From 1837 onwards he filled a series of private notebooks with a riot of evolutionary speculations. He was alert to the subtle balances and relationships between organisms, and between organisms and their environment, seeking an alternative explanation for what was seen by others as 'perfect adaptation'. Some eighteen months later, in September 1838, he took the idea of competitive struggle and differential survival rates from Malthus' *Principle of Population* (1798) as the foundation of his ideas, calling it 'natural selection'. This provided him with a naturalistic mechanism for change and adaptation that did not involve any form of divine action. By 1844 he felt sufficiently confident in his ideas to compose a short essay which he again kept private, although leaving instructions for his wife, Emma Wedgwood, to publish it in the event of his death. From this time, too, he began experiencing the protracted bouts of ill health that dogged his life.

Single-mindedly Darwin set out to provide the exhaustive documentation that he believed he would need in order to convince his contemporaries of evolution by natural selection. He corresponded prolifically with colleagues from all over the world, making effective use of the British colonial system and his contacts in London's premier scientific societies, while also carrying out natural history experiments in his own home, and requesting other men and women from a wide range of backgrounds to help on particular points. If nothing else, Darwin's work represents an astonishing example of cooperative endeavour across the nineteenth-century natural history sciences. His growing fame from his natural history publications and high position in scientific society facilitated these inquiries. At last, after a long study of barnacles, in which he demonstrated evolutionary relationships to his own satisfaction, he decided he was ready to write up his theories in full.

While writing, he received from Alfred Russel Wallace, who was collecting natural history specimens in Malaysia, a short essay containing identical ideas. A joint announcement was arranged at the Linnean Society of London, July 1858, followed by a very brief joint publication of the text in the *Journal of the Linnean Society*.[2] Both Darwin and Wallace were absent when the theory was announced: Darwin's youngest child was dangerously ill, and Wallace was still in Malaysia. Darwin then published *On the Origin of Species by Means of Natural Selection, or the Preservation of Favoured Races in the Struggle for Life* in November 1859. In this he proposed that every animal and plant is variable. Individuals that are best adapted to the surrounding conditions will be the ones that survive and reproduce. Over aeons of time, and with gradually changing conditions, organisms evolve. He included many instances of environmental change. A masterpiece of interconnected reasoning, the book was also impressive for the mass of detailed factual information that Darwin used to support his argument.

Although not the first to propose evolution, the book aroused intense controversy. On the one hand, Victorians found it hard to accept his (and Wallace's) mechanism of natural selection. They were reluctant to regard organisms as being governed solely by chance, and saw little evidence of transitional forms in nature or of intermediate stages leading to complex organs like the eye. Philosophically minded biologists further argued that Darwin could not prove his hypothesis in the conventional way, for he depended on inference and probability to an uncomfortable degree. Others pointed out that he could explain neither the origin of variations nor how they were passed to succeeding generations.

Yet the primary reason for the controversy was that many were deeply unwilling to remove God from the creative process. The issue was at its most intense when the origin of human beings was considered. Darwin did not speak directly about human ancestry in the *Origin*, saying only that 'Light will be thrown on the origin of man and his history'.[3] Nevertheless, human ancestry was the primary focus of the heated debate that followed publication. If Darwin's proposals were accepted as true, then human beings were not specially created by God, as in the Biblical story, and had instead descended from animal ancestors, probably apes. For this reason the *Origin* was frequently perceived as a dangerously atheistic tract. It was vigorously defended by T.H. Huxley, 'Darwin's bulldog', and by his friends Charles Lyell, Joseph Dalton Hooker, Asa Gray and John Lubbock, as well as others. To some degree, however, each of these men had minor reservations about one or another aspect of Darwin's scheme. When Wallace returned from Malaysia in 1862, he too defended the theory vigorously, generously acknowledging that Darwin had produced far more evidence in its favour than he would have been able. Ultimately, however, Wallace and Darwin diverged on their accounts of the origin of the mental life of mankind.

Darwin spent the rest of his life expanding on different aspects of problems raised in the *Origin*. His later books, including *The Variation of Animals and Plants under Domestication* (1868), *The Descent of Man* (1871) and *The Expression of the Emotions in Animals and Man* (1872), were detailed expositions of topics that had been confined to small sections of the *Origin*. He also investigated plant physiology, especially fertilization, making many experiments in his home and garden, relishing his return to practical natural history work after a long period of writing. The first book he published after the *Origin* was a close examination of orchid fertilization expressly intended to show that these intricate flowers were not the result of divine design but merely a remarkable collection of adaptations to ensure insect fertilization. He had always been interested in the wider implications of botany and considered plants as significant evidence for his theories: many of the key arguments for adaptation, variation and descent in the *Origin* hinged on his botanical work, particularly on his innovative ideas about plant geography.

Darwin incorporated some of this botanical research into *Variation of Animals and Plants under Domestication*, which progressed slowly from 1860 to its publication in 1868. In this work he attempted to fill the one major gap left by the *Origin* as to the origin and transmission of variations. He gave numerous instances of the different causes of variability, including reversion, the effects of use and disuse, correlation, monstrosities and the direct influence of the environment and conditions of life. In this last, he was accused of giving undue weight to Lamarckian influences, although in actual fact he had always allowed for the transmission of some changes acquired by parents during their lifetime. The issue became critical when the key concepts of genetics were being worked out in the decades after his death. The extent to which biologists should admit any inheritable effects of the environment became a hotly debated feature of the biological sciences at the end of the nineteenth century.

Throughout the six editions of the *Origin* produced in his lifetime (1859, 1860, 1861, 1866, 1869, 1872), the main thesis stood firm. But

Darwin made considerable changes to the detail: by degrees he broadened his view of the inheritance of acquired variations; he tried to speed up the rate of evolutionary change to account for William Thomson's calculation of a much shortened time span for the age of the earth; and he included answers to many criticisms, especially those of St George Mivart in the sixth edition. He defended his use of the term 'natural selection' while admitting that he probably personified it too much. At Wallace's suggestion, he introduced Herbert Spencer's expression 'survival of the fittest' in the fifth edition (1869).

His last few books were on botanical topics, assisted by his son Francis, who also acted as his secretary. His final work was on earthworms, a return to a subject that interested him as a young man, and reflected his life-long belief that the accumulation of many small actions, or changes, could produce large effects. He proposed that earthworms, by bringing fresh earth to the surface every night, could slowly bury objects and regenerate the surface of the earth. By the end of his life he was happy to let younger men push forward with evolutionary ideas and was content to work on the smaller, more practical natural history questions that intrigued him. His entire intellectual life had, in this regard, been firmly rooted in the real world of natural history. Much of his greatness lay in his ability to move freely between these small details and the expansive vistas opened up by his theories.

Darwin's religious views have naturally been the object of much inquiry. These seem to have waxed and waned. He was brought up as an Anglican, with Unitarian family influence. As a young man and for much of his time on the *Beagle* he was a traditional, if occasionally sceptical, believer. Yet he examined his religious beliefs very closely while working on evolutionary theory, and from 1837 onwards sometimes revealed a fierce, materialistic bent. Even so, he said that when he published the *Origin* he believed in a non-interventionist deity and claimed in his *Autobiography* that he never considered himself an atheist, saying that Huxley's term 'agnostic' was a far better description of his state of mind. A humane and good-natured man, he believed mostly in the Victorian concept of doing one's duty. In politics, he was a liberal.

In later life, Darwin was revered as a grand old man of science. He died on 19 April 1882, at Down House, in Kent, in the house where he had lived since 1842. He was buried in Westminster Abbey.

Darwin's impact upon environmental thinking and practice has been profound but also ambiguous. For some writers, both followers and critics, the theory of natural selection continued the Enlightenment process of the 'disenchantment' of nature, with the effect that notions like 'respect' for nature as the Book of God or as a purposeful organism could be dismissed as merely 'romantic'. Among so-called 'Social Darwinists', the theory was taken to endorse a view of nature and human relations as 'red in tooth and claw' – a view then employed to justify both the economic exploitation of nature and the colonial subjection of less 'fit', 'primitive' peoples. For other writers, however, Darwin's theory, by emphasizing the integral place of human beings in the natural world, served to demolish that 'Cartesian' picture of human beings as 'intellects' set over against the natural world which, in the view of these writers, has been largely responsible for treating the environment as an 'object' or resource to be used in whatever ways people like.

Notes

1   *On the Origin of Species by Means of Natural Selection, or the Preservation of Favoured Races in the Struggle for Life*, London: John Murray, p. 489, 1859.
2   Charles Darwin and Alfred Russel Wallace, 'On the Tendency of Species to Form Varieties, and on the Perpetuation of Varieties and Species by Natural Means of Selection', *Journal of the Proceedings of the Linnean Society of London*, 3 (9), pp. 1–62, 1858.
3   Ibid., p. 488.

See also in this book

Carson, Malthus

Darwin's major writings

Journal of Researches into the Geology and Natural History of the Various Countries Visited by H.M.S. Beagle, 1839, ed. J. Browne and M. Neve, Harmondsworth: Penguin Classics, 1989.
On the Origin of Species by Means of Natural Selection, or the Preservation of Favoured Races in the Struggle for Life, 1859, facs. edn; with an introduction by E. Mayr, Cambridge, MA: Harvard University Press, 1964.

The standard edition of Darwin's writings is *The Works of Charles Darwin*, 29 vols, ed. by P.H. Barrett, P.J. Gautrey and R.B. Freeman, London: Pickering & Chatto, 1986–9. His articles are reprinted in *The Collected Papers of Charles Darwin*, ed. P.H. Barrett, Chicago, IL: University of Chicago Press, 1977. His correspondence has been listed with helpful summaries in *Calendar of the Correspondence of Charles Darwin, 1821–1882*, rev. edn, F.H. Burkhardt *et al.*, Cambridge: Cambridge University Press, 1994, and is also being published volume by volume in *The Correspondence of Charles Darwin*, ed. F.H. Burkhardt *et al.*, Cambridge: Cambridge University Press, 1985–.

Further reading

Barlow, N. (ed.), *The Autobiography of Charles Darwin. With original omissions restored*, London: Collins, 1958.

Barrett, P.H. *et al.* (eds), *Charles Darwin's Notebooks, 1836–1844: Geology, Transmutation of Species, Metaphysical Enquiries*, London: British Museum (Natural History) and Cambridge University Press, 1987.

Browne, E.J., *Voyaging. A Biography of Charles Darwin*, London: Jonathan Cape, 1995.

Colp, R., *To Be an Invalid: The Illness of Charles Darwin*, Chicago, IL: University of Chicago Press, 1977.

Desmond, A.J. and Moore, J.R., *Darwin*, London: Michael Joseph, 1992.

Ellegard, A., *Darwin and the General Reader*, rev. edn, Chicago, IL: University of Chicago Press, 1990.

Keynes, R.D. (ed.), *Charles Darwin's Beagle Diary*, Cambridge: Cambridge University Press, 1988.

JANET BROWNE

HENRY DAVID THOREAU 1817–62

> The West of which I speak is but another name for the Wild; and what I have been preparing to say is, that in Wildness is the preservation of the World.[1]

No doubt there would have been an environmental movement without Thoreau, but it is hard to imagine such a movement without the rhetorical fire of his words or the inspirational force of his actions. It was Thoreau's ability to embody his actions in powerful and incisive language that made them resonate so widely: most famously, his one-night stand in a Concord jail, the consequence of his non-payment of the tax which supported war in Mexico and slavery in the South; and his residence for two years, two months and two days at Walden Pond, a deep glacier-cut lake just a mile from town. The writings that resulted crystallized concepts that helped shape the actions of generations of successors: anger over his night in jail kindled Thoreau's protest essay 'Resistance to Civil Government', which gave Mohandas Gandhi the term 'civil disobedience';[2] and joy in his sojourn at Walden Pond suffused *Walden* with poetic energy, making this the defining event of

Thoreau's life and career as a writer. In *Walden*, Thoreau moves from caustic criticism of American society to a lyrical intimacy with nature, teaching him, and us, how the spirit of the one can redeem us from the evils of the other. Thoreau's writings became the touchstone for a new and deeper valuation of nature which led, in the decades after his death, to the beginnings of the environmental movement in the USA, starting with Ralph Waldo Emerson and John Muir. As Lawrence Buell writes, thousands of devotees have made pilgrimages to Walden Pond and Thoreau has become our 'environmental hero',[3] the father of American nature writing.

Thoreau was hardly born a naturalist. As a child he joined in family outings into the countryside around Concord, Massachusetts, a small farming village and county seat, set in a rolling landscape of farms, lakes, rivers and second-growth woodlands. Apart from these rambles, Thoreau showed no special disposition towards nature study. His education at Harvard turned him into an accomplished scholar of Greek and Latin, well prepared for his intended profession of schoolteaching. When their notions proved too progressive for the established schools, Henry and his elder brother John opened a school of their own, which flourished briefly until John's ill health forced them to close it in 1841. Henry's life took a further turn when John died suddenly of lockjaw, in January 1842. In the years that followed, Henry tried various ways of making a living: as a tutor, a handyman, assistant in his father's pencil factory and surveyor; but with the encouragement of his friend Ralph Waldo Emerson and the 'Transcendental' movement he inspired, Thoreau set his sights on literature as his true vocation. In 1844, Emerson bought land on Walden Pond, and in 1845, with Emerson's blessing, Thoreau began to build his cabin. When he moved in – on Independence Day, 4 July 1845 – Thoreau took with him the materials for his first major writing project, *A Week on the Concord and Merrimack Rivers* (1849), a meditative re-telling of a two-week journey he and John had taken in 1839. While at the Pond, though, Thoreau began gathering materials for his next project, *Walden*. At first he merely sought to explain his unusual actions to his curious fellow-townsmen, but over the years the project grew to encompass the events of his stay at the pond and the philosophy of living he learned to practise on its shores.

It was while he was living at the Pond that Thoreau was seized and jailed, one afternoon in July 1846 when he was running errands in town. The controversy that ensued sharpened his political thought; already a vocal abolitionist and a modest success on the lecture circuit, from the 1840s onwards Thoreau was increasingly prominent as an anti-slavery speaker and activist. Two other events at the Pond also shaped his future career. First, a few weeks after his arrest Thoreau travelled to Maine, where on Mount Katahdin he first encountered true wilderness. The experience, as he narrated it in 'Ktaadn', shattered his image of nature as a safe and nurturing mother: here, 'Vast, Titanic, inhuman Nature' seemed to corner him and query, 'why came ye here before your time? This ground is not prepared for you.' It was difficult, Thoreau pondered, 'to conceive of a region uninhabited by man', for we presume our presence 'everywhere. And yet we have not seen pure Nature, unless we have seen her thus vast, and drear, and inhuman ... Here was no man's garden, but the unhandselled globe.'[4] After this revelation, Thoreau could see that even Walden's peaceful landscape held its terrors, for some element in nature was always and irreducibly Other: or, as he would soon call it, Wild.

The second event suggested one way in which that otherness might be approached, if not fully comprehended. As Thoreau increasingly turned to nature, he also turned to writings about nature, especially to works of natural history. But the arrival in Boston of Louis Agassiz, the famous Swiss natural scientist, turned Thoreau from observer to participant. Agassiz soon organized a collecting network, and by April 1847 Thoreau was shipping specimens of fish, turtles, and even a fox, to Agassiz, who declared some of the species Thoreau collected new to science. Soon afterwards, Thoreau came to the writings of Agassiz's mentor, Alexander von Humboldt, and of Charles Darwin and Charles Lyell, also deeply influenced by Humboldt. Thoreau was critical of natural history surveys, which he condemned as 'inventories of God's property, by some clerk'[5] – but here was something else again, a cosmic vision of nature as one great whole to be approached through the loving and exacting study of its myriads of details, not in the laboratory but out in the wild, through exploration and collection. Thoreau caught the Humboldtian wave just as it was cresting, not only in Europe but in America, where Humboldtian science was stimulating the organization and funding of govern-ment-sponsored Exploring Expeditions to the American West and along the coastlines of North and South America. Humboldt promoted a science that included organism and environment in one interconnected web, a synthesis that decades later would be named 'ecology'. Thoreau's discovery of proto-ecological science was of tremendous importance to his development as a thinker, for in it he found tools and models for conducting his own 'ecological' studies of the Concord environment. By the early 1850s, this new vocation absorbed most of his productive hours, including the records in his *Journal*, which eventually totalled over two million words. Under the excitement of his emerging passion, *Walden* – which had languished in manuscript form since the commercial failure of *A Week* – grew to maturity.

Published at last in 1854, *Walden* remains the classic text at the head of all American nature writing since. It is directed to all those who recognize that, like the 'mass of men', they too 'lead lives of quiet desperation'.[6] Thoreau's 'experiment' at Walden Pond sheds all but the essential trappings of 'civilized' life to reveal a more truly civil life of the mind, lived close to nature's rhythms and attentive to her creatures, of whom we are, of course, one. 'Not till we are lost, in other words, not till we have lost the world, do we begin to find ourselves, and realize where we are and the infinite extent of our relations', Thoreau wrote.[7] Walden is above all a place to dwell and 'find' oneself, and so the emphasis in *Walden* is on domestic nature. Two other works, which overlap in the time of their composition but which were not published in final form until after Thoreau's death, take up the nature of wilderness and of those whose lives border civilized and wild. *The Maine Woods* (1864) collects the narratives of Thoreau's three trips to Maine: 'Ktaadn' was followed by 'Chesuncook', in which Thoreau joins a moose hunt, and 'The Allegash and East Branch', in which he considers the mind and life of the Indian through his friendship with the Penobscot guide Joe Polis. In *Cape Cod* (1865), Thoreau visits the men and women who live in the dunes with the sea at their backs, and here, facing that sea, Thoreau considers that 'wilderness reaching round the globe, wilder than a Bengal jungle, and fuller of monsters'. Thoreau's beach delineates, like Mount Katahdin, the outermost edge of humanity and holds similar terrors: 'It is a wild rank place, and there is no flattery in it ... There is naked Nature, – inhumanly sincere, wasting no thought on man, nibbling at the cliffy shore where gulls wheel amid the spray.'[8]

Thoreau's early death, at age 44, cut short the developing projects of what should have been his middle years. Thoreau was well on his way to a unique synthesis of scientific precision with a poet's love of metaphor. Most notably, 'The Succession of Forest Trees' (1860) presents both a scientific theory accounting for patterns of forest succession and a passionate argument for intelligent forest manage-ment.[9] The need for such an argument reminds us that Thoreau's home landscape was hardly pristine: already in the 1840s it had been worn down by 200 years of European use. Furthermore, the onset of the Industrial Revolution alerted Thoreau to its long-term consequences: the railroad had cut across a corner of Walden just before he moved there, and cutting timber for ties and fuel had by the 1850s nearly levelled the forests he grew up with. Once-familiar species like deer and beaver had long been hunted out of his neighbourhood, and his critique of capitalism included the fear that soon all open land would be fenced and posted against trespassers, outlawing the kind of long cross-country walks he took daily.[10] In another of his late essays, 'Wild Apples', Thoreau warned against the coming of the evil days when 'even all the trees of the field, are withered'.[11] Yet he did not counsel despair. Instead, Thoreau began to work out solutions whereby the community would combine to create 'national preserves',[12] taking selected lands out of the system of private property and holding them in trust for all, 'a common possession forever, for instruction and recreation'. Such land, if forested, was not to be cut but to 'stand and decay for higher uses', suggesting an ethic of preservation;[13] in another late manuscript, Thoreau speculated that 'Forest wardens should be appointed by the town' to oversee the management of private woodlots. Americans had much to learn, Thoreau suggested, from the English, who 'have taken great pains to learn how to create forests', where Americans still bushwack infant forests or foolishly plough them underground.[14] Thus the seeds of the two contending sides of the environmental movement – preservation of resources and their conservation or managed use – may both be found in Thoreau's late writings.

Though he was active in educating his townspeople about better ways to live with the land and the river, Thoreau never sponsored or joined what could be called a 'movement', in environmental activism or anything else. His reasoning is presented in 'Resistance to Civil Government', where political change is shown to emerge from the convergent actions of all persons with a conscience who, based on their independent moral reasoning, refuse to participate in social injustice. As Emerson had written in 'Self-Reliance', the true reformer 'is weaker by every recruit to his banner'.[15] Thoreau pushed Emerson's idea of self-reliant resistance even farther: first, for Thoreau, nature, too, has the power to 'resist' humankind. That is, nature is not plastic in our hands, to mould as we wish; physical nature has the power to push back, against our designs, or is simply indifferent to them, like the Titan of 'Ktaadn' or the world-circling ocean. When Thoreau looked at wild creatures, they looked back at him, and what he saw in their eyes was not his own reflection but something alien, 'wild.' Thus for Thoreau nature had its own moral standing: 'Who hears the fishes when they cry?', he asked of the shad trapped before the Billerica dam; and he went on to warn, 'It will not be forgotten by some memory that we were contemporaries.'[16] Thoreau understood that were humans removed, nature would still exist and she would not mourn. That insight, astonishing for its time, both fascinated and frightened Thoreau, who was fundamentally a humanist in his outlook; that the universe might be *bio*centric was both troubling and exciting to him. As a result, he attended to the relationship between humans and their environment in a way that few were yet capable of imagining.

Second, Thoreau believed that power flowed from the individual to the collective. Emerson had entertained this idea, but like most Romantics he was even more taken by its complement, the way in which power flowed from the whole organization through the individual. Thoreau stubbornly lived his independent convictions in a way that unnerved his friends, but it was in this way that Thoreau joined his political ideals – his vision of the ultimate democracy – with his understanding of how nature worked: through a creative harmony of independent agents, each seeing to their own ends, but in their purposes borrowing each other to combine towards a higher whole. Thoreau's intellectual convictions also shaped his literary style: since the individual initiated social change, Thoreau sought to move each single reader. By turns he shocks, insults, mocks, jokes, disarms, reasons, preaches, contradicts and sings, knowing that while some readers will shake him off, others will be provoked and inspired. Above all, Thoreau knew the power of a good story, and so in *Walden* he tellingly offers a narrative of his own narrow escape from bondage to freedom. Of course, the point is lost if readers could not imagine recreating the story in their own lives, and so Thoreau invites his readers – us – to follow him, not to Walden Pond but to our own 'Walden', from which we might find our way to a life lived not in desperation but in wisdom.

For Thoreau, such a goal was inconceivable apart from nature: 'culture', the definitive characteristic of humanity, was a process of self-growth or 'cultivation' which joined human effort with the unworked natural landscape, changing both together – like Thoreau in his notorious Walden bean-field. We are not set into our environment; rather, we and our environment grow together into an interlinked whole, such that a careful look around will tell who, and what, we are. Thoreau's exacting observation of the landscape of

Concord told him America still had a long way to go, that most human possibility still lay unrealized. If we are a little closer to the civil society he imagined, it is partly because he spoke, in a way that made us listen.

Notes

1 'Walking', in *Natural History Essays*, p. 112.
2 Thoreau's original title is given here, although after his death the second printing of this essay altered the title to 'Civil Disobedience'. Many reprints since have used the later, but non-authorial, title.
3 Lawrence Buell, *The Environmental Imagination: Thoreau, Nature Writing, and the Formation of American Culture*, pp. 315–16.
4 *Maine Woods*, pp. 64, 70.
5 *A Week on the Concord and Merrimack Rivers*, p. 97.
6 *Walden*, p. 8.
7 Ibid., p. 171.
8 *Cape Cod*, pp. 148, 147.
9 The ideas in 'The Succession of Forest Trees' were developed at much greater length in Thoreau's unfinished manuscript, 'The Dispersion of Seeds', recently published in *Faith in a Seed*, pp. 23–173. Other previously unpublished late natural history writings appear in *Wild Fruits*.
10 See Thoreau's essay, 'Walking', in *Natural History Essays*, pp. 93–136, or in any of its many reprints.
11 *Natural History Essays*, p. 210.
12 *Maine Woods*, p. 156.
13 'Huckleberries', in *Natural History Essays*, p. 259.
14 *Faith in a Seed*, pp. 173, 172.
15 Ralph Waldo Emerson, 'Self-Reliance', in *The Collected Works of Ralph Waldo Emerson*, vol. II, Cambridge, MA: Harvard University Press, p. 50, 1979.
16 *A Week on the Concord and Merrimack Rivers*, p. 37.

See also in this book

Darwin, Emerson, Jeffers, Lovelock, Muir

Thoreau's major writings

The standard edition of Thoreau's writings is now being issued by Princeton University Press as *The Writings of Henry D. Thoreau*, ed. Walter Harding *et al.*, 12 vols to date, 1971–. This is gradually replacing the earlier edition, *The Writings of Henry David Thoreau*, 20 vols, Boston, MA: Houghton, Mifflin & Co., 1906.

A Week on the Concord and Merrimack Rivers, 1849, ed. Carl Hovde *et al.*, Princeton, NJ: Princeton University Press, 1980.

Walden, 1854, ed. J. Lyndon Shanley, Princeton, NJ: Princeton University Press, 1971.
Excursions, Boston, MA: Ticknor & Fields, 1863.
The Maine Woods, 1864, ed. Joseph J. Moldenhauer, Princeton, NJ: Princeton University Press, 1972.
Cape Cod, 1865, ed. Joseph J. Moldenhauer, Princeton, NJ: Princeton University Press, 1988.
Natural History Essays, ed. Robert Sattelmeyer, Salt Lake City, UT: Peregrine Smith, 1980.
Faith in a Seed: The Dispersion of Seeds and Other Late Natural History Writings, ed. Bradley P. Dean, Washington, DC: Island Press, 1993.
Wild Fruits, ed. Bradley P. Dean, New York: Norton, 1999.

Further reading

Buell, L., *The Environmental Imagination: Thoreau, Nature Writing, and the Formation of American Culture*, Cambridge, MA: Harvard University Press, 1995.
McGregor, R.K., *A Wider View of the Universe: Henry Thoreau's Study of Nature*, Urbana, IL: University of Illinois Press, 1997.
Richardson, R.D., Jr, *Henry David Thoreau: A Life of the Mind*, Berkeley, CA: University of California Press, 1986.
Walls, L.D., *Seeing New Worlds: Henry David Thoreau and Nineteenth-Century Natural Science*, Madison, WI: University of Wisconsin Press, 1995.

LAURA DASSOW WALLS

KARL MARX 1818–83

> The bourgeoisie, during its rule of scarce one hundred years, has created more massive and more colossal productive forces than have all preceding generations together. Subjection of nature's forces to man, machinery, application of chemistry to industry and agriculture, steam navigation, railways, electric telegraphs, clearing of whole continents for cultivation, canalisation of rivers, whole populations conjured out of the ground – what earlier century had even a presentiment that such productive forces slumbered in the lap of social labour?
>
> Karl Marx and Friedrich Engels,
> *The Communist Manifesto* (1848)

Karl Marx, economist and philosopher, is generally regarded as the founder of modern communism as well as a major influence on socialist theory. He was born in Trier, the son of a lawyer, and studied law and philosophy at Bonn and Berlin. After a lively and short-lived career in political journalism he sought refuge first in Paris and then in London, where he was supported financially in a life of impoverished scholarship by Friedrich Engels, with whom he collaborated extensively in his writings. He worked in the British Museum on his great study of the principles of capitalism, *Das Kapital*; it was unfinished at his death and completed by Engels from the notes that Marx left.

At the heart of Marx's thinking lies an acute sense of the damage done to human life and the human spirit by social and economic conditions, conditions which were not new but which had been exacerbated by the Industrial Revolution, as the quotation above indicates. Marx saw the rapid growth of capitalist economy as achieved by exploitation: the exploitation of one social class (the proletariat, roughly the 'working classes') by another (the bourgeoisie or owners of capital, such as the owners of mills and factories). Under these conditions all values and relations, including environmental ones, become subordinated to monetary or commercial ones: there occurs what we would now call the triumph of market values. Marx regarded this as the cause of *alienation*, of a great gulf that estranges man from nature, from himself and his own vitality, and from his fellow-man. His ambition was to free humankind from narrow utilitarian and commercially inspired desires and help us to 're-humanise' our senses.

This acute sense of the alienating properties of capitalism, which requires us to engage in what Marx called *labour*, as opposed to productive and fulfilling *work*, remains one of his most enduring achievements. The connection he drew between these properties and the estrangement of humankind from the natural world is the principal reason for his continuing importance to thinking about the environment.

Marx's view of the moral standing of the natural world (a concept that would probably have struck him as wholly obscure), and of our relationship to it, is equivocal. In places he directly criticizes the exploitation of nature by humankind. In one essay[1] he writes that 'The view of nature attained under the domination of private property and money is a real contempt for, and practical debasement of, nature', and in the same essay he approvingly quotes Thomas Munzer as declaring it intolerable 'that all creatures have been turned into property, the fishes in the water, the birds in the air, the plants on the earth; the creatures, too, must become free'. It is usual to attribute this kind of view exclusively to the early Marx; nevertheless, in the relatively late third volume of *Das Kapital*, written between 1863 and 1883, he is still insisting that we are not the owners of the planet, whether 'we' here are construed as a society or a nation or 'even all simultaneously existing societies taken together'. We are only 'its possessors, its usufructuaries', and 'must hand it down to succeeding generations in an improved condition'.

When he writes like this Marx can appear to hold a 'stewardship' view of our responsibilities to the ecosphere. Sometimes he sounds as holistic as any modern Green could wish: 'Man lives from nature – i.e., nature is his body – and he must maintain a continuing dialogue with it if he is not to die. To say that man's physical and mental life is linked to nature simply means that nature is linked to itself, for man is a part of nature.'[2] The method of 'dialectical materialism' typical of later Marxist thought (it is necessary to be cautious here: Marx never used the phrase, though he often wrote of 'dialectics') also appears to promise a kind of holism. Thinking that is dialectical, in Marx's sense, is impressed by the non-static nature of things and the propensity of any state of affairs to generate contradictions and opposite states, a tension out of which new and often better conditions emerge. Hence later Marxists often welcome contradictions and conflicts as a sign that social evolution is occurring; sometimes their welcome extends to denying that it is logically impossible to maintain directly contradictory propositions (for example, that nothing can be both completely white and completely black at the same time). There is a clear connection here with the modern complaint that binary thinking (yes *or* no, black *or* white, 1 *or* 0) is at the root of the techno-rationalism that fuels our ecological ills. Engels, rather than Marx, makes explicit the relationship between dialectics and respect for nature, meaning here by 'metaphysics' roughly what we would now call binary thinking: 'Dialectics, on the other hand, comprehends things and their representations, ideas, in their essential connection, concatenation, motion, origin and ending ... Nature is the proof of dialectics ... Nature works dialectically and not metaphysically.'[3]

Marx himself sometimes appears to regret the 'disenchantment' of the world that comes from the increasing gulf between the natural world and humankind: it is this gulf that he believes communism will bridge, this conflict (among many others) that it will resolve. 'This communism, as fully developed naturalism, equals humanism, and as fully developed humanism equals naturalism; *it is the genuine resolution of the conflict between man and nature*, and between man and man, the true resolution of the conflict between existence and being, between objectification and self-affirmation, between freedom and necessity, between individual and species. It is the solution of the riddle of history ... '[4]

It would however be a mistake to attribute to Marx on the basis of remarks such as these any great degree of environmental sensitivity as we would now understand it. First, this almost romantic strain in his thinking is at odds with the far more central and dominant materialist strain. It is more typical of the mature Marx to repudiate any notion of mystical or spiritual unity between humankind and nature as an expression of false consciousness, a manifestation of the 'superstructure' put in place by priests and others in order to secure their own power base. He writes that nature 'first appears to men as a completely alien, all-powerful and unassailable force, with which men's relations are purely animal and by which they are overawed like beasts':[5] thus we need to be liberated from such a superstitious view of nature as much as from any other kind of mystification. Disenchantment then is the name of our cure, not of our disease.

Second, Marx's labour theory of value makes it clear that nature is not to be understood as having any intrinsic worth: nature acquires worth insofar as it is transformed by human work. It is, otherwise, simply nothing: 'nature, too, taken abstractly, for itself, and fixed in its separation from man, is nothing for man'.[6] Although Marx comments here and there on the importance of respecting nature and not 'appropriating' it, the importance lies in the benefits for humankind and not in any sense for nature itself. The fundamental outlook is thoroughly anthropocentric and often Marx writes as if nature exists simply in order to be used: 'The worker can create nothing without nature, without the sensuous external world. It is the material in which his labour realizes itself, in which it is active and from which, and by means of which, it produces.'[7] The danger of mastering nature is not simply that we shall lose our awe for the natural world, but that we shall do so only to replace it with awe for the man-made one. 'What a paradox it would be', he writes, 'if the more man subjugates nature through his labour and the more divine miracles are made superfluous by the miracles of industry, the more he is forced to forgo the joy of production and the enjoyment of the product out of deference to the power of technology and those miracles of the industrial process.'[8]

Third, Marx has a pronounced tendency to deprecate peasant communities and those who work on the land in traditional ways – people who are, we might think, significantly in touch with nature – as reactionary and superstitious. He placed his hopes of progress in the urban proletariat, whom he expected to form the backbone of revolution. He saw industrial capitalism, whose effects he criticized so eloquently, as nevertheless a necessary phase in sweeping away the old peasant economies and moving humankind forward into a new age when the limitations of capitalism would be transcended in turn.

What can we say about the environmental legacy of Marx and Marxism? Certainly the architects of Soviet Marxism showed no sign of nostalgia for the peasant way of life: they collectivized it at enormous human cost, introducing a form of factory-farming that mirrored in the countryside the industrialization of the cities. But it is simplistic to blame Marxism for the environmental shortcomings of the few socialist republics that have taken Marx's writings as doctrine. For example, the devastating pollution of parts of the former Soviet bloc, or the disaster of Chernobyl, can probably be traced to a significant degree to the over-rapid industrialization of backward economies and to a host of other factors, including the reluctance of many Western countries to share the benefits of advanced technology with regimes to which they are ideologically opposed. It is no more sensible to make a direct, causal connection between Marxism and Chernobyl than between capitalism and global warming or the *Exxon Valdez*.

Notes

1 *On the Jewish Question*, 1844.
2 *Economic and Philosophical Manuscripts*, first manuscript.
3 *Socialism: Utopian and Scientific*, 1892.
4 *Economic and Philosophical Manuscripts*, third manuscript, emphasis added.
5 *The German Ideology*, 1846.
6 *Economic and Philosophical Manuscripts*, first manuscript.
7 Ibid.
8 Ibid.

See also in this book

Bahro, Bookchin, Malthus, Passmore

Marx's major writings

Economic and Philosophical Manuscripts, 1844.
The Communist Manifesto, with Friedrich Engels, 1848.
The Eighteenth Brumaire of Louis Bonaparte, 1852.
Grundrisse, 1857.
A Contribution to the Critique of Political Economy, 1859.

Kapital, 1867–94.

These texts can be found in numerous editions. They are conveniently available on the world-wide web at the Marx/Engels Library, whose home-page can be found at http://www.marxists.org/archive/marx/works/index.htm. All quotations in the article above are taken from this electronic edition.

Further reading

Fromm, E., *Marx's Concept of Man*, New York: Frederick Ungar, 1973.
McLellan, D., *The Thought of Karl Marx*, London: Macmillan, 1980.
Parson, H.L. (ed.), *Marx and Engels on Ecology*, Westport, CT: Greenwood Press, 1977.
Schmidt, A., *The Concept of Nature in Marx*, London: New Left Books, 1971.
Singer, P., *Marx*, Oxford: Oxford University Press, 1980.

RICHARD SMITH

JOHN RUSKIN 1819–1900

> Building a city fit for people to live in [Ruskin wrote] means 'remedial action in the houses that we have; and then the building of more, strongly, beautifully, and in groups of limited extent, kept in proportion to their streams, and walled round, so that there may be no festering and wretched suburb anywhere, but clean and busy street within, and the open country without, with a belt of beautiful garden and orchard round the walls, so that from any part of the city perfectly fresh air and grass, and sight of far horizon, might be reachable in a few minutes' walk'.[1]

John Ruskin was born of a possessive mother and wine-merchant father. Instead of being sent away to school he was tutored at home; at Oxford University he won the prestigious Newdigate prize for poetry in 1839. He published the first volume of his series *Modern Painters,* establishing the importance of the painter Joseph Turner, when he was 24 years of age. He rapidly established a reputation as the foremost art critic of his time, later holding a Chair of Fine Arts at Oxford. His personal life was not wholly happy: his wife Euphemia ('Effie') Gray divorced him, on the grounds that the marriage had not been consummated after five years, to marry the painter Millais who had been an intimate friend of the couple. Three years later Ruskin met and fell in love with the young Rose La Touche; when he eventually proposed to her, Rose's parents opposed the marriage. In late life he experienced periods of mental illness, exacerbated perhaps partly by this disappointment and partly by the strain of the libel action brought against him by the painter Whistler (whose *Nocturn in Black and Gold* he had accused of 'flinging a pot of paint in the public's face'). He resigned his Oxford Chair in 1879, four years after Rose la Touche's death and two years after Whistler's technical victory in his lawsuit (Ruskin was ordered to pay damages of one farthing). A connection between his resignation and Oxford University's proposal to sponsor vivisection is unproven. His house, Brantwood, overlooking Lake Coniston in England's Lake District, still stands as a memorial to many of his artistic and environmental ideals.

By the 1860s Ruskin was drawing significant connections between art and architecture on the one hand, and the natural world and social and economic conditions on the other. He criticized the economic thinking of his day for emphasizing material wealth at the expense of social welfare, and insisted on the moral basis of any true economics: 'The idea that directions can be given for the gaining of wealth, irrespectively of the consideration of its moral sources, or that any general and technical law of purchase and gain can be set down for national practice, is perhaps the most insolently futile of all that ever beguiled men through their vices'.[2] The conditions of industrial mass production, he argued, were destructive of human sensibility and of a harmonious relationship with nature. They involved making the worker into a tool, his fingers like cog-wheels and his arms like compasses. Demanding 'engine-turned precision' of human beings is a degradation of them.

More than that, to demand precision or perfection goes against what we understand of the natural world of which we are part. Nature teaches us that imprecision and imperfection are essential if anything is to be good. This is what might be called 'The Foxglove Principle':

Nothing that lives is, or can be, rigidly perfect: part of it is decaying, part nascent. The foxglove blossom, – a third part bud, a third part past, a third part in full bloom, – is a type of the life of this world.[3]

It is this principle that Ruskin believed was enshrined in the 'rude and wild' Gothic architecture that he revered. It displayed a 'look of mountain brotherhood between the cathedral and the Alp', its ruggedness and even crudeness paying homage to its models in nature. Since no truly great man stops working until he has reached his point of failure, Ruskin notes, it follows that 'no good work whatever can be perfect, and *the demand for perfection is always a sign of a misunderstanding of the ends of art*' (ibid., emphasis in original). He is prepared to point up the paradox: 'Of human work none but what is bad can be perfect' (ibid.). It would be interesting to hear his comments on that tawdry educational slogan of our times, 'Excellence'.

Gothic architecture, furthermore, is the product of the medieval guild system, which Ruskin viewed with romantic eyes as embodying 'healthy and ennobling labour'. It rejected the idea of the division of labour with its excessive specialization and repetition, and thus it involved work which was intrinsically satisfying, as opposed to work which merely makes possible the acquisition of satisfactions through the wages it commands. It fostered creativity, or *Invention*, as Ruskin calls it, never demanding exactness for its own sake but only where there is a practical or aesthetic need for exactness; it discouraged mere imitation. The cathedral's gargoyles are 'signs of the life and liberty of every workman who struck the stone' (ibid.). Here Ruskin's ideas are remarkably similar to those of Marx and Engels, who shared his sense of the damage done by industrialization; the language of his denunciation of the evils of industrialization is considerably more vehement and impassioned even than theirs.

Ruskin's reputation now is not primarily that of an environmental thinker. Yet his formulation of the connections between social and economic, artistic and what we would now call environmental questions is important and humane. It is not misleading to call his thinking holistic in its lively sense of those interconnections. At the same time he repudiates any easy distinction between anthropocentrism and the idea of intrinsic value in nature, in a way that many later writers might learn from:

> The desire of the heart is also the light of the eyes. No scene is continually and untiringly loved, but one rich by joyful human labour; smooth in field; fair in garden; full in orchard; trim, sweet, and frequent in homestead; ringing with voices of vivid existence ... As the art of life is learned, it will be found at last that all lovely things are also necessary: – the wild flower by the wayside, as well as the

tended corn; and the wild birds and creatures of the forest, as well as the tended cattle ... [4]

Ruskin was a powerful influence on the development of socialism, on the arts and crafts movement of the later nineteenth century, and on a diverse range of thinkers. For example, Gandhi reported that he discovered some of his deepest and most life-transforming convictions from reading *Unto This Last* on an overnight train from Johannesburg to Durban. William Morris wrote the utopian *News from Nowhere* in 1890: it is a vision of a pastoral and ecologically harmonious England that we would now perhaps call 'ectopian'. It has been claimed that 'the most important period of green politics before 1980 lay between 1880 and 1900'.[5] During this twenty-year period were founded many environmental and conservation groups, such as the Edinburgh Environment Society and the Coal Smoke Abatement Society. Ruskin's ideas inspired many of these, as well as the founding of the National Trust and the Society for the Protection of Ancient Buildings.

Notes

1   *The Mystery of Life and its Arts*, 1868.
2   *The Veins of Wealth*, 1862.
3   *The Stones of Venice*, 1851–3.
4   *Unto This Last*, 1887.
5   P. Gould, *Early Green Politics*, Brighton: Harvester, 1988.

See also in this book

Marx

Ruskin's major writings

The Stones of Venice, Orpington, Kent: G. Allen, 1886.
Unto This Last and Other Writings, ed. C. Wilmer, Harmondsworth: Penguin, 1985.
Seven Lamps of Architecture, New York: Dover Publications, 1995.

Further reading

Anthony, P., *John Ruskin's Labour: A Study of Ruskin's Social Theory*, Cambridge: Cambridge University Press, 1983.
Landow, G.P., *Ruskin*, Oxford: Oxford University Press, 1985.

Wheeler, M. (ed.), *Ruskin and Environment: The Storm-cloud of the Nineteenth Century*, Manchester: Manchester University Press, 1995.

RICHARD SMITH

FREDERICK LAW OLMSTED 1822–1903

> The dominant and justifying purpose of Central Park was conceived to be that of permanently affording, in the densely populated central portion of an immense metropolis, a means to certain kinds of REFRESHMENT OF THE MIND AND NERVES which most city dwellers greatly need and which they are known to derive in large measure from the enjoyment of suitable scenery.[1]

> The special value of the Central Park to the city of New York will lie ... in its comparative largeness. There are certain kinds of beauty possible to be had in it ... because on no other ground of the city is there scope and breadth enough for them.[2]

Among the many fields in which he excelled, Frederick Law Olmsted, as a noted journalist, travelled through the pre-Civil War Southern states from 1852 to 1856, reporting on the social abuses of apartheid to the *New York Daily Times*. In his books, *A Journey in the Seaboard Slave States* (1856), *A Journey through Texas* (1857) and *A Journey in the Back Country* (1860), Olmsted exposed the social and economic deprivation of negroes in America. His writings became rallying documents for the repeal of slavery. In 1855, Olmsted became the managing editor of *Putman's Monthly Magazine*, a journal on social, political, scientific and aesthetic issues. In 1866, he was to become one of the founders of *The Nation*, a national intellectual monthly.

Olmsted also was a 'scientific' farmer, from 1844 to 1852, utilizing new agricultural methods and advanced horticultural cultivars on his successive farms at Hartford, Connecticut, and Staten Island, New York. In his travels, he observed the latest innovations and recorded them in his many writings for the *Horticulturist*, a monthly journal. His book, *Walks and Talks of an American Farmer in England,* was published in 1852.

Olmsted too was a great reformist public administrator. He became the first Superintendent of Central Park in New York City in 1856 and skillfully manoeuvred between Republican and Democratic patronage to prepare the site for the park. During the Civil

War, Olmsted founded and directed the American Sanitation Commission, which became the blueprint for the American Red Cross.

Olmsted was also a social critic of America's cities. Joining the Century Association in New York City in 1856 and while living on the Lower East Side of Manhattan, Olmsted banded together with a group of radical artists, writers and religious leaders – William Cullen Bryant, Jacob Reis, Asher Durhan, Rev. Henry W. Bellows, Washington Irving, Peter Cooper and Andrew Jackson Downing – to discuss strategies to alleviate poverty, poor sanitation and lack of organization in services to the poor. He was an early champion of providing large-scale 'pleasure ground for all citizens' which would become the central park of the city.

As significant as these achievements were, Frederick Law Olmsted's most noted accomplishments were the creation of public parks and the establishment of a new profession – landscape architecture. Landscape architecture, which he founded with Calvin Vaux, a collaborating architect who had trained under the landscape artist A.J. Downing, was founded for the purpose of creating a specific type of urban open space. Olmsted and Vaux, in 1857, entered the design competition for the new park for New York City. They named the project by its advocacy and location, 'The Central Park of New York'. Of thirty-two entrants, their 'Greensward Plan' won. The first 'Commission of the Park' which voted for Olmsted and Vaux's design included such noted reformers as William Cullen Bryant, David Dudley Field, Parke Godwin, Cornelius Grinnell, Charles H. Marshall, Henry Jay Raymond and Russell Sturgis. Without the unflagging support of these literary and artistic reformers, the park would never have been realized.

The design of Central Park was uniquely American. Central Park broke all precedents. It was revolutionary in social response, power and control, in layout and organization and in emotional content. Until then, no other city had such a park. In Europe, parks were either remnants of royal preserves or parks built for the privileged few, with limited access. Central Park broke all traditions in size alone. Its 770-acre expanse was enormous, greater than any park that had ever been proposed and was an enormous undertaking in terms of expenditure of money and manpower. It was promoted for many reasons: scientifically, for the prevention of malaria and for clean air; economically, to provide employment at low wages at a time of recession; and to increase land values for real estate profit. More practically, it provided the city with a new reservoir and water delivery system. It converted polluted and derelict pig farms with clean fill and erosion control vegetation. It provided improved positive drainage and storm water management through new streams and seemingly natural water courses. And it provided much needed public infrastructure for the future growth of the city. In its infrastructure, it was visionary, providing for grade-separated cross-town through traffic and grade-separated internal park circulation for carriages, pedestrian and equestrian traffic. Like all parks since Central Park, the project was proffered to the public on issues of health, safety and welfare.

So revolutionary was the design that it was often criticized by public officials as being too ambitious, but as it was built, enthusiastic approval attracted unprecedented numbers in great social and economic diversity. Its landscape character replicated, in well-defined areas, the very landscapes that the Hudson River School painters had captured on their canvases. The painters, Frederick E. Church, Asher B. Durand, John Frederick Kensett and George Innes, were exhibited in the Century Association's galleries in New York, one of the few venues for American artists at that time – works depicting nature in just this manner. Through this new park development, and for the very first time, a park was designed for the average citizen. It was democratic. Even the most indigent of New York City's citizens were able to experience the beauty and pleasure of the scenic natural settings available previously to only the most wealthy. The park would provide activities for citizens of all classes and it would be open and accessible to all, by design and location, in the centre of the city. Central Park was dubbed the People's Park. Unlike any European model, this was a park for a democratic society, a truly American Park.

> Citizens who would take excursions in carriages, or on horseback, could have the substantial delights of country roads and country scenery, and forget for a time the rattle of the pavements and the glare of brick walls.
>
> William Cullen Bryant

The construction of Central Park proved enormously beneficial. New York gained in reputation in Europe where, previously, the city was thought to be foul and unrefined. The park proved very useful to politicians as well, as the recession of 1857 had left many unemployed and park building provided employment for many at low wages. The park provided an upgraded image and viewing space for the city's cultural institutions, which competed for a place within, or on its perimeter. It accreted New York's most significant civic institutions: the Metropolitan, Guggenheim and Frick Museums, the Museum of Natural History, the City of New York Museum and many others. And, it proved very successful for the property owners adjacent to the park, as their property values escalated overnight.

Central Park's reputation quickly spread world-wide. Every American city wanted a park of this type. The American Park Movement was born. The city of Brooklyn was the second to commission Olmsted and Vaux to design a new public park on an abandoned brick quarry. 'Surely nothing will grow here' many politicians said of the idea. Frederick Law Olmsted brought to the task his farming skills, combining them with techniques of large-scale earth moving that he learned in the building of Central Park. He refashioned the central part of the quarry into the Long Meadow, an undulating sloping green expanse with an axis of a curvilinear valley similar to those of the Hudson River paintings. All walkways were designed along the perimeter of the space. Plantings were added for depth and layering of distant views. It was a magnificent composition in total. Years later, after completing hundreds of parks, Frederick Law Olmsted Sr stated that his singular most successful landscape space was that of the Long Meadow of Prospect Park, as it encompassed all the attributes of the picturesque style.

Olmsted's ability to make every natural feature a design asset enabled him to produce brilliant regional parks and solve many urban problems at once. As his public works commissions grew larger and more extensive, they became park systems organizing an entire city. Boston's 'Emerald Necklace' of 1875 became a planned park system that would organize the existing city and connect it to several of its affluent small suburban neighbouring communities such as Brookline and Newton. Here Olmsted's plan strung together widely differing parks. They were: (1) the Commons, a traditional New England pastoral space; (2) the Garden, a Victorian public park; (3) Massachusetts Avenue, a new tree-lined, median-divided boulevard; (4) the Fens, a degraded marshland that was bio-regeneratively reconstituted as a spatial centrepiece for cultural institutions such as the Boston Museum of Fine Arts, Isabel Gardener Museum, etc.; (5) the Muddy River Run, a transportation corridor and linear park along an unsightly urban stream – this would contain five differing circulation types within its narrow

100 foot width; (6) the Arnold Arboretum site, a hilltop site cleared for research; (7) Jamaica Pond, a public reservoir for drinking water; and (8) Franklin Park, a large regional park at the terminus, modelled after Prospect Park in Brooklyn. Olmsted, no longer in partnership with Calvin Vaux, brought in the noted Boston architect, Henry Hobson Richardson, to design the park's numerous bridges and crossings. The Emerald Necklace established America's first green corridor, created a unified park space – a serpentine connection from natural feature to natural feature, through existing regional settlements, and tying Boston proper with its newly annexed suburbs. It has guided all urban development throughout Boston's hundred years of growth. During the 1890s, Olmsted's disciple and partner, Charles Elliot, expanded the park system as proposed by Olmsted and developed new parkland acquisition criteria based on five new scientific principles: safeguarding drinking watershed; providing tidal estuaries to protect the urban populace from diseases; preserving unique scenic resources; designing for river flood planes; and establishing barrier beaches. The latter two were especially important to prevent massive flood damage to property. The full concept of a uniquely American urban park system was formulated around naturalistic, scenic and conservation design parameters.

From this beginning, the profession of landscape architecture grew in multiple directions, justifying its broad definition as 'architecture of the land', or as the 'design of land and the objects placed upon it' or as the 'design of all exterior spaces'. The design competition for Central Park drew thirty-two entrants. Among the others, there were a number of entrants who were architects, planners, engineers and landscape engineers. For some, it was the beginning of a new career in a subject and profession that had no name, definition or direction for future growth. Olmsted himself groped for a name before combining two words, prevalent at the time, to best describe the profession – 'architecture' from the art of building, and 'landscape' from the art of painting. Olmsted thought that building parks was closely related to the art of building the landscape scenes of paintings and gave the profession its name, 'landscape architecture'. Through park projects in other American cities, still others became acquainted with the new potential of this profession.

Frederick Law Olmsted, ... this man who designed Central Park in New York, Riverside Drive, Rockaway, Morningside

Heights, the Arnold Arboretum, the Boston Parkways, Fairmount Park in Philadelphia, the Chicago Parks, the Brooklyn Parks, the National Cemeteries, Mount Royal Park at Montreal, the grounds of Yale, Princeton, Lawrenceville, the University of California, Groton and the National Zoo, [also designed] twenty-five hundred other parks.[3]

The most written about project in America after Central Park must surely have been the great Chicago's World Columbian Exposition. It was a watershed for architecture, planning and landscape architecture. In his 1894 Chicago Fair plan, Olmsted managed to create a master plan that responded perfectly to the widely divergent design philosophies of the Prairie and Classical schools of design. For the Classical, eclectic buildings of the east coast's group, which included McKim, Mead and White and Daniel Burnham, Olmsted proposed a central formal water basin around which the building would form an imposing unified urban/civic space. For the organic American Prairie Architecture School of Louis Sullivan, Frank Lloyd Wright and Daniel Burley Griffen, Olmsted proposed the Lagoon area, with softened edges and romantic islands. And for his own naturalistic and American park landscape, he proposed the lake-front barrier beach areas. The Midway was an unusual development linking the main Fair site, Jackson Park, with a Fair expansion site, Washington Park. Together, the Fair's landscape composition embodied not just answers to specific problems Olmsted encountered, but it created a system of connected park types which had applications to many varied urban conditions. The principal positive contribution of the Fair was that it represented a total cohesiveness of design from section to section in spite of varying architectural styles, uses, commercial enterprises and land forms. The Fair was distinctly urban and presented a new urban ideal of beauty, codified in America as the City Beautiful Movement.

In the City Beautiful philosophy, formal landscape architecture became a powerful counterpoint to the informal pictorial, naturalistic landscapes. Its design philosophy drew from the architecture and art of the classical periods and used art construction terms as the basis for its rational design methodology. The rational aesthetic was firmly established among eastern architects, especially the most influential firms of New York City. It also captivated many landscape architects who collaborated with these architects or with landscape architects who were engaged in planning the urban expansions of America's cities. The resultant combination of Olmsted Sr's urban parks combined with City Beautiful boulevards, local parks and natural riverway conservation corridors have given these cities a uniquely American green infrastructure.

The designs of all American parks were a distinct break from the European parks of England, France and Italy. Olmsted loved the natural scenery and began to introduce systematically ecological processes within his parks. Olmsted's parks were works of nature. In place of water basins, there were natural water bodies such as lakes, ponds, meandering streams and cascades – all representative of natural water courses. Instead of flat planes of grass or manicured rolling hills, one discovered rugged ravines, buttes, rock outcrops of all geologic forms – all representative of the surrounding regions. The trees of Central Park were not trimmed bosques of singular types of trees nor ornamentals as in European parks, but were complete collections of species, of differing vegetation types and differing ecological associations, all representative of the region's ecology. An Olmsted park always provided a wilderness area. In Central Park it was the Ramble, an area where the forces of nature were left to define the parkland in complete, perfect representation of the natural environment. Brooklyn's Prospect Park contained the Bramble; Boston's Franklin Park, the Wilderness; and Boston's Emerald Necklace, the Fens.

Olmsted, in the prime of his professional life, approached the problem of the conservation of scenic areas. He realized the extreme uniqueness of America's majestic landscapes and witnessed the encroachment by commercial developments. In 1865, while in California working as manager of the Mariposa mines, Olmsted visited the scenic valley of Yosemite Falls. Included in the tour was the stunning Mariposa Big Tree Grove featuring some of the most mature Giant Sequoia trees in America. Enthralled by the majestic scale of the valley and its delicate waterfall suspended high above the valley floor, he envisioned its despoilment by commercial loggers, miners and other resource-extracting enterprises. With support from leading American conservationists, he successfully petitioned the United States Congress to set these lands aside, 'granting the Yosemite Valley to the State of California as a public park' and to create a commission to manage this 'land grant'. In turn, he became the preserve's first Commissioner and set in course the concepts and the basis for America's National Park System. This action is considered to be the centrepiece of the American Conservation

Movement, which is generally placed in the period 1850 to 1920. This movement, led by such American conservation notables as Henry David Thoreau, Asher Durand, Samuel H. Hammonds, James Russell Lowell, Albert Bierstadt, John Muir and George Perkins Marsh, began the unprecedented public and private initiatives intended to insure the wise and scientific use of natural resources, and the preservation of wildlife, forestry and landscapes of great natural beauty.

In 1880, Olmsted visited Niagara Falls with the intent of rekindling boyhood memories. He was shocked and disheartened by the rampant commercialization of both the American and Canadian sides. Armed with the support of Canadian colleagues, he strove for the first international park to organize the visitor experience of this scenic wonder. After many years of political battles in the State of New York legislature, a limited conservation 'land reserve' was formed. In 1887, Olmsted submitted his plan to remove all commercial enterprises from the reserve and provide visitor facilities open to the general public. His plans for parklands on the American embankment and on Goat Island were carried out and the natural ecology restored. While the Canadian side was approved quickly and the removal of commercial enterprises swift, their landscape development was that of the formal European Park. Olmsted's effort at Niagara Falls was the first park of a state-wide park system established for a state, New York. It was also instrumental in establishing the need for regional and state parklands across America.

To Frederick Law Olmsted Sr no land-oriented problem seemed out of bounds for his professional interest. Olmsted, the social reformer, took on the design of numerous new social institutions. In Buffalo, he planned the State Asylum, a mental health facility; in Hartford, the Insane Retreat; in Waverly, Massachusetts, the McLean Asylum; and in Boston, Massachusetts General Hospital. He worked on America's most distinguished universities' campuses. He also remodelled many, such as Yale University, planned the expansion of others, and designed whole new institutions, including Stanford University in California, the University of California at Berkeley, the University of Massachusetts at Amherst and the University of Florida at Gainesville. Olmsted rethought the typical American residential subdivision, giving new order to commercial centres, street patterns, housing mixes by densities and, most importantly and unique for its time, Commons as public open space. These projects, along with parks and park systems, city plans and private estates, formed the scope of landscape architecture for the professionals of his day.

Notes

1   F.L. Olmsted and C. Vaux, *The Conception of the Winning Plan Explained by its Authors, Part Two; The Greensward Plan*, Central Park Competition, New York: New York, pp. 1–6, 1856.
2   Olmsted, *Letter to Mr. Ignaz A. Pilat, Chief Landscape Gardener of Central Park*, Panama: 26 September 1863.
3   Mrs Luther P. Eisenhart, *Frederick Law Olmsted: Landscape Architect, Bulletin of the Garden Club of America*, 11, September, pp. 15–20, 1938.

See also in this book

Muir, Thoreau

Olmsted's major writings

Walks and Talks of an American Farmer in England, 2 vols, New York: G.P. Putnam, 1852.
A Journey in the Seaboard Slave States: With Remarks on Their Economy, New York: Dix & Edwards, 1856.
A Journey through Texas: or a Saddle-Trip on the South-western Frontier, with a Statistical Appendix, New York: Dix & Edwards, 1857.
A Journey in the Back Country, New York: Dix & Edwards, 1860.

There are numerous professional reports published for each of the parks. Olmsted was a prolific writer of articles for numerous journals and newspapers. Additionally, Olmsted corresponded extensively with colleagues and friends. These writings are too numerous to present here.

Further reading

Beveridge, Charles E. and Rocheleau, Paul, *Frederick Law Olmsted: Designing the American Landscape*, New York: Rizzoli, 1995.
Fabos, Julius, Milde, Gordon T. and Weinmayer, V. Michael, *Frederick Law Olmsted, Sr.: Founder of Landscape Architecture in America*, Amherst, MA: University of Massachusetts Press, 1968.
Hall, Lee, *Olmsted's America: An 'Unpractical' Man and His Vision of Civilization*, Boston, MA: Bulfinch Press, 1995.
Newton, Norman T., *Design on the Land: The Development of Landscape Architecture*, Cambridge, MA: Harvard University Press, 1971.
Roper, Laura Wood, *FLO: A Biography of Frederick Law Olmsted*, Baltimore, MD: Johns Hopkins University Press, 1973.

R. TERRY SCHNADELBACH

JOHN MUIR 1838–1914

> In God's wildness lies the hope of the world ... The great
> fresh unblighted, unredeemed wilderness. The galling
> harness of civilization drops off, and words heal ere we
> are aware.[1]

Of his beloved wild Sierra, John Muir wrote, 'mountains as holy as
Sinai ... they are given, like the gospel, without money and without
price. 'Tis heaven alone that is given away.'[2] Like the mountain
creatures he so admired, ranging from prophets of old to grizzly
bears, Muir was the mountain embodied: 'I am hopelessly and
forever a mountaineer',[3] he wrote, and it was in mountains that he
found meaning and metaphor, glory and imaginative possibility.

'The mountains are fountains of men as well as of rivers, of
glaciers, of fertile soil. The great poets, philosophers, prophets, able
men whose thoughts and deeds have moved the world have come
down from the mountains.'[4] Like Moses and visionaries of ancient
Christianity such as Augustine and John of Damascus, Muir
delivered his message from the mountains with prophetic purity
and power. Key events in Muir's life, documented in his voluminous
journals and recollections and in public records of his fame, have
explanatory power in the raising of this mighty, righteous voice of
the mountains.

Born to the family Muir, meaning 'a wild stretch of land', in
Dunbar, Scotland on 21 April 1838, son of Daniel and his second
wife Anne Gilrye, John Muir spent a lifetime living up to the name
and to his father's stern expectations. Daniel was a convert to
evangelical Presbyterianism, and a strict, dour man who beat John
throughout his childhood. Biographer Stephen Fox writes: 'John
read his Bible and grew pious beyond his years, but he could never
please his father. The endless scoldings and beatings made his
adolescence a grimly unequal contest of wills with a tyrant blinded
by his own righteousness.'[5]

According to Edwin Way Teale, young Muir was

> repelled by the harsh fanaticism of his father's religion ...
> he affiliated himself with no formal creed. Yet he was
> intensely religious. The forests and the mountains formed his
> temple. His approach to all nature was worshipper. He saw
> everything evolving yet everything the direct handiwork of

God. There was a spiritual and religious exaltation in his experiences with nature.[6]

Muir's dutiful and passionate engagement with learning led him from a Wisconsin farm to which his family had emigrated, to the University of Wisconsin where he took no degree but took the courses he felt he needed. Avoiding the American Civil War and often depressed and lonely, Muir wandered and worked in Ontario and Wisconsin.

Muir came to a turning point in his life when, while working at a wagon factory, he was blinded by a file flying into his right eye and by a sympathetic reaction in his left eye. Struck into abject fear at the prospect of never again seeing natural beauty, he later wrote 'my days were terrible beyond what I can tell, and my nights were if possible more terrible. Frightful dreams exhausted and terrified me every night without exception.'[7]

Recovering his vision, Muir determined to have a three-year-long 'sabbatical' to store, he wrote, 'a stock of wild beauty sufficient to lighten and brighten my after life in the shadow'.[8] Muir's own Sierran baptism and mountain enlightenment climaxed this long search for self-understanding – an odyssey of spiritual and intellectual searching that took place largely out of doors across the North American continent.

In 'First Glimpse of the Sierra' he begins:

[W]hen I set out on the long excursion that finally led to California, I wandered, afoot and alone, from Indiana to the Gulf of Mexico, with a plant-press on my back ... I crossed the Gulf to Cuba, enjoyed the rich tropical flora there for a few months ... but I was unable to find a ship bound for South America ... therefore I decided to visit California for a year or two.[9]

Arriving in San Francisco by steamer on 1 April 1868, he set out to meet his destiny in the Yosemite Valley:

A landscape was displayed that after all my wanderings still appears as the most beautiful I have ever beheld. At my feet lay the Great Central Valley of California, level and flowery, like a lake of pure sunshine, forty or fifty miles wide, five hundred miles long ... from the eastern boundary of this vast golden flowerbed rose the mighty Sierra, miles in

height, and so gloriously colored and so radiant, it seemed
not clothed in light but wholly composed of it, like the wall
of some celestial city.[10]

Muir saw in such wilderness the source of humanity's spiritual
health and wholeness. His philosophy of nature as the glorious
handiwork of a God who created a democracy of life forms has
inspired the post-modern deep ecology movement. Muir was keenly
aware of the anthropocentric character of human attitudes towards
nature, including the values embedded in utilitarian conservation. In
his mind, a different ethic was at work – one which was to inspire
Aldo Leopold, Arne Naess, John Seed and contemporary deep
ecologists.

> The world we are told was made especially for man – a
> presumption not supported by the facts ... why should man
> value himself as more than a small part of the one great unit
> of creation? And what creature of all the Lord has taken the
> pains to make is not essential to the completeness of that
> unit – the cosmos? The universe would be incomplete
> without the smallest transmicroscopic creature that dwells
> beyond our conceitful eyes and knowledge ... [P]lants are
> credited with but dim and uncertain sensation, and minerals
> with positively none at all. But why may not even a mineral
> arrangement of matter be endowed with sensation of a kind
> that we in our blind exclusive perfection can have no matter
> manner of communication with? ...
>   [B]ut glad to leave these ecclesiastical fires and blinders,
> I joyfully return to the immortal truth and immortal beauty
> of Nature.[11]

For him such truth and beauty as one can know in nature answered
his questions. Through immersion in wild nature one could know
how best to live. As Michael P. Cohen puts it, 'ecological
consciousness would generate an ecological conscience'.[12] Muir
moved from his own profound spiritual experiences in wilderness to
preaching action to a nation. According to Cohen: 'His vision, he
now felt, must lead to concrete action, and the result was a
protracted campaign that stressed the ecological education of the
American public, government protection of natural resources, the
establishment of National Parks, and the encouragement of

tourism.'[13] He was much ahead of his time in promoting action based on ecological responsibility. Many have called Muir the voice of the wilderness and his passion to protect it from destruction gave birth to the popular conservation movement. In 1898 he founded the Sierra Club for these purposes.

Within his own historical context Muir had remarkable influence – literary, political and philosophical – on those who were to follow him in environmental ethics and environmental education. Inspired from an early age by the Bible, Shakespeare, Milton, Scott and Burns, he later discovered Thoreau and Emerson.[14] He kept journals with no intent to publish and his first book was not printed until he was aged 56. Literary fame came fast though – the result of a turn-of-the-century love of nature and the urgent need to conserve America's vast natural resources from the unbridled rapaciousness of her maturing capitalism.

His political influence grew as he devoted himself to proselytizing the grandeur of the American West and the vital importance of protecting it. He led an array of important figures from Ralph Waldo Emerson to Theodore Roosevelt on excursions in the Sierra. Some of these camping trips had an enormous effect, such as that upon Robert Underwood Johnson, editor of the influential *Century* magazine, who subsequently launched a campaign to create Yosemite National Park, while President Roosevelt ordered his Secretary of the Interior to extend the Sierra Reserve one day after emerging from his sojourn with Muir.

For generations his work inspired not only the movement to conserve nature but the impetus to appreciate it. His journals brim with the power of his experiences which could and ought to be accessible to all. He thought if only people would save the land and take the time to saunter on it, then would come wisdom. His encounter with the rare orchid *Calypso Borealis*, later famous as marking the beginning of his evolution into pantheism, is recorded in such a journal entry. The entry was written in 1864 near Lake Huron. Muir was in Canada to avoid being drafted into the American Civil War:

> I never before saw a plant so full of life; so perfectly spiritual. It seemed pure enough for the throne of its Creator. I felt as if I were in the presence of superior beings who loved me and beckoned me to come. I sat down beside them and wept for joy.[15]

His philosophical contributions to the conception of wilderness, to the democratic ethical responsibility of humans towards all life forms, and to the ecological consciousness of a vast eternal unity are immense. Earlier, among Americans, only Thoreau spoke with such moral authority; later only Carson had such an influence on environmental thinking.

Notes

1 'Alaska Fragments, June–July (1890)', in *John of the Mountains*, p. 317.
2 Quoted in 'Chronology' by William Cronon, *John Muir: Nature Writings*, New York: The Library of America, p. 839, 1997.
3 Edwin Way Teale, *The Wilderness World of John Muir*, Boston, MA: Houghton-Mifflin Company, p. 143, 1954.
4 'The Philosophy of John Muir', in Teale, op. cit., p. 321.
5 Stephen Fox, *John Muir and His Legacy: The American Conservation Movement*, Boston: Little, Brown & Co., p. 31, 1981.
6 Teale, op. cit., p. xiii.
7 Fox, op. cit., p. 48.
8 Ibid.
9 Teale, op. cit., p. 99.
10 Ibid, p. 100.
11 Ibid, p 318.
12 Michael P. Cohen, *The Pathless Way: John Muir and the American Wilderness*, Madison, WI: University of Wisconsin Press, 1984, quote appears on the dust jacket.
13 Ibid.
14 Cronon, op. cit., p. 836.
15 Fox, op. cit., p. 43.

See also in this book

Carson, Leopold, Naess, Thoreau

Muir's major writings

The Mountains of California, New York: Century, 1894.
Stickeen, Boston, MA: Houghton-Mifflin Company, 1909. This book is in the public domain and can be found on the web at the following URL: http://www.sierraclub.org/history/muir/stickeen.html.
The Story of My Boyhood and Youth, Boston, MA: Houghton-Mifflin Company, 1913.
Letters to a Friend, Boston, MA: Houghton-Mifflin Company, 1915.
Travels in Alaska, Boston, MA: Houghton-Mifflin Company, 1915.
A Thousand Mile Walk to the Gulf, Boston, MA: Houghton-Mifflin Company, 1916.
John of the Mountains: The Unpublished Journals of John Muir, ed. Linnie Marsh Wolfe, Boston, MA: Houghton-Mifflin Company, 1938.

John Muir in His Own Words: A Book of Quotations, compiled and ed. Peter Brown, Lafayette, CA: Great West Books, 1988.

For more writings by John Muir visit the WWW John Muir exhibit: http://icr.ucdavis.edu/John_Muir/.

On-line text versions of many of Muir's works are available at the URL: http://www.sierraclub.org/john-muir-exhibit/index/html

Further reading

Kimes, William F. and Kimes, Maymie B., *John Muir: A Reading Bibliography*, Fresno, CA: Panorama West Books, 1986.
Miller, Sally M. (ed.), *John Muir: Life and Work*, Albuquerque: University of New Mexico Press, 1995.
Nash, Frederick, *Wilderness and the American Mind*, New Haven, CT: Yale University Press, 1967.
Wolfe, Linnie Marsh, *Son of the Wilderness*, New York: Alfred A. Knopf, Inc., 1945.

PETER BLAZE CORCORAN

ANNA BOTSFORD COMSTOCK 1854–1930

In order to appreciate truly his farm, the farmer must needs begin as a child with nature-study; in order to be successful and make the farm pay, he must needs continue in nature-study; and to make his declining years happy, content, full of wide sympathies and profitable thought, he must needs conclude with nature-study; for nature-study is the alphabet of agriculture and no word in that great vocation may be spelled without it.[1]

A serious agricultural depression in the north-eastern United States drove people from rural landscapes to burgeoning cities in the late nineteenth century. Such a migration took place in New York 1891 to 1893. Anna Botsford Comstock wrote in the Preface to her *Handbook of Nature-Study*:

the charities of New York City found it necessary to help many people who had come from the rural districts – a condition hitherto unknown. The philanthropists managing the Association for Improving the Condition of the Poor asked 'What is the matter with the land of New York state that it cannot support its own population?'[2]

In response, a movement was created to interest 'the children of the country in farming as a remedial measure', being that 'the first step toward agriculture was nature-study'.[3]

From such a utilitarian concern for the future of rural life grew the American nature-study movement. The centre of the movement towards reiteration of the importance of agriculture and country values was Cornell University in Ithaca, New York, which from its founding in 1865 had been committed to the problems of agricultural extension.

The leader of this movement was Liberty Hyde Bailey (1858–1954), the great communicator of the idealistic, progressive, romantic beliefs of the Cornell school of thought. The practical purpose of this effort was 'making children sympathetic with nature-study so that they would truly enjoy rural life and be happy on the farm'.[4]

Working with Bailey at Cornell, and the spiritual leader of the nature-study movement, was the great proselytizer of happy, intimate contact with the earth, Anna Botsford Comstock. Born into a Quaker family in rural Cattaraugus County of upstate New York in 1854, she lived until she was 3 years old in a log cabin which she remembers well enough to describe in her autobiography, *The Comstocks of Cornell*. Farm life and a mother named Phoebe who loved nature made indelible impressions on the young Anna. She quotes her mother as saying one day at sunset: 'Anna, heaven may be a happier place than earth, but it cannot be more beautiful'.[5]

An educated female neighbour, Mrs Ann French Allen, was an important influence in directing Anna Botsford towards higher education. She chose new, nearby Cornell, which had opened its doors to women. Zoology study with Professor John Henry Comstock led to long walks and courtship. Marriage interrupted her formal education but led to a decades-long partnership in scientific research, teaching and entomology illustration.

Her path of inquiry seems to have been selected by both cultural limits and personal choices. Biographer Pamela Henson writes that she 'entered science through the "back door" as many female relatives of scientists did, and she always worked on the "peripheries" of science in art, popularization, and children's education.'[6] Henson cites Evelyn Fox Keller and the gendered theory of masculinist objective science in explaining Comstock's choices.

> Comstock found it more comfortable to incorporate her aesthetic appreciation of nature into scientific interpretation

for children and a popular audience. This appreciation was part of her overall sense of subjective connectedness to the world around her. Anna Comstock experienced the natural world in emotional terms and felt a sense of personal relationship and responsibility to living things around her.[7]

Comstock also faced sexist societal barriers as a member of the first generation of American women with university educations. Often cited as the first woman professor at Cornell, appointed in 1898, it is less often noted that the Board of Trustees revoked the title and did not allow women professors until 1911, and only in home economics. Comstock was finally reappointed in 1915.

During her Cornell career Comstock worked with other founders of the nature-study movement. She called Wilbur Samuel Jackman of Chicago the father of nature-study. Jackman's belief that children derived intellectual benefit, as well as personal satisfaction, from the formal study of their immediate environments seems to have been one of the most influential ideas in the history of nature-study. Comstock carried this idea in the period from 1900 to 1920 – the zenith of the nature-study movement and the era of Anna's leadership. She edited the *Nature-Study Review* and served as president of the American Nature Study Society, now more than a century old.

Nature-study was to a large extent a reform movement which rejected the methodologies of schools in the late nineteenth century. Although the movement was to become fractured at a later date by conflicting purposes, there was a common purpose at the outset. According to Richard Raymond Olmsted in his dissertation 'The Nature-Study Movement in American Education', 'like most curriculum movements, the nature study agitation developed into a complex phenomenon. The leaders of this movement found initial agreement, however, in the assumption that elementary school children should be taught about nature, defined usually as the immediate countryside, through field trips and other direct experiences.'[8]

The relationship of events in the nature-study movement to social conditions is vital to an understanding of the controversy which surrounded its introduction into the schools. The historical period from the mid-nineteenth century to 1880 was one of signal change. The Civil War, westward expansion, immigration of millions of new citizens and rapid industrial growth altered the nature of American society. Education was influenced by the introduction of universal schooling, the publication of Charles Darwin's *The Origin of Species* and the growth of child psychology.

Comstock and her heroes Jackman and Bailey saw nature-study as a pedagogical ideal and social reform initiative with roots in the work of Johan Amos Comenius, Heinrich Pestalozzi, Jean-Jacques Rousseau and Friedrich Froebel.[9] She was able to channel her own feeling for nature and her progressive social ideals into an educational and environmental philosophy much needed in her cultural period in America.

Her philosophy was that at the heart of a fully human existence is the cultivated imagination and insight for truth and beauty, as found in nature. In her seminal essay 'The Teaching of Nature-Study' she wrote:

> nature-study cultivates ... a perception and a regard for what is true, and the power to express it. All things seem possible in nature; yet this seeming is always guarded by the quest of what is true. Perhaps half the falsehood in the world is due to lack of power to detect the truth and to express it. Nature-study aids both in discernment and in expression of things as they are. Nature-study cultivates in the child a love of the beautiful; it brings to him early a perception of color, form, and music. He sees whatever there is in his environment, whether it be the thunder-head piled up in the western sky, or the golden flash of the oriole in the elm, whether it be the purple of the shadows on the snow, or the azure glint on the wing of the little butterfly. Also, what there is of sound, he hears; he reads the music score of the bird orchestra, separating each part and knowing which bird sings it. And the patter of the rain, the gurgle of the brook, the sighing of the wind in the pine he notes and love becomes enriched thereby.[10]

She also believed nature was a nurse for human health, an elixir of youth for the teacher, and a cure for problems of school discipline. Her reverence for the power of nature in strengthening human nature was reiterated throughout her writing.

A respected scientist, she published in 1911 what was to become the classic, *Handbook of Nature-Study*; since reprinted in many editions. She advocated direct observation and contact and made great, even extravagant, claims for the mental and physical wellbeing of students and teachers.

'Nature-study is nature love taught in the schools',[11] she wrote. She advocated, without apology, love of the world through harmonious relationship with it. Nature-study is the vehicle for such love – for student and teacher, in school and out. In a speech in Philadelphia in 1914 she said:

> If nature-study as taught does not make the child love nature and the out-of-doors, then it should cease. Let us not inflict permanent injury on the child by turning him away from nature instead of toward it. However, if the love of nature is in the teaching heart, there is no danger; such a teacher, no matter by what method takes the child gently by the hand and walks with him in paths that lead to the seeing and comprehending of what he may find beneath his feet or above his head. And these paths, whether they lead among the lowliest plants, or whether to the stars, finally converge and bring the wanderer to that serene peace and hopeful faith that is the sure inheritance of all those who realize fully that they are working units of this wonderful universe.[12]

In her retirement speech as president of the American Nature Study Society, she said:

> the nature-study idea almost from the first overflowed the school boundaries to enrich and make happier the lives of those who loved the life of the woods and fields, and who would fain know something of the mysteries and wonders therein hidden.[13]

The advocacy of nature-study made her well read and well regarded. She lectured widely in the Chautauquan movement and published science writing for the public. According to Pamela Henson, 'Comstock's popularity was built on a melding of accurate science with popular sentimentality and her aesthetic talents ... '[14]

Called the Dean of American Nature-Study and finally promoted to full professor in Entomology and Nature-Study, she was admitted to Phi Kappa Phi, the honorary society. In 1923 the League of Women Voters elected her one of the twelve greatest women in America. She remained energetic over a long productive career. She was not, however, tireless. When asked why she did not actively fight for women's suffrage, she said: 'I had been using all of my strength to fight narrowness, prejudice, and injustice, in the curriculum of the common schools, and I was weary with fighting'.[15]

She was at her best, as she humbly proclaimed herself, as an interpreter of science. Keller and others have said this was so that she need not assume the objectivist perspective of Western male science. She was an artist *and* a scientist. She educated about the complex power of nature in symbolic forms.

This style also enabled her to advocate for educational reform and nature conservation. She saw the power of the human spirit and of love of nature as the best motivator, putting aside the more typical American concern with practical benefit. In 1914, she said:

> With a fatuity that our descendants of three centuries hence will characterize a criminal stupidity we have exterminated many species of birds, destroyed many interesting and harmless wild animals, hacked down our trees ruthlessly and cleared our streams of valuable fish. Men of science had remonstrated in vain. It was not until the nature-study movement permeated the people throughout the land that they came to resent this extermination; and not until then was there a sufficiently strong popular opinion created to establish and carry out protective laws ... It should be remembered that in all history crusades have been born and led of the spirit.[16]

Her own leadership in science education and in environmental thinking was an inspiration to the American conservation movement. Also important was her gender. She helped make possible the later environmental leadership of many American women from Alice Rich Northrup to Edith M. Patch, from Rosalie Edge to Rachel Carson. And she made legitimate advocacy for nature on spiritual and emotional grounds by both women and men.

Notes

1 *Handbook of Nature-Study*, p. ix.
2 Ibid.
3 Ibid.
4 Leo E. Klopper and Audrey B. Champagne, 'Six Pioneers of Elementary School Science', University of Pittsburgh, Manuscript Draft, p. 299, 1975.
5 *The Comstocks of Cornell*, p. 57.
6 Pamela M. Henson, 'Through Books to Nature: Anna Botsford Comstock and the Nature Study Movement', in T. Gates and Ann B.

Shteir, *Natural Eloquence: Women Reinscribe Science*, Madison, WI: University of Wisconsin Press, p. 116, 1997.

7 Ibid., pp. 118–19.

8 Richard Raymond Olmsted, 'The Nature-Study Movement in American Education', Indiana University, Dissertation, p. 2, 1967.

9 Liberty Hyde Bailey, *The Nature-Study Idea: Being an Interpretation of the New School-Movement to Put the Child in Sympathy with Nature*, New York: Doubleday, Page & Company, p. 7, 1903.

10 *Handbook of Nature-Study*, p. 4.

11 Ibid., p. 3.

12 Speech delivered at Philadelphia, 30 December 1914, entitled 'The Growth and Influence of the Nature-Study Idea'.

13 Comstock, 'The Attitude of the Nature-Study Teacher toward Life and Death', *Nature-Study Review*, 5 (May), p. 121, 1909.

14 Henson, op. cit., p. 128.

15 Marcia Myers Bonta, *Women in the Field: America's Pioneering Women Naturalists*, College Station, TX: Texas A & M University Press, p. 164, 1991.

16 Speech delivered at Philadelphia, 30 December 1914.

See also in this book

Darwin, Emerson, Rousseau

Comstock's major writings

Manual for the Study of Insects, Ithaca, NY: Comstock Publishing Company, 1895.

Ways of the Six-Footed, Ithaca, NY: Cornell University Press, 1903.

How to Know the Butterflies: A Manual of the Butterflies of the Eastern United States, with John Henry Comstock, New York: D. Appleton Publishing Company, 1904.

Confessions of a Heathen Idol, originally published under the pseudonym Marian Lee, New York: Doubleday, Page, & Company, 1906.

Handbook of Nature-Study, Ithaca, NY: Comstock Publishing Associates, 1911.

The Comstocks of Cornell, with John Henry Comstock, Ithaca, NY: Comstock Publishing Associates, 1953.

Comstock also wrote many essays for *The Nature-Study Review*, 1904–23.

Further reading

National Society for the Scientific Study of Education, *The Third Yearbook, Part III, Nature Study*, Chicago, IL: University of Chicago Press, 1904.

PETER BLAZE CORCORAN

RABINDRANATH TAGORE 1861–1941

> I still remember the very moment, one afternoon, when I
> ... suddenly saw in the sky ... an exuberance of deep,
> dark clouds lavishing rich, cool shadows on the atmo-
> sphere. The marvel of it ... gave me a joy which was
> freedom, the freedom we feel in the love of our friend.[1]

Rabindranath Tagore was a great poet and profound thinker. He was
born in Calcutta on 6 May 1861. He belonged to a family which is
the most gifted in Bengal in the realm of religion, philosophy,
literature, music and painting. Although he was not educated in any
college or university, he was clearly a man of learning. He had his
own original ideas about education which led him to establish an
educational institution at Shantiniketan in December 1901 following
the model of the forest hermitages of ancient India. He named it
Viswa Bharati with the intention of re-opening the channel of
communication between the East and the West. He was a versatile
genius. There is no aspect of literature – poetry, short story, novel,
drama – which he has not enriched. He was awarded a Nobel Prize
in 1913 in recognition of his outstanding literary activities. Equally
important are his innumerable essays and many books which reveal
his deep socio-political as well as spiritual commitments. He was
also a most original composer of music. He travelled extensively in
different countries of the world, and was a successful mediator
between Western and Eastern cultures. He died on 7 August 1941.

Crucially, Tagore's poems, short stories and novels, as well as
books and essays, exhibit his love and concern for nature, for land,
sea, air, plants and animals that constitute the 'environment' around
us. His concern or thinking about the environment is not, however,
activated by any pragmatic or utilitarian consideration. Rather it
grows on a different – non-utilitarian – ground. And here we may
profitably utilize his idea of 'surplus'. The surplus in man which,
according to Tagore, constitutes his spiritual make up, overflows
pragmatic need, the stage of pure utility, and 'extends beyond the
reservation plots of our daily life'.[2] This surplus indicates an aspect
of human being, 'a fund of emotional energy' which is 'useless' or
'superfluous' in the sense that it is not regulated by self-interest, by
any moral or other practical ends. Thus the point is that we no doubt
have one side which is governed by pragmatic necessity, but parallel
to it we have also another side – a spiritual one – which requires
fulfilment of our creative urge, our capacity to appreciate and enjoy.
And our life cannot be meaningful in the strict sense of the term by

pragmatic fulfilment alone, without this spiritual fulfilment. This is what Tagore wants to convey by his notion of 'surplus'.

This insistence on surplus gives us the clue as to why the environment matters to Rabindranath; why he would want to see it defended against any unnecessary tampering . Nature is dear to him, since with all its enthralling beauty it can evoke our appreciation, and thus fulfil the demand of the surplus in us. To put it in a different way, he entertains nature in terms of the aesthetic appreciation or delight that it prompts. 'Would they not attract me from all sides – / These trees, creepers, rivers, mountains and woods / The deep blue eternal sky?'[3] This explains clearly why natural environment with its 'special harmony of lines, colours and life and movement' should be preserved.[4] It should be preserved, for it gives us aesthetic joy, and thereby a bond of love is established between it and us.

But this defence of the environment on aesthetic ground will not enjoy the approval of all ecologists even in India. Some will condemn it as an anthropocentric denial of the intrinsic value of nature. Let us ponder how far it is fair to bring this charge against Rabindranath. Strictly speaking, the use of the words 'intrinsic' and 'anthropocentric' is infected by ambiguity: 'An object X has intrinsic value' may be understood in at least two senses:

1   'X has intrinsic value' may be understood to mean that 'X has non-instrumental value', i.e. the value of X does not consist in its being a means to some end. So 'X has intrinsic value' will denote that X is an end-in-itself. This is the sense in which many environmentalists consider the value of nature. Hence it will be wrong, in their opinion, to view nature only as instrument for serving some end of man. That would be anthropocentric imperialism. I call this anthropocentrism in the first sense.
2   'X has intrinsic value' refers to what may be designated as 'objective' value. An objective value is that which X possesses independently of any human evaluation. The denial of objective value in this sense will amount to what I call anthropocentrism in the second sense.

The view of Rabindranath indeed has an anthropocentric flavour at least in the second sense, since he thinks that no account of value can be isolated from all relations to human being. That is why he observes, 'What we call nature is what is revealed to *man* as nature.'[5] Or, 'Reality is ... [that] by which we are affected, that which we express.'[6] Evidently Rabindranath wants to emphasize that even to say that nature has value must involve some reference to man, to his being affected by it. There is nothing wrong in highlighting this human reference. It does not mean that values are conferred on things by man. What it implies is the crucial truth that even if, like the ecologists, we grant values to nature on account of the qualities it has, this has no real sense or bearing unless we are able to understand 'why something with those qualities should matter to us, how it might fit into the orbit of our concerns'.[7]

But to hold that Rabindranath takes an anthropocentric attitude in the second sense to nature (which sounds quite reasonable) is not to hold that he is inclined towards anthropocentrism in the first sense, towards the stronger claim that nature is only a means for satisfying human purpose. Rabindranath's point that value pre-supposes human evaluation is only a 'formal' one about how value is to be understood; but from this does not follow the stronger claim, which is a 'substantial' one about what makes something valuable.[8]

That Tagore would not endorse any instrumentalism is strength-ened by another consideration of his when, like Kant, he employs, as already suggested, the concept of 'disinterestedness'. He talks about aesthetic enjoyment – 'the enjoyment which is disinterested'.[9] The disinterestedness of aesthetic contemplation can be made explicit by the idea of an 'alternative world'.[10] The same forest which is the source of one's livelihood can open a different horizon – an alternative world – which is unconnected with any question of livelihood, with any pragmatic concern or interest. Then the smell of grass, the graceful movement of boughs of trees, the sweet melody of birdsongs begin to move us in a new way. Thus emerges the aesthetic moment when the forest is imaginatively explored and when any thought of using it for our interest or personal benefit becomes completely redundant.

This comes out more clearly from Tagore's insistence, as indicated in the opening quotation, on the relation of love we enter into with nature in our aesthetic contemplation of it. Inspired by the teachings of the *Upanishads*, he holds that when I love anyone, I cannot think of seeing my beloved in the light of any usefulness. On the contrary, I find in my beloved an extension of my own being which gives me the feeling of real freedom. It is this relation of love or of heart that we have with nature in our aesthetic experience of it. Hence this relation must be 'superfluous', i.e. beyond the bounds of any interest or satisfaction of practical purpose. 'There is an element of [the] superfluous in our heart's relation with the world.'[11]

Incidentally, but very crucially, this disinterestedness will also enable Rabindranath to meet the challenge often made by ecologists that aesthetic appreciation, since it admits of variations, cannot be effectively utilized as the ground for environmental preservation. Even if we concede that aesthetic appreciation is variable, there is yet a very good sense of it that we can hopefully attend to in the context of environment protection. The concept of disinterestedness helps us extract this good sense. As Kant puts it: 'where anyone is conscious that his delight in an object is with him independent of interest, it is inevitable that he should look on the object as one containing a ground of delight for all men'.[12] In other words, if aesthetic appreciation is based on disinterestedness, as Tagore thinks it is, we can very reasonably be assured that nature can give rise to the same appreciation or delight in others as it does in my case. And then it can well provide a formidable reason in favour of environment preservation.

I have tried to explain and defend Tagore's thinking about the environment on aesthetic and spiritual grounds. True, some environmental thinkers would not receive him well. Yet it is also true that his emphasis on the beauty of nature endeared him to many of his eminent contemporaries both in India and abroad. Note how D.R. Bhandarkar, a great Indian thinker, approves of and admires his sensitivity to nature: 'Everywhere in his poems and songs you see sunshine ... still night and various aspects of nature ... His is a mind most responsive to nature.'[13] Similarly, another eminent writer, Lim Boon Keng from the University of Amoy, China, writes that 'His soul seems at once to vibrate in full harmony with the orchestra of melodies and echoes reflected from the sound of rushing waters, from the songs of birds, from the rustling of leaves ...'[14] And it cannot be denied that caring for nature on aesthetic grounds, as Tagore did, has now become one of the major environmental concerns in the developed countries of the world.

Notes

1   *The Religion of an Artist*, pp. 16–17.
2   *The Religion of Man*, p. 33.
3   Tagore, 'Vasundhara', trans. in Rabindranath Choudhury, *Love Poems of Tagore*, Delhi: Orient Paperbacks, p. 55, 1975.
4   *The Religion of Man*, p. 85.
5   Ibid., p. 72, original emphasis.

6 Ibid., p. 83.
7 D.E. Cooper, 'Aestheticism and Environmentalism', in D.E.Cooper and J.A.Palmer (eds), *Spirit of the Environment*, London: Routledge, p. 103, 1998.
8 Ibid., p. 102.
9 *Lectures and Addresses*, p. 79.
10 D.E. Cooper, op. cit., p. 109.
11 *Lectures and Addresses*, p. 93.
12 Immanuel Kant, *The Critique of Judgement*, trans. J.C. Meredith, Oxford: Clarendon Press, p. 50, 1928.
13 D.R. Bhandarkar, 'My Impressions about the Poet', in R. Chatterjee (ed.), *The Golden Book of Tagore*, p. 36.
14 Lim Boon Keng, 'The Beauty and Value of Tagore's Thoughts', in Chatterjee (ed.), op. cit., p. 125.

See also in this book

Gandhi

Tagore's major writings

Creative Unity, 1922, New Delhi: The Macmillan Company of India Limited, 1980.
The Religion of Man, 1970, paperback edn, London: Allen & Unwin, 1970.
The Religion of an Artist, 1936, Calcutta: Viswa-Bharati, 1988.

The standard edition of Tagore's writings is *The Works of Tagore* in 15 volumes, Calcutta: West Bengal Government, 1961. A one-volume selection of Tagore's important lectures is *Lectures and Addresses*, ed. Anthony X. Soares, New Delhi: Macmillan Pocket Tagore Edition, 1995.

Further reading

Chatterjee, Ramananda (ed.), *The Golden Book of Tagore*, Calcutta: The Golden Book of Tagore Committee, 1990.
May, Larry and Sharratt, Shari Collins (eds), *Applied Ethics: A Multicultural Approach*, Englewood Cliffs, NJ: Prentice-Hall, 1994.

KALYAN SEN GUPTA

BLACK ELK 1862–1950

Birds make their nests in circles, for theirs is the same religion as ours.[1]

Black Elk was born in 1862 on the banks of the Little Powder River, a tributary of the Yellowstone River in what is now the state of

Wyoming. Then it was in the westernmost territory of the Lakota. Black Elk belonged to the Oglala Band. His father and grandfather – both also named Black Elk – were medicine men. He followed them in this calling. Black Elk was born into a world radically different from the one in which he would die. It was a sacred world in which 'the two-leggeds and the four-leggeds lived together like relatives, and there was plenty for them and plenty for us'.[2] By the time of his death, at the age of 88, on the Pine Ridge Indian Reservation in South Dakota, the vast herds of game, especially bison, that his people hunted for their subsistence were a fading memory; the faces of four United States presidents had defaced Mount Rushmore in the Black Hills, which were sacred to the Lakotas; the Yellowstone Plateau was a National Park; and the prophecy of Drinks Water, a contemporary of Black Elk's grandfather, had been fulfilled: 'you shall live in square gray houses, in a barren land, and beside those square gray houses you shall starve'.[3]

Trouble began the year after Black Elk was born. As a young child, he never saw a 'Wasichuö' (the name means not 'white', but 'too-many-to-count'), but he grew up hearing of them. Black Elk's mother would invoke the name as a bugbear: 'If you are not good the Wasichus will get you'.[4] His father was wounded fighting the Wasichus when Black Elk was only three. Black Elk later fought for his people and saw their defeat and dispossession. He was a cousin of the great Lakota warrior, Crazy Horse. He was an eye witness of Custer's Last Stand at the Battle of Little Big Horn: 'These Wasichus wanted it, they came to get it, and we gave it to them'.[5] Black Elk participated in the Ghost Dance, a millenarian pan-Indian religious revival. Although at first sceptical, it was, indeed, he who dreamed of and reproduced the famous Ghost Shirt that was supposed to protect its wearer from bullets. Black Elk was present at the slaughter of more than 300 Lakota men, women and children at Wounded Knee Creek, the last 'battle' of the 'Indian wars' in the USA. He travelled to England and France as a dancer in Buffalo Bill's Wild West show. In short, Black Elk lived through the transformation of the central plains of North America from its aboriginal condition inhabited by indigenous peoples to a land of Wasichu farms, ranches, railroads, highways, power lines, towns, motels, monuments, parks, diners, movie theatres, and all the other trappings of modern American civilization. And he participated in some of the most legendary events in the history of the American West.

After the murder of Crazy Horse and the pacification and

reservationization of the Plains Indians, Black Elk undertook a vision quest, and began his career as a Thunder-Being medicine man, age 17. As a condition of employment in Buffalo Bill's troupe, he converted to Christianity in his mid-20s, and seems, during his three years abroad (1886–9), to have been a sincere and devout convert. Then, the fervour of the Ghost Dance, which swept the country in 1889 and 1990, encouraged him to return to his native religious beliefs. Wounded Knee ended the Ghost Dance episode in American history on 29 December 1990, and embittered and demoralized the Lakota, who were the sole victims of the massacre. Afterwards, Black Elk, like most of the Lakota, turned his back on European-American culture, and defiantly continued to practise traditional medicine, which put him in conflict with the missionaries on his reservation. As the psychic and spiritual wounds of the tragedy at Wounded Knee scarred over and the nineteenth century gave way to the twentieth, Black Elk slowly abandoned his traditional medical practice and the religious world-view in which it was embedded in favour of Catholicism and modernity. In this transformation, he may been encouraged by his first wife, Katie War Bonnet. He was baptized in 1904 and given the name Nicholas.

Above all else, Black Elk was a religious genius, and he turned this genius into a career as a catechist in the Catholic Church's St Joseph Society, spreading the gospel to other Lakotas in their own language. For the next ten years, he travelled the Great Plains as something of a Native evangelist. With fragile health (he suffered from tuberculosis) and failing eyesight, Black Elk quit travelling and settled on the Pine Ridge reservation, the head of a large family, a pillar of the Church. His humble home was a centre of Catholic social life, and he was a man to whom the missionaries pointed with pride as a model of their success in leading the Lakota from the darkness of heathenism into the light of Christianity and civilization.

In August 1930 John G. Neihardt came to Pine Ridge looking for informants on the Ghost Dance and the massacre at Wounded Knee for the final volume of his epic poem *Cycle of the West*. He was directed to Black Elk, who seemed to be expecting him, in the traditional manner of a prescient shaman recruiting a spirit-designated apprentice. The two immediately discovered they had an extraordinary rapport. At the end of the day Black Elk said: 'There is so much to teach you. What I know was given to me for men and it is true and it is beautiful. Soon I shall be under the grass and it will be lost. You were sent to save it, and you must come back, so that I can teach you.'[6] Neihardt did return the following spring, not for the purpose of fulfilling his own agenda, but Black Elk's. A special teepee was erected. In it, Black Elk spoke for many days to Neihardt in Lakota; Black Elk's son Benjamin interpreted; and Neihardt's daughters, Enid and Hilda, recorded the translation, from which they later made typescripts. Neihardt then drew upon his literary skills to craft these interviews into *Black Elk Speaks*, one of the greatest achievements of American letters, and a genre exemplar in post-colonial American Indian literature. According to Vine Deloria, Jr, a Lakota philosopher and activist, the book has realized Black Elk's intent and more: 'The most important aspect of the book ... is not its effect on the non-Indian populace who wished to learn something of the beliefs of the Plains Indians, but upon the contemporary generation of young Indians who have been aggressively searching for roots of the structure of universal reality. To them the book has become a North American bible of all tribes ... So important has this book become that one cannot today attend a meeting on Indian religion and hear a series of Indian speakers without recalling the exact parts of the book that lie behind contemporary efforts to inspire and clarify those beliefs that are "truly Indian".'[7]

It is a mistake to suspect that *Black Elk Speaks* is solely the product of Neihardt's romantic imagination. The typescripts of the 1931 interview were preserved among Neihardt's papers in the archives of the University of Missouri and were published in 1985. Comparison with these shows the book to be a faithful rendition. Neihardt's contribution was in fact purely literary, editing the narrative, simplifying and stylizing the prose. Indeed, in Neihardt's own estimation, *Black Elk Speaks* is 'the first absolutely Indian book thus far written ... all out of the Indian consciousness'.[8] How Black Elk's poignant account of the 'truth' and 'power' of his Great Vision – vouchsafed to him when he was only a 9-year-old boy, as innocent of missionary propaganda as he was of all things Wasichu – may be reconciled with his later and never recanted devotion to Christianity remains unclear. In response to Neihardt's question about that, he said simply, 'My children had to live in this world'.[9] *Black Elk Speaks* should, therefore, be taken at face value – as an authentic window into the traditional Lakota world-view (if not that 'of all tribes').

And when we look through that window, what do we see? Many wonderful things, including a powerful environmental ethic.

The Lakota world-view, although thoroughly indigenous, is hardly aboriginal. As late as the eighteenth century, the Lakota were a woodland people living in the region of the western Great Lakes. They were pushed out onto the plains by the Algonkian-speaking Ojibwa in a kind of domino-effect of expanding European settlement of the American Eastern Seaboard. They quickly adopted the mounted bison-hunting plains culture that was already established, and which was itself a post-Columbian phenomenon. Although evolved in North America, the horse, upon which reliable bison hunting depended, had been extinct in the Western hemisphere for ten thousand years. It was reintroduced by the Spanish, and the domesticated species re-established feral populations on the vast grasslands of North America. It was welcomed by the Indians of the interior, not, as formerly, a game animal, but as a beast of burden and a companion in war and in the chase. Further, the Lakota themselves recognized that their sacred-pipe religion is of recent historical origin in the myth of White Buffalo Cow Woman, who gave it to them.

The Lakota world-view grew out of and reflected the relatively featureless, open spaces of the Great Plains. Its parameters are six in number – sky, earth, and the cardinal directions: west, north, east and south – each personified as a 'power'. *Black Elk Speaks* opens with an invocation and an explanation of the symbolism of the sacred pipe:

> These four ribbons hanging here on the stem are the four corners of the universe. The black one is for the west where the thunder beings live to send us rain; the white one for the north, whence comes the great white cleansing wind; the red one for the east, whence springs the light and where the morning star lives to give men wisdom; the yellow for the south, whence come the summer and the power to grow.[10]

Either the traditional collective Lakota world-view is very abstract and sophisticated or Black Elk's own personal version of it is, for there is a unity within this multiplicity that one scholar compares to the concept of *Brahman* in Vedic Hindu philosophy, to the mystery of the Trinity in Christian theology (one God, three persons), and to the monism of the early modern European philosopher Benedict Spinoza.[11] The unifying concept is *Wakan Tanka*, the 'Great Spirit', whom Black Elk often refers to as 'Grandfather':

> But these four spirits are only one Spirit after all, and this

eagle feather here is for that One ... Is not the sky a father
and the earth a mother, and are not all living things with feet
or wings or roots their children? And this hide upon the
mouthpiece here, which should be bison hide, is for the
earth, from whence we came and at whose breast we suck as
babies all our lives, along with all the animals and birds and
trees and grasses.[12]

So, in brief, the sky is a universal father; the earth, a universal
mother; each of the four quarters (sometimes also called winds and
each associated with its distinctive colour) is a spirit with a peculiar
power. All are united, however, in the Grandfather (as distinct from
the Father) Spirit, *Wakan Tanka*, the Great Spirit.

This world-view is the foundation of an environmental ethic,
which is quite expressly stated in *Black Elk Speaks*, albeit with
characteristic simplicity and brevity: after invoking each of these
spirits individually and the Great Spirit, of which they are all
particular manifestations, Black Elk prays: 'Give me the strength to
walk the soft earth, a relative to all that is!'[13] Black Elk's rhetoric
routinely implies a familial egalitarianism among all the children of
Father Sky and Mother Earth – human animal, non-human animal
or plant. Human beings differ from other living beings only in
number of legs, or the absence of wings or roots. Again, this
egalitarianism is expressly stated briefly and simply: 'all over the
earth the faces of living things are all alike'.[14] In bad things as well as
good, the native two-leggeds and four-leggeds share a common
destiny: 'the Wasichus came, and they have made little islands for us
and other little islands for the four-leggeds, and always these islands
are becoming smaller, for around them surges the gnawing flood of
the Wasichu'.[15]

The Lakota environmental ethic is similar to, but, in important
ways, also differs from the familiar 'land ethic' formulated by Aldo
Leopold in 1949. The Leopold land ethic is based on a social model
of nature, which is similarly egalitarian – in which a human being is
but a 'plain member and citizen' of the 'biotic community'.[16] But
nature in the land ethic is represented as one big *society*, while in the
Lakota environmental ethic nature is portrayed as one big *family*.
According to the ecological 'community concept', each species
occupies a niche, role or profession in the economy of nature. Just as
in the human social microcosm there are farmers, truckers and
doctors, each specializing in a particular task, so in the natural
macrocosm there are producers (the green plants), consumers

(animals of all sorts), and decomposers (fungi, bacteria and the like). And just as our non-privileged membership in human communities generates our human-to-human ethics, so our 'plain' membership in biotic communities generates land ethics, according to Leopold. In the Lakota environmental ethic, however, the relationship of human beings to nature seems closer, warmer – just as our relationship to a family member is more intimate and our obligations more compelling than to a fellow citizen of our municipality or country. Instead of a 'land' environmental ethic, perhaps we could call Black Elk's a 'family' environmental ethic.

Notes

1 *Black Elk Speaks, Being the Life Story of a Holy Man of the Oglala Sioux*, p. 199.
2 Ibid., p. 9.
3 Ibid., p. 10.
4 Ibid., p. 13.
5 Ibid., p. 127.
6 Ibid., p. 10.
7 Vine Deloria, Jr, 'Introduction', *Black Elk Speaks, Being the Life Story of a Holy Man of the Oglala Sioux*, Lincoln, NE: University of Nebraska Press, pp. xii–xiii, 1979.
8 John G. Neihardt to Julius T. House, 3 June 1931, *The Sixth Grandfather*, p. 49.
9 Ibid., p. 47.
10 *Black Elk Speaks*, p. 2.
11 *The Sacred Pipe.*
12 Ibid., pp. 2–3.
13 Ibid., p. 6.
14 Ibid.
15 Ibid., p. 9.
16 Aldo Leopold, *A Sand County Almanac and Sketches Here and There*, New York: Oxford University Press, p. 204, 1949.

See also in this book

Leopold

Black Elk's major writings

Neihardt, John G., *Black Elk Speaks, Being the Life Story of a Holy Man of the Oglala Sioux*, New York: Morrow, 1932.
—— *When the Tree Flowered: An Authentic Tale of the Old Sioux World*, New York: Macmillan, 1951.
Brown, Joseph Epes, *The Sacred Pipe: Black Elk's Account of the Seven Rites of the Oglala Sioux*, Norman, OK: University of Oklahoma Press, 1953.
DeMallie, Raymond J. (ed.), *The Sixth Grandfather: Black Elk's Teachings*

Given to John G. Neihardt, Lincoln, NE: University of Nebraska Press, 1984.

Further reading

Deloria, Ella C., *Dakota Texts*, Publications of the American Ethnological Society, no. 14, New York: G.E. Stechert, 1932.

Lame Deer, John (Fire) and Erdoes, Richard, *Lame Deer: Seeker of Visions*, New York: Simon & Schuster, 1972.

Luther Standing Bear, *Land of the Spotted Eagle*, Boston: Houghton Mifflin, 1933.

Rice, Julian, *Black Elk's Story: 'Distinguishing Its Lakota Purpose'*, Albuquerque, NM: University of New Mexico Press, 1991.

Walker, James R., *Lakota Belief and Ritual*, ed. Raymond J. DeMallie, Lincoln, NE: University of Nebraska Press, 1980.

J. BAIRD CALLICOTT

FRANK LLOYD WRIGHT 1867–1959

> What, then, is architecture? It is man in possession of his earth. It is the only true record of him ... While he was true to earth his architecture was creative.[1]

Frank Lloyd Wright was an American architect whose early designs were the catalyst for the emergence of Modern architecture around 1900, and whose seventy-two-year career has been the single greatest influence on the architecture of the twentieth century. Today, forty years after his death, Wright is the most famous architect in the world, and his designs, including Unity Temple, Fallingwater and the Guggenheim Museum, are among the most well-known works of architecture built in the twentieth century.

Wright was born in Richland Center, Wisconsin, in 1867, and raised in a family where the study of nature, the Unitarian faith and the ideas of American transcendental philosophy were all powerfully present. Aged 20, without any formal university training, Wright moved to Chicago and entered the practice of architecture. After five years in the office of Louis Sullivan, leader in the development of 'organic' architecture and the skyscraper, Wright opened his own practice in 1893. During the next sixty-six years, Wright designed over six hundred built works, revolutionizing architecture as we understand it in the modern world.

Wright idealized Nature (which he spelled with a capital N) as the absolute reference and evaluative measure for the works of man.

Nature was the source of both ethical principles, for the living of life, and formal principles, for the design of architecture. Wright based this interpretation of nature on the writings of the American transcendental thinkers, Walt Whitman, Henry Thoreau, Horatio Greenough and, most importantly, his beloved Ralph Waldo Emerson. The transcendentalists held as fundamental the fact that the material and spiritual worlds were inseparable, being in fact one and the same. Nature was the ideal manifestation of divine order, and Emerson called on his readers to 'esteem nature a perpetual counselor, and her perfections the exact measure of our deviations'.[2] Every physical thing, natural or man-made, was the consequence of, and had consequences for, spiritual thought – all form had moral meaning. 'All form is an effect of character',[3] Emerson said, and Wright believed that a person's character was an effect of the form and construction of the place in which they dwelled: 'Whether people are fully conscious of this or not, they actually derive countenance and sustenance from the "atmosphere" of the things they live in or with. They are rooted in them just as a plant is in the soil in which it is planted.'[4]

Wright's formal principles of architectural design were also drawn from the natural world. The formative experience of working on his uncle's farm during the summers of his childhood established Wright's great love and respect for nature. The Friedrich Froebel kindergarten training Wright received transformed this naïve love of nature into a precise method of making form. Based on learning from nature, Froebel training taught the child to seek the fundamental geometries underlying all natural forms. From this training, reinforced by his later studies of nature-based ornament with Sullivan, Wright would develop his definition of architectural design: discovering the underlying geometric structure of nature and building with it. Wright believed that man does not learn from nature by merely copying its surface effects – the underlying structure and geometry of nature were nature's true gifts for the architect, to be discovered only through close analysis of both natural forms and their determining functions. The ideal of an 'organic' architecture, first proposed by Horatio Greenough in his 1852 essay 'Form and Function', and defined thirty-five years later by Wright's mentor Louis Sullivan as 'form follows function', was redefined by Wright as 'form and function are one'. Wright sought to build an architecture that attained the perfect fusion of geometric form and life-giving function he found in his studies of nature.

Yet for Wright, nature as the ideal source of geometric order for design (Whitman's 'the square deific') was not to be confused with the particular building site or landscape in which he was called upon to work. While idealized Nature was sacred, the inhabited landscape was always in need of the redemptive power of design. Wright believed that no site selected for building was ever untouched by the hand of man. 'Fallingwater', the most famous modern house in the world, was designed by Wright in 1935 on what most visitors today assume was a 'wild, natural' site, but which in fact had been inhabited for more than forty years – this most 'natural' house itself sits in the hill cut of a pre-existing road. Wright believed that humans never built in and inhabited the natural world without fundamentally changing it, but he felt that, if the architect worked with the underlying geometric order of nature, it was possible to make the built landscape as beautiful, in its own way, as wild nature.

The vast majority of Wright's buildings were built in the American suburbs, where the original landscape had been sub-divided into lots served by street grids and utilities, and where often most of the original trees and vegetation had been removed before any houses were built – these suburbs were far indeed from being 'natural' places. From the very beginning of his career in Oak Park, a suburb of Chicago where Wright built his house in 1889, Wright conceived of the architect's task in designing houses for the American suburbs to be one of reconstitution of a lost natural balance, a nature now fundamentally changed through the inhabitation of man. For Wright, man was an integral part of nature; 'Man takes a positive hand in creation whenever he puts a building on the earth beneath the sun. If he has birthright at all, it must consist in this: that he too is no less a feature of the landscape than the rocks, trees, bears, or bees of that nature to which he owes his being.'[5]

Wright thus conceived of architectural design as encompassing both the landscape and the architecture that engaged it. The 'Prairie Houses' of 1900–15, Wright's first important domestic design innovation, also involved an equally innovative (if rarely noted) strategy of relating to the landscape. Wright's Prairie Houses were often located at the edge of their suburban lots, allowing their gardens to occupy the geometric centre of the sites (usually reserved for the house itself), and weaving together interior and exterior spaces so that the house and landscape were inextricably bound to one another. Rather than the free-standing object in the landscape, so typical of much later Modern architecture, Wright from the very beginning of his career constructed a remarkable interdependence between house and landscape, such that neither appears complete without the other.

Wright's was a truly 'organic' design ethic, embracing both architecture and landscape, and all that takes place within them: 'buildings are the background or framework for the human life within their walls and the natural efflorescence without; and to develop and maintain the harmony of a true chord between them ... These ideals take the buildings out of school and marry them to the ground.'[6] Wright began each design by incorporating the formative power of the landscape as the primal place of inhabitation – the building literally began with the ground on which it was to stand: 'It is in the nature of any organic building to grow from its site, come out of the ground into the light – the ground itself held always as a component basic part of the building itself'.[7]

Wright believed that architecture was determined by 'the nature of materials' of which it was constructed. He believed that the way a space was experienced was directly related to the way it was made or constructed. Wright built with both the underlying structures of nature (the cantilevered skyscraper based upon the tree) and the actual materials of nature. Wright employed each material in its natural state, displaying its inherent colours and texture, whether it was stone mined from a nearby quarry, concrete cast into ornamented block, or wood cut in the mill, and exposing the marks of cutting and shaping inevitably involved in taking materials from nature and preparing them for use in construction. Wright employed each material so that it contributed its own unique character to the spatial experience of inhabitation – the 'natural house' was literally made from nature. In this way, Wright believed his buildings were natural places within and without, where man could truly be at home in nature.

For Wright, architecture was literally mankind's place in nature, our particular manner of dwelling on the earth, under the sky. Whatever the commission, Wright always designed for a balanced condition – man in nature and nature in man. In the public urban building, such as Unity Temple, Johnson Wax and the Guggenheim Museum, vertical sunlight fell from above, filtered through the 'natural' geometric forms of skylights, bringing nature deep into the very heart of the city. In the private suburban house, such as the Coonley House, the Robie House and the Jacobs House, horizontal views, sheltered by the brow of the broad overhanging roof, opened to the surrounding landscape, bringing nature all the way into the hearth at the centre of the house. The public urban building was given the arc of the sun, and the private suburban house was given the line of the land – sky and horizon, as respective boundaries of the natural world, brought by Wright into the spaces of daily life.

Wright held that it was essential for daily life to be lived in direct communion with nature, and that architecture should be designed as a place in nature. Wright believed, following Emerson and Thoreau, that because man was a product of nature, he was only able to learn about his own essential nature through regular and intimate contact with the natural landscape. Wright felt that the American democratic experiment would ultimately fail unless all its citizens had the opportunity to live intimately in nature. In 1935, the same year he designed Fallingwater, Wright designed the first of his 'Usonian' Houses, modestly priced prototype homes for the growing American middle class. The Usonian Houses were L-shaped in plan, framing two edges of their suburban sites in such a way that the garden was the centre of both the site and the spatial composition of the house itself. Flooded with light, these gardens became the focus of the house and the life that took place within it; Wright strove to 'make the garden be the building as much as the building will be the garden, the sky as treasured a feature of daily indoor life as the ground itself'.[8]

A life taking place in nature was what Wright sought to make possible through his house designs, and thus his opposition to the flattening of landscape contours or the mechanical control of climate: 'To me air conditioning is a dangerous circumstance ... I think it far better to go with the natural climate than to try to fix a special artificial climate of your own. Climate means something to man. It means something in relation to one's life in it.'[9] The remarkable energy-efficiency and unerring solar orientation of Wright's houses from the very beginning of his career, though unprecedented in architectural practice, is entirely consistent with his vision of architecture's harmony with nature. While often considered 'ahead of his time' in his willingness to embrace and employ technical developments, Wright remained absolutely opposed to the instrumental aspects of the modern industrial era that in any way diminished mankind's experience of being at home in nature. Primary among these were land speculation and speculative building, which Wright believed were inherently evil and unnatural, noting that in the typical American suburb 'architecture and its kindred, as a matter of course, are divorced from nature in order to make [architecture] the merchantable thing ... It is a speculative commodity.'[10]

Wright's designs engaged both the natural land form and the history of human occupation of the site. He believed agriculture (to care for and cultivate) and architecture (to build and to edify) were related human activities on the earth – the tending and transforming of the landscape. Broadacre City, designed in 1935, was Wright's greatest and most comprehensive counter-proposal to the crowding of the traditional city, but also to the isolation of both agrarian life and the developer's speculative suburb. For Wright, culture and cultivation were closely related, and the level of culture of a society was directly indicated in the level of cultivation of its landscape: 'You will find the environment reflecting unerringly the society'.[11] At the most fundamental level, Wright believed that the natural environment should be integrated into daily domestic life: each of his designs was intended 'to be a natural performance, one that is integral to site, integral to environment, integral to the life of the inhabitants'.[12]

Notes

1 'Architecture and Modern Life', *Collected Writings*, vol. 3, p. 222, 1937.
2 Ralph Waldo Emerson, 'Prudence', *Emerson's Essays*, New York: Harper & Row, p. 166, 1926, 1951.
3 Emerson, 'The Poet', ibid., p. 269.
4 Wright, *The Natural House*, New York: Horizon Press, p. 135, 1958.
5 'Architecture and Modern Life', p. 223.
6 'In the Cause of Architecture', *Collected Writings*, vol. 1, p. 95, 1908.
7 *The Natural House*, p. 50.
8 Ibid., p. 53.
9 Ibid., pp. 175–8.
10 'Architecture and Modern Life', p. 237.
11 'Concerning Landscape Architecture', *Collected Writings*, vol. 1, p. 57, 1900.
12 *The Natural House*, p. 134.

See also in this book

Emerson, Ruskin, Thoreau

Frank Lloyd Wright's major built works

Larkin Company Building (demolished), Buffalo, NY, 1903.
Darwin Martin House, Buffalo, NY, 1904.
Unity Temple, Oak Park, IL, 1905.
Avery Coonley House, Riverside, IL, 1907.
Frederick Robie House, Chicago, IL, 1907.
Frank Lloyd Wright House/Studio, 'Taliesin', Spring Green, WI, 1911–25.

Midway Gardens (demolished), Chicago, IL, 1913.
Imperial Hotel (demolished), Tokyo, Japan, 1914–22.
Aline Barnsdall 'Hollyhock' House, Los Angeles, CA, 1919.
Samuel Freeman House, Los Angeles, CA, 1923.
Edgar Kaufmann House, 'Fallingwater', Mill Run, PA, 1935.
Johnson Wax Buildings, Racine, WI, 1935, 1944.
Herbert Jacobs House, Madison, WI, 1936.
Frank Lloyd Wright House/Studio, 'Taliesin West', Scottsdale, AZ, 1937.
Florida Southern College, Lakeland, FL, 1938–59.
Guggenheim Museum, New York, 1943–59.
H.C. Price Company Tower, Bartlesville, OK, 1952.
Beth Sholom Synagogue, Elkins Park, PA, 1954.
Marin County Civic Center, San Rafael, CA, 1957.

Frank Lloyd Wright's major writings

Frank Lloyd Wright: Collected Writings, ed. Bruce Brooks Pfieffer,
New York: Rizzoli, 1992–5:

Volume 1: 1894–1930 (1992)
Volume 2: 1930–1932 (1992)
Volume 3: 1937–1939 (1993)
Volume 4: 1939–1949 (1994)
Volume 5: 1949–1959 (1995)

Further reading

Hoffmann, Donald, *Frank Lloyd Wright: Architecture and Nature*,
New York: Dover, 1995.
Levine, Neil, *The Architecture of Frank Lloyd Wright*, Princeton, NJ:
Princeton University Press, 1996.
McCarter, Robert, *Frank Lloyd Wright*, London: Phaidon Press, 1997.
Riley, Terrance (ed.), *Frank Lloyd Wright, Architect*, New York: Harry
Abrams/Museum of Modern Art, 1994.
Sergeant, John, *Frank Lloyd Wright's Usonian Houses*, New York: Whitney
Library of Design, 1976.

ROBERT McCARTER

MAHATMA GANDHI 1869–1948

the next step should not be destructive agriculture but the
planting of plenty of fruit trees and other vegetation.[1]

Although Gandhi has become a household name, the lean, saintly
looking bespectacled son of India who took on the British Empire

with his sharp wit and prolific pen is better known for his ethics of non-violence and truth-force than for his environmental philosophy. However, just as leaders of non-violent civil rights movements across the globe attribute their inspiration to Gandhi's strategy of making the oppressors confront their own unjust practices, leading environmental theorists and activists in India and other parts of the world defer to Gandhi's insights and practices in the area of ecology as well. While much of what Gandhi said or wrote on ecology is of an anecdotal nature, his criticism of structures antithetical to a healthy ecological life-world ramified into ideas which developed and were put into action in different areas of environmental concern. Gandhi's importance as an environmental thinker may be marked in terms of the strategies and vistas opened up by his pursuits, both public and private, towards a sustained animal and environmental liberation struggle. Looked at another way, Gandhi's environmental thinking is rooted in his larger philosophical and moral thinking.

The Mahatma ('great soul') was born Mohandas Karamchand Gandhi in Porbandar, now in the State of Gujarat, on 2 October 1869. As a child he had learned to appreciate the beauty of the coastal region washed by the Arabian seas and surrounded by temples, churches and mosques. Although by caste the Gandhis were merchants, his family held high office in the sovereign province's court and were devout Hindus. Very early on he came to the realization that morality is an inexorable part of the objective reality he preferred to call Truth rather than God, and that nature was a substance within this reality. Hence, as in traditional wisdom, nature was not there merely for human use or as an appendix to civilization but was a presence, much like one's nourishing nurse, to be respected. Gandhi's Hindu background taught him about the basic elements that constituted the physical and material world, namely, earth, water, fire, ether and space, which he saw ritually invoked in home worship (*puja*) as well as in meditational practices. Indeed, Hindu biocosmology, with its large pantheon of gods and goddesses, appeared to share these elemental constituents in varying measures and permutations.

During his education in England, Gandhi rediscovered the virtues of his family's vegetarianism, albeit on the moral grounding articulated by Henry Salt, and inspired by Shelley, Thoreau, Whitman and Ruskin. At the same time Gandhi sought out theosophists who initiated him into a non-ritual moral reading of the *Bhagavad Gita*; this instilled humanitarian ideals that were to take Gandhi further towards a complete break with Western civilization. In South Africa, where he went to practise as an attorney, Gandhi withdrew from time to time to deepen his understanding of Tolstoy, the Upanishads, Quakerism, the Gospels through contacts with Trappists, Methodists and Jewish acquaintances. He also tried his hand at living in a commune. The influence of Ruskin's *Unto This Last* led Gandhi to write his own treatise on *Sarvodaya* ('welfare for all') which became the basis of the movement of the same name which he launched upon his return to India in 1914. It was part of the larger programme he envisioned for India of *swadeshi* or 'self-sufficiency' and had outlined in the 1908 treatise *Hind Sawaraj*. Both socio-ethical directives, as well as that of non-violent resistance (*ahimsa*), were propelled by a common volitional determination he called '*satyagraha*' or 'truth-force'. Gandhi acknowledges the influence of the Jaina ethical precept of non-injury (which Buddhism and Hinduism also heed and which has its parallel in the Golden Rule of 'turning the other cheek' or 'non-resistance', as Tolstoy had christened this practice). Under Gandhi's impetus, however, this basically passive and individual stance becomes a positively empowering and collective experience with enormous potential for unleashing liberative but, at times, also coercive and indignant energies.[2]

From these general articulations and stances, also sprang the more practical ideal of minimal or 'reactionary' economy and Luddite manufacturing skills, such as the humble spinning wheel (*charkha*) and weaving of yarns (*khadi*), and small-scale farming. Gandhi also experimented extensively with 'earth treatments' and 'dietetics' as means of healing and rejuvenation that did not depend on chemical-based medicines and toxic pollutants. Personal ecology for him was the basis for social and environmental ecologies as well.[3] Traditional methods of farming, husbandry, and irrigation were explored in the Ashrams which Gandhi helped set up in different regions.

Gandhi's overall social and environmental philosophy is based on what human beings need rather than what they want. His early introduction to the teachings of Jains, theosophists, Christian sermons, Ruskin and Tolstoy, and most significantly the *Bhagavad Gita*, were to have profound impact on the development of Gandhi's holistic thinking on humanity, nature and their ecological interrelation. His deep concern for the disadvantaged, the poor and rural population created an ambience for an alternative social thinking that was at once far-sighted, local and immediate. For Gandhi was acutely aware that the demands generated by the need to feed and sustain human life, compounded by the growing industrialization of India, far outstripped the finite resources of nature. This might nowadays appear naïve or commonplace, but such pronouncements were as rare as they were heretical a century ago. Gandhi was also concerned about the destruction, under colonial and modernist designs, of the existing infrastructures which had more potential for keeping a community flourishing within ecologically-sensitive traditional patterns of subsistence, especially in the rural areas, than did the incoming Western alternatives based on nature-blind technology and the enslavement of human spirit and energies.

Perhaps the moral principle for which Gandhi is best known is that of active non-violence, derived from the traditional moral restraint of not injuring another being. The most refined expression of this value is in the great epic of the *Mahabharata*, (*c*.100 BCE to 200 CE), where moral development proceeds through placing constraints on the liberties, desires and acquisitiveness endemic to human life. One's action is judged in terms of consequences and the impact it is likely to have on another. Jainas had generalized this principle to include all sentient creatures and biocommunities alike. Advanced Jaina monks and nuns will sweep their path to avoid harming insects and even bacteria. Non-injury is a non-negotiable universal prescription. Gandhi relates this principle to the value that the *Bhagavad Gita* places on the welfare of all beings:

The one whose self is disciplined by yoga
Sees the self abiding in every being
And sees every being in the self;
He sees the same in all beings.[4]

The transcendence of the self from constricting human conditions of desire and attachment and the prudential ethic of not causing injury to other beings for fear of attracting more karma into one's soul is turned by Gandhi into a categorical value: one does X because X is right and it is also just from the position of the other.

This principle, more than anything else, becomes the foundation-stone for Gandhi's approach to environmental ethics. Much that can be gleaned from Gandhi's own practices, as noted earlier, is of anecdotal value. His obsession with the hygiene of man and animals alike – safer waste disposal systems and cleanliness of both the body and the surrounding environs – have been meticulously noted in the Gandhiana literature and his own writings. Gandhi's weakness, as many writers have pointed out, is that he did not compose a systematic treatise on this subject, nor did he lead a major ecological campaign in the way that he did political campaigns, such as the symbolic 'Salt March', an act of nationalist defiance against the British monopoly over access to sea-salt. His impact, nevertheless, has been tremendous, and Gandhi's visions, if not his words, have certainly left traces in the great works on ecological thinking, especially those of Arne Naess and other 'deep ecology' or pan-eco-theistic thinking in recent decades. Gandhi, with his advocacy of *sarvodaya* and radical empowerment of localized or microecono-culture, was a forerunner of the avant-garde movements nowadays associated with 'deep ecology' and the Greens. But Gandhi went further in some respects with his emphasis on the aboluteness of non-violence and *dharma*.

Gandhi was also adamant about the need for a rigorous ethic of non-injury in our treatment of animals.[5] On active environmental renewal projects, Gandhi wrote in 1926 that for India the next step should not be destructive agriculture but the planting of fruit trees and other vegetation as these provide nourishment, stability in the soil, and attract rainfall as well as provide fodder for the insect and animal world. The implications of such simple ecological wisdom have only just begun to dawn on a tech-fested agricultural economics. Likewise Gandhi's symbolic insistence on *khadi* spinning was instructive for avoidance of factory-emitted pollution, desalina-tion of soil through over-cultivation and dependence on raw materials produced through suffering caused on animals (e.g. silk and wool). Gandhi's advocacy of simple living through the principles of non-violence and holding steadfastly to truth challenge modern-day Hindus to reconsider their lifestyle engendered by pressures of contemporary consumerism. They have had to consider whether social duty can be expanded to include ecological community and whether the Hindu tradition can develop new modalities of caring for the earth.[6] Can *dharma* be re-interpreted in earth-friendly terms to meet the challenges of modern post-industrial 'civilization'?

Gandhian activists have attempted to deal with just these challenges. *Sarvodaya* has increasingly become a basis for a number of *asarkari* or NGO groups across India. Inspired by Gandhi and especially his wife-partner Kasturba's dedicated *sarvodaya seva* or service ideal, these groups regularly travel to remote villages to teach women and youth the virtues and simple practices of hygiene and earth-care. Rural development and alternative technology pro-grammes have been helping villagers to construct *chulas* or smoke-less ovens, mudbrick dwellings, and to utilize non-toxic organic fertilizers. Schools and colleges have been established to explore and promote safe ecological practices. Tribal groups have been encouraged to preserve the wild bushland, to curtail excessive use of wood for cooking, and to develop a technology for dealing with local conditions while resisting the technologies and wares brought in by profit-driven urban and corporate enterprises. Gandhians have not been unanimous on a complete biospheric egalitarianism, and most have come to accept small-scale 'soft' technology supplemented heavily with hand-crafting and local cottage industries.[7]

Active in northern regions of the subcontinent is Sunderlal Bahuguna, best known for his spectacular Himalayan campaign, known as the Chipko ('Hug the Trees') Movement, aimed at resisting environmental destruction, particularly by governmental agencies and corporate interests which, in exploiting the hill regions, leads inexorably to irreparable deforestation.[8] Bahuguna is also a great believer in locally renewable 'sustainable economy'; hence, he has been one of the leading critics of India's current policy of economic liberalization which has allowed the influx of multinational companies and unilateral concessions on produce and plant variety rights forced upon India by WTO treaties.

Another scene which has been drawing world-wide attention where similar non-violent resistance tactics have been used to raise awareness of environmental concerns is the Narmada Bacho Andolan in southern Gujarat. Environmentalists led by the veteran Medha Patkar have ceaselessly argued that the 3,200 dams planned on the Narmada and tributary rivers would cause immense damage to surrounding land mass which would also lead to the dislocation of 2,500 families in nearly 60 villages of tribal people who have lived along the river basins and maintained a healthy eco-community for countless generations. The Gandhian spirit lives on. There are numerous other grassroots groups and movements that invoke traditional wisdom and practical ethics in their expression of resistance to and concerns for radical transformations of the local environment. The supply of safer drinking water to rural areas, conserving rain water and utilizing dead water from hydro-electric dams, have become joint initiatives of NGOs, religious leaders and some State governments as well (e.g. the southern taluks around Puttaparthi in Andhra Pradesh).

The Bhopal incident in 1984 where the ill-maintained Union Carbide chemical plant unleashed thousands of tons of poisonous chemical fumes into the atmosphere, killing and disabling thousands of people, perhaps highlighted a particular kind of challenge facing

Gandhian environmentalists. The challenges of industrialization, modernity, globalization and a rapidly expanding liberal economy present Gandhians with a very different set of circumstances and contexts from those that Gandhi could have foreseen. These call for quite different sorts of responses on the environmental front, and they can only be forthcoming case by case. Still, there are a number of Gandhian followers who are prepared to 'risk their all' in order to meet these challenges for the sake of non-violent truth and to bring greater welfare to all beings on the planet Earth.

Notes

1   Letter to Kaka Kaleklar, 1926, quoted in I. Harris, *Gandhians in Contemporary India: The Vision and the Visionaries*, Lewiston, NY: Edward Mellon, p. 274, 1998.
2   Joan V. Bondurant, *Conquests of Violence*, Princeton, NJ: Princeton University Press, 1985.
3   *An Autobiography*, p. 271.
4   *The Teaching of the Gita*, VI.29.
5   Gandhi, *My Socialism*, Ahmedabad: Navajivan, pp. 34–5, 1959.
6   Christopher Key Chapple, 'Hinduism, Jainism and Ecology', *Earth Ethics*, Fall, pp. 16–18, 1998.
7   P. Bilimoria, 'Indian Religious Traditions', in D.E. Cooper and J.A. Palmer (eds), *Spirit of the Environment: Religion, Value and Environmental Concern*, London and New York: Routledge, pp. 1–14, p. 12, 1998.
8   Harris, op. cit., p. 265.

See also in this book

Buddha, Naess, Ruskin, Tagore

Gandhi's major writings

An Autobiography: The Story of My Experiments with Truth, Boston, MA: Beacon Press, 1957.
Collected Works of Gandhi, 90 vols, New Delhi: Ministry of Information and Broadcasting, India,1958–94.
The Essential Gandhi, ed. L. Fischer, New York: Vintage Books, 1962.
The Teaching of the Gita, Bombay: Bharatiya Vidya Bhavan, 1962.

Further reading

Bilimoria, P. and McCulloch, J., *Environmental Ethics*, Victoria: Deakin University, 1992.
Brown, Judith, *Gandhi: Prisoner of Hope*, New Haven, CT: Yale University Press, 1989.

Chapple, C.K., *Nonviolence to Animals, Earth, and Self in Asian Thought*, Albany, NY: State University of New York Press, 1994.

Joshi, Nandini, *Development Without Destruction*, Ahmedabad: Navajivan, 1992.

Macy, Joanna, *Dharma and Development: Religion as Resource in the Sarvodaya Self-Help Movement*, West Hartford, CT: Kumarian Press, 1983.

Naess, Arne, *Gandhi and the Nuclear Age*, trans. A. Hannay, Totowa, NJ: Bedminster Press, 1965.

Weber, Thomas, *Hugging the Trees: The Story of the Chipko Movement*, Delhi: Viking Press, 1988.

PURUSHOTTAMA BILIMORIA

ALBERT SCHWEITZER 1875–1965

> Man has lost the capacity to foresee and to forestall. He will end by destroying the earth.[1]

The publication of Rachel Carson's *Silent Spring* in 1962 is frequently regarded as the beginning of the modern environmental movement. It was to Albert Schweitzer that Carson dedicated the work, and she opened her text using his above words.

'In terms of intellectual achievement and practical morality', Schweitzer has been described as 'probably the noblest figure of the twentieth century.'[2] Born in 1875, he was brought up at Gunsbach in Alsace. His intellectual achievements span four major disciplines. He learnt the organ under Widor in Paris and eventually published *J.S. Bach, le musicien-poète* in 1905. He studied theology and philosophy at Strasbourg, Paris, and Berlin, and published major works of New Testament scholarship, most notably *The Quest of the Historical Jesus* (English translation 1910). In 1896 he made his famous decision to live for science and art until age 30 and then devote his life to serving humanity. Accordingly, despite his international reputation as a musician and theologian, he turned to medicine and qualified as a physician. In 1905 he resigned as principal of the theological college in Strasbourg and founded the hospital at Lambaréné in the heart of what was French Equatorial Africa.

By 1962 Schweitzer had already become a legend in his lifetime. Although his work in Lambaréné captured the public imagination, earning him the Nobel Prize for peace in 1952, Schweitzer considered that his most meaningful contribution, the one for which he most wished to be remembered, was his ethic of 'reverence for life'. Travelling slowly upstream in a tug-steamer – amidst the panorama of the tropical forest – on Gabon's Ogowe River, the 'unforeseen and unsought' phrase, 'reverence for life', 'flashed' into his mind. The phrase, simple as it is profound, unlocked for him the 'iron door' of ethical thought.

Although the concept of 'reverence for life' is now well known, it has been subject to a range of distortions, and it is important that we confront these in order to understand what Schweitzer meant by this term.

The first distorting lens is *legalism*. Contrary to many commentators, Schweitzer does not propound reverence as a new moral law but rather as 'ethical mysticism'. Ethical mysticism emerges out of reflection upon the 'will-to-life' (*Wille zum Leben*). 'The essential thing to realise about ethics', he writes, 'is that it is the very manifestation of our will-to-live.'[3] His use of the term 'will-to-live' is derived from Arthur Schopenhauer, the principal advocate of the German Voluntarist school, who articulated the phrase in *The World as Will and Idea* (1819). Schweitzer follows Schopenhauer's conviction that 'the essence of things in themselves, which is to be accepted as underlying all phenomena', is 'will-to-live'.[4] Whereas Immanuel Kant denied that the 'thing-in-itself' (his term for an 'object considered as it is independently of its cognitive relation to the human mind'[5]) was knowable, Schweitzer believed that the 'thing-in-itself' was the 'will-to-live' and readily ascertainable through the physiological make-up of animate phenomena. That which underlies all life – actually its very essence – is the will-to-live.

Schweitzer's metaphysics begins with the supposition that despite the diversity of individual things in the world, they all manifest the same inner essence. From a comprehension of oneself (the microcosm), one is able to acquire knowledge of the world (the macrocosm); the key to understanding the world is proper self-understanding. Schweitzer's argument largely rests on whether knowledge that originates from the inner experience of the will-to-live is more reliable than knowledge derived from empirical examination of the outer, physical world. The non-empirical quality of the will-to-live as the core self is a presupposition of his work. His view is that all empirical reality must, like himself, have an inner nature (will-to-live), and he uses this notion to offer a new account of the relationship between the self, the natural world and God.

It is from this reflection on the will-to-live that Schweitzer derives the ethic of reverence for life. Though he starts from the personal ('I am life which wills-to-live'), he goes on to assert the radical interdependence of all life. Each life 'wills-to-live' not in isolation, but 'in the midst of other wills-to-live'. This assertion is not as an ingenious dogmatic formula but rather a personal revelation:

> Day by day, hour by hour, I live and move in it. At every moment of reflection it stands fresh before me ... A mysticism of ethical union with Being grows out of it.[6]

This immediate, experiential identification of one's individual will-to-live (or life) with other life, and through life with Being, is the foundation of his ethical mysticism. Indeed, the mystical nature of the experience of reverence is implicit in the very word: 'reverence' (*Ehrfurcht*) implies 'awe', 'wonder' and 'mystery'.

The second distorting lens is *inviolability*. Many commentators have assumed that Schweitzer is proposing the moral inviolability of all life of whatever kind. It is true that he sometimes writes in such a way as to invite this misunderstanding. The ethical person, he maintains:

> tears no leaf from a tree, plucks no flower, and takes care to crush no insect. If in the summer he is working by lamplight, he prefers to keep the window shut and breathe a stuffy atmosphere rather than see one insect after another fall with singed wings upon his table.
>
> If he walks on the road after a shower and sees an earthworm which has strayed on to it, he bethinks himself that it must get dried up in the sun, if it does not return soon enough to ground into which it can burrow, so he lifts it from the deadly stone surface, and puts it on grass. If he comes across an insect which has fallen into a puddle, he stops a moment in order to hold out a leaf or a stalk on which it can save itself.[7]

At first sight the sheer practical impossibility of these injunctions presents itself. But what Schweitzer offers here are not *rules* but rather *examples* of what reverence for life may require in a given situation. Schweitzer's basic definition of the moral is that 'it is good to maintain and to encourage life, it is bad to destroy life or obstruct it'.[8] Beyond this statement, he affords the reader only instances of the kind of action expected from one who upholds this ethic.

The third distorting lens is *inconsistency*. Since Schweitzer defines reverence as an 'absolute' ethic which enjoins 'responsibility without limit towards all that lives',[9] it is perhaps not surprising that reverence is judged to entail inconsistency in practice. And Schweitzer himself has not escaped this charge. He notoriously captured fish to feed his sick pet pelican, engaged in a pre-emptive strike against poisonous spiders, and did not fully embrace vegetarianism until later in life. These apparent inconsistencies are made more glaring by his rejection of any moral hierarchy:

> The ethics of reverence for life makes no distinction between higher and lower, more precious and less precious lives. It has good reasons for this omission. For what are we doing, when we establish hard and fast gradations in value between living organisms, but judging them in relation to ourselves, by whether they seem to stand closer to us or farther from us? This is a wholly subjective standard. How can we know the importance other living organisms have in themselves and in terms of the universe?[10]

Some commentators have interpreted Schweitzer at this point as suggesting that no form of life should ever be destroyed and that all creatures from human beings to microbes should have the same moral standing. It is doubtful whether this was Schweitzer's intention. Rather what he is doing is rejecting here the long tradition of moral hierarchy which places humanity at the top of the pyramid of descending moral worth. Schweitzer would have admitted (as his personal examples demonstrate) that it is sometimes necessary to make choices between one form of life and another, but what he wanted to emphasize was the essentially *subjective* and *arbitrary* nature of these declarations.

Any time life is sacrificed or injured, either 'for the sake of maintaining [one's] own existence or welfare' or 'for the sake of maintaining a greater number of other existences or their welfare', one is no longer wholly 'within the sphere of the ethical'.[11] In other words, killing may be *'necessary'* but it can never be *'ethical'* as such. When one is constrained by 'necessity', one must bear the 'responsibility' and 'guilt' of having injured life. 'Whenever I injure life of any sort', wrote Schweitzer, 'I must be quite clear whether it is necessary. Beyond the unavoidable, I must never go, not even with what seems insignificant.'[12]

Having clarified aspects of Schweitzer's thought, it is now

possible to indicate some of his main contributions to the development of ecological consciousness.

The first and most important contribution concerns *the mystical apprehension of the value of life*. At the heart of many environmental controversies is the issue of value: whether beings outside of ourselves have value, of what kind, and why. What Schweitzer emphasizes is that the recognition and appreciation of the value of life is actually a mystical apprehension. This apprehension is 'primary' because all subsequent decisions and choices depend upon it. To understand Schweitzer at this point we do best perhaps to make a comparison with Plato. Plato describes philosophers in a democratic state as those who 'wrangle over notions of right in the minds of men who have never beheld *Justice* itself'.[13] Likewise, Schweitzer would maintain that one can have no proper sense of oneself and others in the world unless, first and foremost, one has a sufficient sense of the value of *Life* itself. Everything depends practically upon this prior recognition of value.

The second contribution concerns *service to life as practical mysticism*. In contrast to most mystics, Schweitzer maintains that the goal of union with the Divine is achieved not through contemplation, but primarily through service to other life:

> Ethics alone can put me in [a] true relationship with the universe by my serving it, co-operating with it; not by trying to understand it ... Only by serving every kind of life do I enter the service of that Creative Will whence all life emanates ... It is through the community of life, not community of thought, that I abide in harmony with that Will. This is the mystical experience of ethics.[14]

The phenomenon we call 'life', in short, is not something put here for our use or pleasure; we are part of 'life' (or as Schweitzer would say 'the will-to-live') and our role is to enhance and serve each and every manifestation of it.

The third contribution concerns the recognition of *the tragedy of life in conflict with itself*. Schweitzer is not a pantheist – that is, someone who thinks that the world is God or co-terminos with God. Indeed he is sharply critical of those who seek to deify the natural world as it is instead of recognizing its essentially tragic and incomplete nature. Schweitzer writes movingly of the world as 'the ghastly drama of the will-to-live divided against itself'.[15] To affirm life and the value of life is not to affirm the parasitical and predatory aspect of nature itself. Schweitzer's own preaching is clearly eschatological – that is, he looks forward to a time when creation will be renewed and redeemed. His pioneering work in the field of New Testament scholarship – especially on the teaching of Jesus and Paul – emphasizes Jesus as an eschatological figure who will inaugurate 'the Kingdom', understood as the liberation of all creation from its present predation and suffering. Reverence for life was for Schweitzer 'practical eschatology'.

The fourth contribution concerns *non-injury to life as the central ethical imperative.* 'A man is truly ethical', Schweitzer writes, 'only when he obeys the compulsion to help all life which he is able to assist, and shrinks from injuring anything that lives.'[16] 'The time is coming ... when people will be astonished that humankind need so long a time to learn to regard thoughtless injury to life as incompatible with ethics.'[17]

Schweitzer regarded traditional philosophy which restricted ethics to human-to-human relations as spiritually impoverished. He was deeply critical of animal experimentation, opposed hunting for sport and eventually embraced a vegetarian diet. His hospital at Lambaréné was a model of ecological responsibility: he went out of his way to preserve trees and flora, re-used every piece of wood, string and glass, and rejected modern technological developments which would have resulted in environmental degradation.

It is unsurprising then that Rachel Carson, and others, have found in Schweitzer an inspiration for a wider ecological ethic. When Carson received the Schweitzer Medal from the Animal Welfare Institute in 1963, she summed up her work in Schweitzerian-like terms: 'What is important is the relation of man to all life'.[18] An inscribed photograph of Schweitzer (together with a letter of thanks for the dedication of *Silent Spring*) were encased, centre-stage, in her study. According to Carson's housekeeper, Ida Sprow, it was 'her most cherished possession'.[19]

Notes

1 Rachel Carson, *Silent Spring*, New York: Houghton-Mifflin Company, p. v, 1962.
2 Magnus Magnusson and Rosemary Goring (eds), Chambers Biographical Dictionary, London: Chambers Harrap Publishers, p. 1314, 1990.
3 'The Ethics of Reverence for Life', p. 229.
4 *The Philosophy of Civilisation*, p. 236.
5 Ted Honderich (ed.), *Oxford Companion to Philosophy*, Oxford: Oxford University Press, p. 871, 1995.
6 *The Philosophy of Civilisation*, p. 310.

7   Ibid.
8   Ibid., p. 309.
9   Ibid., p. 311.
10 *The Teaching of Reverence for Life*, p. 47, 1965.
11 *The Philosophy of Civilisation*, p. 325, reprint of New York: Macmillan, 1949.
12 Ibid., p. 318.
13 Plato, *The Republic*, trans. F.M. Cornford, Oxford: Oxford University Press, VII.518, p. 232, 1969; emphasis added.
14 'The Ethics of Reverence for Life', p. 239.
15 *The Philosophy of Civilisation*, p. 312.
16 Ibid., p. 310.
17 Ibid., pp. 310–11.
18 Linda Lear, *Rachel Carson: Witness for Nature*, London: Allen Lane, p. 440, 1997.
19 Ibid., p. 438.

See also in this book

Carson

Schweitzer's major writings

Kulturphilosophie I: Verfall und Wiederaufbau and *Kulturphilosophie II: Kultur und Ethik*, Bern: Paul Haupt, 1923. Both volumes published together as *The Philosophy of Civilisation*, trans C.T. Campion, New York: Prometheus Books, 1987.
Aus Meinem Leben und Denken, Leipzig: Felix Meiner, 1931; *Out of My Life and Thought: An Autobiography*, trans. A.B. Lemke, New York: Henry & Company, 1990.
'The Ethics of Reverence for Life', *Christendom*, I (winter), 1936.
The Teaching of Reverence for Life, trans. Richard and Clara Winston, New York: Holt, Rinehart & Winston, 1965.
A Place for Revelation: Sermons on Reverence for Life, trans. David Larrimore Holland, ed. Martin Strege and Lothar Stiehm, New York: Macmillan Press, 1988.

Further reading

Brabazon, James, *Albert Schweitzer: A Biography*, London: Gollancz, 1976.
Clark, Henry, *The Ethical Mysticism of Albert Schweitzer*, Boston, MA: Beacon Press, 1962.
Ice, Jackson Lee, *Schweitzer: Prophet of Radical Theology*, Philadelphia, PA: Westminster Press, 1977.
Joy, Charles, *Schweitzer: An Anthology*, Boston, MA: Beacon Press, 1956.
Linzey, Andrew, *Animal Theology*, London: SCM Press, 1994.
Seaver, George, *Albert Schweitzer: The Man and His Mind*, London: A. & C. Black, 1947.

ARA BARSAM AND ANDREW LINZEY

ALDO LEOPOLD 1887–1948

> If the individual has a warm personal understanding of
> the land, he will perceive of his own accord that it is
> something other than a breadbasket. He will see land as a
> community of which he is only a member ... He will see
> the beauty as well as the utility of the whole, and know
> that the two cannot be separated. We love (and make
> intelligent use of) what we have learned to understand.[1]

Aldo Leopold was born in Burlington, Iowa, in 1887, the eldest of
Carl and Clara's four children, an American family of German
ancestry. Aldo had two brothers and one sister. His father was fond
of hunting and introduced his sons to it at an early age; and before
the advent of game laws, he imposed upon himself and his sons a
sporting ethic, including closed seasons and bag limits. Leopold
dedicated his first published book, *Game Management*, to his father,
'a pioneer in sportsmanship'.[2] His mother was interested in music,
especially opera; from her he acquired a keen aesthetic sensibility.
Leopold was educated in Burlington public schools, the Lawrenceville Preparatory School in New Jersey, and the Sheffield Scientific
School and Forest School of Yale University, from which he was
graduated in 1909 with a master's degree. He immediately joined the
US Forest Service and was posted to District 3, the Arizona and
New Mexico territories. After a shaky start, Leopold advanced
through the ranks to become Assistant District Forester in Charge
of Operations, the second highest position in the unit. He married
Estella Bergere in 1912, and together they reared five children,
Starker, Luna, Nina, Carl and Estella – all of whom have gone on to
distinguished careers in the geo-biological sciences or conservation.
In 1924, Leopold accepted a transfer to a comparable position at the
Forest Products Laboratory in Madison, Wisconsin. During his
fifteen-year tenure in the southwest, he was primarily interested in
'secondary' forest uses, especially recreational hunting. After four
years in his job as Assistant Director of the Forest Products lab,
Leopold resigned to pursue his vocation full time. Supported by the
Sporting Arms and Ammunition Manufacturers' Institute, he
conducted game surveys of eight midwestern states and worked on
a textbook of game management. In 1933, after several months of
unemployment during the depths of the Great Depression, Leopold
joined the University of Wisconsin as the nation's first professor of
game management. He spent the rest of his life conducting research,
teaching, writing and shaping conservation policy in this capacity.

Leopold died suddenly of a heart attack in 1948, just a week after learning that his new book manuscript, which would become *A Sand County Almanac*, had been accepted for publication by Oxford University Press.

Leopold was an innovator and a visionary. He was indeed a founder of a number of environmental fields. First, and most obviously, Leopold was a founder of game management, which became wildlife management in his own lifetime, then wildlife ecology, and finally now conservation biology. This, Leopold's central and eventually professional interest, grew directly out of his life-long passion for hunting, which he acquired in boyhood. The original idea was simple: for a variety of reasons American game was growing scarce in the first quarter of the twentieth century, and game management was essential for 'producing something to shoot'.[3] At the end of his life, Leopold envisioned the desire to 'seek, find, capture, and carry away' game animals becoming transformed into the desire to seek, find, capture, and carry away knowledge about animals of all kinds; that is, the transformation of the consumptive sport of hunting into the non-consumptive 'sport' of wildlife research.[4]

Leopold was a founder of the North American Wilderness Movement. In the 1920s he argued passionately and voluminously for a system of wilderness areas in the national forests, primarily for purposes of primitive and virile kinds of recreation. The first National Forest Wilderness Area, surrounding the headwaters of the Gila River in Arizona, was designated the year he left the region. Leopold helped form the Wilderness Society in 1935. His understanding of the importance of wilderness shifted, during his university years, from an emphasis on recreation to biological conservation. Designated wilderness areas were important to conservation for two reasons. First, they afforded a vital habitat for some 'threatened species' – those that, for whatever reason, do not co-exist well with human beings, our cities, suburbs, factories, dwellings, farms, ranches and mines.[5] Second, wilderness areas provide 'a base-datum of normality, a picture of how healthy land maintains itself as an organism'.[6] By reference to such base-data in 'each biotic province' we can measure the health of similar areas (which should also be conserved to the extent possible) that are used for timber extraction, grazing and farming.[7]

Leopold was a founder of ecological restoration, another very recently emerged formal conservation discipline. In his view, the main purpose of the University of Wisconsin Arboretum and

Wildlife Refuge in Madison was 'to construct ... a sample of original Wisconsin, a sample of what Dane County looked like when our ancestors arrived here in the 1840s'.[8] In addition to his restoration work at the Arboretum, Leopold spent leisure hours during the last thirteen years of his life restoring a property that he bought in 1935 on the banks of the Wisconsin River. The first part of his chief work, *A Sand County Almanac*, is devoted to literary sketches of this place. In the Foreword, Leopold describes this section of the book in terms of ecological restoration. 'Part I tells what my family sees and does at its week-end refuge from too much modernity: "the shack." On this sand farm in Wisconsin, first worn out and then abandoned by our bigger and better society, we try to rebuild, with shovel and axe, what we are losing elsewhere.'[9]

Leopold was a founder of ecosystem-management forestry, to which the US Forest Service has been converting since 1992. For most of the twentieth century, the Forest Service was devoted to an agronomic model of forestry, the purpose of which was, in Leopold's words, 'to grow trees like cabbages, with cellulose as the basic forest commodity'.[10] The alternative that Leopold envisioned 'sees forestry as fundamentally different from agronomy because it employs natural species and manages a natural environment rather than an artificial one'.[11] The current policy of ecosystem management adopts another conservation concept that Leopold formulated – 'land health' – as its norm. The basic idea is to manage forest ecosystems with the primary goal of restoring or maintaining their health, with commodity extraction an ancillary or subordinate goal. Land health is a concept that Leopold struggled to articulate during the last years of his life. He most frequently characterized it as 'the capacity for self-renewal in the biota'.[12] Today the concept is called 'ecosystem health', and is defined in terms of normal ecosystem processes and functions. Beginning in the 1990s, there is an International Society for Ecosystem Health, which has convened several global congresses, and an academic journal, *Ecosystem Health*, which has been in publication since 1994.

Leopold contributed foundationally to conservation philosophy. He himself closely associated land health with land integrity, the full complement of the native species of a biotic province in their characteristic numbers. He believed that preserving its integrity was a necessary and sufficient condition for preserving a particular piece of land's health. Today, ecosystem health and biological integrity are not so tightly coupled. The biological integrity of an area is a sufficient, but not a necessary, condition for its ecosystem health.

Certainly, that is, an ecosystem containing the full complement of its native species populations in their characteristic numbers will be healthy, but an ecosystem with a simplified biota, including non-native species, may also exhibit normal processes and functions. Leopold himself was especially interested in promoting land health as the conservation norm for the extensively modified farmscapes of southern Wisconsin. To this central conservation concern of his latter years, he linked both his concern for wilderness preservation and ecological restoration. Wilderness provided 'the most perfect norm' of ecosystem health; less perfect, but still useful is 'a reconstructed sample of old Wisconsin to serve as a benchmark ... in the long and laborious job of building a permanent and mutually beneficial relationship between civilized men and a civilized landscape'.[13] Leopold frequently characterized this relationship as 'a state of harmony between men and land'.[14]

Contemporary practitioners in many other environmental fields can (and do) legitimately claim Leopold as an important figure in its development. Take range management: while with the Forest Service, Leopold discovered connections between over-grazing, fire suppression and the disastrous shift from grassy forage to unpalatable brush in the southwest, and recommended management strategies for range recovery in the region. Take erosion control: Leopold was alarmed by the extensive grazing-related erosion he encountered in the southwest, and continued to be concerned about it in the midwest; and he worked in both regions to stanch it. So great, indeed, was his concern about erosion that it may give a more literal sense to his 'land ethic'. Take sustainable agriculture: Leopold's Chair of Game Management at the University of Wisconsin was at first located in the Department of Agricultural Economics and much of his work on the ground was with farmers, first to encourage them to 'grow' a 'crop' of wild game, and later to practise methods of farming that are more accommodating to wildlife of every kind. Shortly after assuming his academic duties, Leopold began monthly broadcasts over the University's extension radio station, addressed to farmers; the next year he began offering a Farmer's Short Course in Game Management; and, between 1938 and 1942, he published a series of thirty-four short 'how-to' pieces in the *Wisconsin Agriculturist and Farmer*. Leopold was among the first to observe and decry 'the tremendous momentum of industrialization ... spread to farm life' and to conceive an alternative 'new vision of "biotic farming"'.[15] Take environmental history: Leopold's essay 'Good

Oak' in *Sand County* is a pioneering contribution to the field. Take environmental policy and law: Leopold chaired a blue-ribbon American Game Association Committee on Game Policy and was the senior author of its influential 1930 *Report*. He was offered the job of Chief of the United States Bureau of Biological Survey (forerunner of the present US Fish and Wildlife Service) in 1934, but turned it down. He was also appointed to the Wisconsin Conservation Commission in 1943 and was embroiled for the rest of his life in bitter controversy over his recommended state deer management policy. His conservation policy advice was sought on every scale, from the local and private to the public and national. Take environmental education: in addition to training graduate students for careers in wildlife management, Leopold offered an undergraduate Wildlife-Ecology course open to any University of Wisconsin student. He published a paper addressed to fellow academics in the field titled 'The Role of Wildlife in a Liberal Education', the most important advice of which was 'to use wildlife ecology to teach the student how to put the sciences together' – because 'all the sciences and arts are [conventionally] taught as if they were separate', but 'they are separate only in the classroom'; all one need do is 'step out on the campus and they are immediately fused'.[16] Take nature writing: *A Sand County Almanac* has become more than a classic in the field; it is a genre exemplar.

Of all the environmental fields that Leopold either founded or that his genius shaped, none is of more lasting significance than environmental ethics. The climactic essay of the *Almanac*, 'The Land Ethic', is the seminal text in this new field of philosophy. After all of his years working for a *public* conservation agency, the US Forest Service, and helping to formulate policy and law for such newer agencies as the National Park Service and the Bureau of Biological Survey, Leopold came to believe that conservation would never succeed without a land ethic on the part of individual, *private* landowners. Government alone could not do the job.

Leopold based his proposed land ethic on two scientific cornerstones: evolution and ecology. From Charles Darwin he borrowed an account of ethics as a necessary condition for human social organization. 'All ethics so far evolved', Leopold wrote, 'rest upon a single premise: that the individual is a member of a community of interdependent parts. His instincts prompt him to compete for his place in that community, but his ethics prompt him also to co-operate (perhaps in order that there may be a place to

compete for).'[17] From Charles Elton, he borrowed the concept of a
'biotic community', a social model of the inter-relationships of
plants and animals studied in ecology. Ecology, Leopold wrote,
'simply enlarges the boundaries of the community to include soils
and waters, plants, and animals, or collectively: the land'.[18] Putting
these two elements together, he formulated 'a land ethic', which
'changes the role of Homo sapiens from conqueror of the land
community to plain member and citizen of it. It implies respect for
his fellow-members and also respect for the community as such.'[19]
The golden rule of the land of the land ethic is this: 'A thing is right
when it tends to preserve the integrity, stability, and beauty of the
biotic community. It is wrong when it tends otherwise.'[20] Leopold, of
course, intended for the land ethic to supplement our human-to-
human ethics, not replace them. And in light of subsequent
developments in ecology, in which 'stability' is down-played, the
golden rule of the land ethic may have to be revised. Nevertheless,
the very idea of a land or environmental ethic, and Leopold's sketch
of its contours, has taken the contemporary environmental movement out of the domain of mere utility and into that of morality. If
for no other reason, then for this one Leopold would deserve the
frequently conferred metonym of 'prophet' and his masterpiece that
of 'the bible of the contemporary environmental movement'.

Notes

1 'Wherefore Wildlife Ecology?', in *The River of the Mother of God*,
  pp. 336–7, p. 337.
2 *Game Management*, p. v.
3 'The State of the Profession', in *The River of the Mother of God*,
  pp. 276–80, p. 280.
4 *A Sand County Almanac and Sketches Here and There*, p. 168.
5 'Threatened Species', in *The River of the Mother of God*, pp. 230–4.
6 'Wilderness as a Land Laboratory', in *The River of the Mother of God*,
  pp. 287–9, p. 288.
7 Ibid., p. 289.
8 J. Baird Callicott, '"The Arboretum and the University: The Speech
  and the Essay", Appendix A: The Speech "What Is the University of
  Wisconsin Arboretum and Wild Life Refuge, and Forest Experiment
  Preserve?" by Aldo Leopold', *Transactions of the Wisconsin Academy of
  Sciences, Arts and Letters*, 87: p. 15, 1999.
9 *Sand County*, pp. vii–viii.
10 Ibid., p. 221.
11 Ibid.
12 'The Land-Health Concept and Conservation', in *For the Health of the
  Land*, pp. 218–26, p. 219.
13 'Wilderness as a Land Laboratory', p. 288; J. Baird Callicott, '"The

Arboretum and the University: The Speech and the Essay", Appendix
A', p. 17.
14 *Sand County*, p. 207.
15 'The Outlook for Farm Wildlife', in *The River of the Mother of God*,
pp. 323–6, p. 326; *Sand County*, p. 222.
16 'The Role of Wildlife in a Liberal Education', in *The River of the
Mother of God*, pp. 301–5, p. 302.
17 *Sand County*, pp. 203–4.
18 Ibid., p. 204.
19 Ibid.
20 Ibid., pp. 224–5

See also in this book

Darwin

Leopold's major writings

Game Management, New York: Charles Scribner's Sons, 1933.
A Sand County Almanac and Sketches Here and There, New York: Oxford
University Press, 1949.
Round River: From the Journals of Aldo Leopold, ed. Luna B. Leopold,
New York: Oxford University Press, 1953.
The River of the Mother of God and Other Essays by Aldo Leopold, ed. Susan
L. Flader and J. Baird Callicott, Madison, WI: University of Wisconsin
Press, 1991.
*For the Health of the Land: Previously Unpublished Essays and Other
Writings by Aldo Leopold*, ed. J. Baird Callicott and Eric T. Freyfogle,
Washington, DC: Island Press, 1999.

Further reading

Callicott, J. Baird (ed.), *Companion to A Sand County Almanac: Interpretive
and Critical Essays*, Madison, WI: University of Wisconsin Press, 1987.
—— *In Defense of the Land Ethic: Essays in Environmental Philosophy*,
Albany, NY: State University of New York Press, 1989.
—— *Beyond the Land Ethic: More Essays in Environmental Philosophy*,
Albany, NY: State University of New York Press, 1999.
Flader, Susan L., *Thinking Like a Mountain: Aldo Leopold and the Evolution
of an Ecological Attitude Toward Deer, Wolves, and Forests*, Columbia,
MO: University of Missouri Press, 1974.
Meine, Curt, *Aldo Leopold: His Life and Work*, Madison, WI: University of
Wisconsin Press, 1988.

J. BAIRD CALLICOTT

ROBINSON JEFFERS 1887–1962

Robinson Jeffers' statement in 1928 that 'I'd sooner, except for the penalties, kill a man than a hawk' startled the reading public with a different understanding of human significance in the world.[1] Twenty years later, Jeffers labelled his philosophy 'inhumanism', which he defined as 'a shifting of emphasis and significance from man to not-man; the rejection of human solipsism and recognition of the transhuman magnificence'.[2]

After stunning initial success, Jeffers found that his high regard for the natural world and low regard for humans combined with his isolationist political stance had earned him the scorn of public taste-makers, especially during the Great Depression and the Second World War. Since his death in 1962, however, he has been hailed as the foremost American poet of the ecological movement and a philosopher-poet who, in giving voice to the coastal landscape of the Big Sur area, has set a pattern followed by such nature-writers as Gary Snyder. He has also taught to a consistently broad-based readership a different understanding of human relationships to the natural world.

Jeffers was born in 1887 in a suburb of Pittsburgh, Pennsylvania. His father, a stern and scholarly professor of Old Testament literature and history, began his son's study of Greek at age 5. For much of his childhood, Jeffers attended schools in Europe. He and his younger brother lived with their mother and were visited regularly by their father. Jeffers' summary of his early years suggests the importance of intellectual development:

> When I was nine years old my father began to slap Latin into me, literally, with his hands; and when I was eleven he put me in a boarding-school in Switzerland – a new one every year for four years – Vevey, Lausanne, Geneva, Zurich. Then he brought me home and put me in college as a sophomore. I graduated accordingly at eighteen, not that I was intelligent but by sporting my languages and avoiding mathematics.[3]

After graduation from Occidental College, Jeffers entered graduate school in literature at the University of Southern California. A year later he returned to Europe and at the University of Zurich began studying philosophy, Old English, French literary history, Dante, Spanish literature and the history of the Roman Empire. He returned to Los Angeles to enter the USC Medical School, where he ranked highest in his class. He briefly taught physiology at the USC Dental College. In 1910, deciding that medicine would leave him too little time to write, he entered the University of Washington, where he studied forestry until 1913.

While a graduate student at USC, Jeffers met Una Call Kuster, who was to become, along with the Big Sur landscape, the greatest influence on his life. Married to a well-to-do lawyer, Una carried on a love affair with Jeffers until, after seven years of rumour and scandal, she divorced her husband and married Jeffers. The two planned to move to Europe where Jeffers could write, but the outbreak of the First World War led them to move up the coast to Carmel instead. As soon as they arrived, Jeffers felt he had found the place about which he would write.

On their land overlooking the Pacific Ocean at Carmel Point, Jeffers began a daily pattern which would hardly vary for many years: he wrote every morning, quarried stone in the afternoon for their granite house and forty-foot stone tower, took walks with Una and their twin boys in the late afternoon, and read Shakespeare and other literature aloud to the family in the evening. Jeffers lived with no telephone, no electricity and no heat but for the heat from a Franklin stove and fireplaces. Not until the 1940s did Jeffers have electricity brought to the house.

Jeffers' first volumes to gain widespread attention were *Tamar and Other Poems*, published in 1924, and *Roan Stallion*, published in 1925. The *New York Herald Tribune* enthusiastically reviewed Jeffers' work, critic Mark Van Doren praised it in the *Nation*, calling Jeffers 'a major poet' and critic Babette Deutsch compared reading *Tamar* to Keats looking into Chapman's Homer.[4] Leonard and Virginia Woolf's Hogarth Press published editions of Jeffers' work in Great Britain, and a French edition went through five printings. Within three years, his work appeared in eight anthologies. Jeffers' work was widely discussed, his books were bestsellers, and he was well paid for his writing. Soon Jeffers was even on the cover of *Time* magazine.

Jeffers' time at the pinnacle of American letters did not last long. His relegating of humanity and human consciousness to the importance of basalt and lichen offended socially progressive sensibilities of the 1930s. Jeffers' opposition to American involvement in the Second World War (or in any war) and his portrayal of all the leaders – Stalin, Roosevelt, Churchill, Hitler, Mussolini – as equally evil in leading their people to war offended many. Religious conservatives condemned Jeffers' portrayal of Jesus and his anti-Christian stance; for Jeffers, Jesus and Christianity turn people away from the beauty of the physical world, which is wrong because the beauty of the physical world is God, or at least the manifestation of God, and deserves our worship. Moralists condemned Jeffers' acceptance of violence as an essential aspect of life, and they condemned Jeffers' use of sexual acts, especially his use of incest to illustrate the human obsession with humans and human things to the exclusion of the beauty of the greater outside world.

Yet Jeffers has always had fervent admirers and a general readership. Jeffers' *Selected Poetry* of 1938 has sold continuously in many reprintings; his *Selected Poems* of 1965 continues to sell well. Jeffers articulates ideas that readers outside the academy and outside literary circles know to be vitally important: the non-human world is complex, interactive, conscious, a whole; every aspect of the non-human world is beautiful, and can lead people to a greater understanding of God and of our temporary and insignificant position in the cosmos; our scientific and our religious ways of knowing have serious flaws. Jeffers presents his ideas in memorable narratives, characterizations, images and metaphors which anyone, not just experienced readers of poetry, can understand.

What is the right relationship between nature and humans? Deriving ideas from Lucretius, Herodotus, Nietzsche and Schopenhauer, his four main philosophical pillars, Jeffers offers a series of answers in direct opposition to the prevailing Western belief that 'no man is an island, entire of itself': we should turn from our 'incestuous' involvement with each other in our corrupt 'communal' life, and pay attention to nature. The problem with humanity, Jeffers says, is our self-absorption. He describes how we might look to future ages:

> ... we shall seem a race of cheap Fausts, vulgar magicians.
> What men have we to show them? but inventions and appliances.
> Not men but populations, mass-men; not life
> But amusements; not health but medicines.[5]

The solution for Jeffers is to turn to permanent, natural things. It doesn't matter where we turn to nature for instruction: 'The ocean will show us / The inhuman road',[6] and 'there are left the mountains'.[7] In 'A Little Scraping', Jeffers says to 'Shake the dust from your hair', and he lists various elements of the landscape that are 'real' and more worthy of observation: a mountain sea-coast, lean cows which 'drift high up the bronze hill', a 'heavy-necked plough team', gulls, rock, 'two riders of tired horses' on a cloudy ridge, topaz-eyed hawks, and more.[8]

Though city dwellers can't very easily lean on rocks and contemplate hawks soaring, they still partake of the permanent reality because we all have bodies, and we eat: 'Broad wagons before sunrise bring food into the city from the open farms, and the people are fed. / They import and they consume reality. Before sunrise a hawk in the desert made them their thoughts.'[9] The landscape in Jeffers, then, is bodily consumed and afterwards influences everyone's thoughts; we must pay attention to the values the landscape might bring us. We should even attempt to imitate the landscape, for it provides examples of how humans should live: 'The beauty of things is the face of God: worship it; / Give your hearts to it; labor to be like it.'[10]

If we labour enough to be like the beauty of the natural world, we might experience the feeling of union with the landscape of the sort one of Jeffers' characters describes: ' ... I entered the life of the brown forest ... / ... and I was the stream / Draining the mountain wood; and I the stag drinking; and I was the stars, / Boiling with light ... / ... I was mankind also, a moving lichen / On the cheek of the round stone'.[11] The speaker is one with the universe, experiencing a feeling of union with all creation more common to mystics than to heroes of Euro-American narratives.

This religious feeling in his poetry, Jeffers says, 'is the feeling – I will say the certainty – that the universe is one being, a single organism, one great life that includes all life and all things; and is so beautiful that it must be loved and reverenced; and in moments of mystical vision we identify ourselves with it'.[12] Jeffers continues his explanation with a contrast to the kind of mysticism we might be more familiar with: 'This is, in a way, the exact opposite of Oriental pantheism. The Hindu mystic finds God in his own soul, and the outer world is illusion. To this other way of feeling, the outer world is real and divine; one's own soul might be called an illusion, it is so slight and so transitory.'[13] Jeffers' character who becomes the stag and the stars emphasizes the importance of the outer world to Jeffers when he concludes, 'I have fallen in love outward.'[14]

The importance of a holistic understanding of reality permeates Jeffers' writing. In a response to a request for a comment on his 'religious attitudes', Jeffers says, 'I believe that the universe is one being, all its parts are different expressions of the same energy, and they are all in communication with each other, influencing each other, therefore parts of one organic whole. (This is physics, I believe, as well as religion.)'[15] This is dialogism, we might add, as well as physics, religion and ecology: all entities are in communication with each other, creating each other by their interaction. In what has become perhaps one of Jeffers' most quoted passages, he summarizes his holistic view and its benefits:

> ... a severed hand
> Is an ugly thing, and man dissevered from the earth and stars
> and his history ...
> Often appears atrociously ugly. Integrity is wholeness, the
> greatest beauty is
> Organic wholeness, the wholeness of life and things, the divine
> beauty of the universe. Love that, not man
> Apart from that, or else you will share man's pitiful confusions,
> or drown in despair when his days darken.[16]

Perhaps the greatest difference between Jeffers' view of landscape and the landscape we are given by those scientifically trained to manage our forests, range lands and waters is this insistence upon seeing things whole. The search for truth is 'foredoomed and frustrate', Jeffers says, 'until the mind has turned its love from itself and man, from parts to the whole'.[17]

Jeffers has been called a prophet in the Old Testament pattern, a mystic, a seer, a religious teacher. Jeffers, ever the son of a Calvinist minister, does not come to know the conventional Christian God through creation; rather, Jeffers asserts that 'Things are the God', and he gives a formula for arriving at this understanding: 'Lean on the silent rock until you feel its divinity'.[18] Other routes to God tend to obscure reality, and throughout his poetry Jeffers aims great invective at saviours of all kinds. Religion apart from the landscape can only lead to disaster.

Jeffers emphasizes the importance of understanding science, too, in relating to the natural world: 'The happiest and freest man is the scientist investigating nature, or the artist admiring it', he tells us.[19] Jeffers considers 'a scientific basis' to be 'an essential condition' for the thinker. Jeffers says, 'We cannot take any philosophy seriously if it ignores or garbles the knowledge and view-points that determine the intellectual life of our time.' While an artist need not know science well, Jeffers says that if an artist has no familiarity with modern science, 'his range and significance would be limited accordingly'.[20]

Jeffers' own background in science, including medical school and forestry school, seems to have given him not only a will and an ability to incorporate a scientific outlook into his poetry, but also a sense of the limits of science as it is practised in the twentieth century: 'Science and mathematics / Run parallel to reality, they symbolize it, they squint at it, / They never touch it', he says.[21]

Not only do scientists miss the truth, but they also apply their efforts to unworthy ends. Science 'has fallen from hope to confusion at her own business / Of understanding the nature of things', Jeffers continues.[22] The echo of *De rerum natura* in the expression 'the nature of things' suggests that Lucretius, for all his pessimism, was on the right track. The methods of science have also gone awry. 'Man, introverted man ... cannot manage his hybrids', Jeffers writes in 'Science'; 'Now he's bred knives on nature turns them also inward'.[23] Science itself is admirable, Jeffers implies throughout his poetry; it is a means to knowledge and hence to truth. But as it is practised in the twentieth century, science needs severe critiquing.

Jeffers' belief in constant change and the need to adapt to change prevented him from pursuing a Luddite path. When asked on a questionnaire in 1926, 'How should the artist adapt himself to the machine age?' Jeffers replies, 'The machine age is only a partial change; the artist should adapt himself to it without ignorance but without excitement. It provides at the most, some shift of scenery for the old actors.'[24] But behavioural adaptation, a necessary characteristic for survival in a changing world, need not entail a whole-hearted embracing of the values that produce change. At roughly the same time, Jeffers writes, 'I don't think industrial civilization is worth the distortion of human nature and the meanness and the loss of contact with the earth, that it entails.'[25]

Like Thoreau, Jeffers returns to the single individual when any question of social reform arises: in one letter he says simply, 'I think that one may contribute (ever so slightly) to the beauty of things by making one's own life and environment beautiful, so far as one's power reaches.'[26] In 'The Answer' Jeffers advises, 'To keep one's own integrity, be merciful and uncorrupted and not wish for evil; and not be duped / By dreams of universal justice or happiness.'[27]

Jeffers achieves a relative complacency about environmental destruction by taking a longer view of reality, in geological and astronomical time rather than human time: 'Man's world puffs up his mind, as a toad / Puffs himself up; the billion light-years cause a serene and wholesome deflation.'[28] Jeffers has a clear idea of the extent of the destruction, as in the depiction of the death of a canyon of redwoods or an abandoned mine, where 'The sweat of men laboring has poisoned the earth'.[29] Jeffers is particularly affected by such abuse of the landscape when it occurs close to home: 'This beautiful place defaced with a crop of suburban houses'.[30] Simply looking harms a landscape as well. In an application of Heisenberg's Uncertainty Principle to everyday life, Jeffers says, 'Whatever we do to a landscape – even to look – damages it.'[31]

Yet change is inevitable, and beautiful places especially call for tragedy involving violence and pain. Jeffers is remarkable in part because he can so easily think beyond the greatest tragedy for the human race – our extinction. The world, he says, will think, 'It was only a moment's accident, / The race that plagued us', and then resume 'the old lonely immortal splendor'.[32]

Notes

1 'Hurt hawks', *Collected Poetry*, vol. 1, p. 377.
2 'Preface', *The Double Axe*, p. xxi.
3 *Selected Letters*, p. 353.
4 Alex A. Vardamis, *The Critical Reputation of Robinson Jeffers: A Bibliographical Study*, Hamden, CT: Archon, p. 9, 1972.
5 'Decaying Lambskins', *Collected Poetry*, vol. 2, p. 604.
6 'The Torch-bearers' Race', ibid., vol. 1, p. 99.
7 'Shine, Perishing Republic', ibid., p. 15.
8 'A Little Scraping', ibid., vol. 2, p. 282.
9 'Meditation on Saviors', ibid., vol. 1, p. 399.
10 'The Inhumanist', ibid., vol. 3, p. 304.
11 'The Tower Beyond Tragedy', ibid., vol. 1, pp. 177–8.
12 *Themes*, p. 182.
13 Ibid.
14 'The Tower Beyond Tragedy', p. 178.
15 *Selected Letters*, p. 221.
16 'The Answer', *Collected Poetry*, vol. 2, p. 536.
17 'Theory of Truth', ibid., p. 610.
18 'Sign-post', ibid., p. 418.
19 *Themes*, p. 184.
20 *Selected Letters*, p. 254.
21 'What's the Best Life', *Collected Poetry*, vol. 3, p. 425.
22 'Triad', ibid., vol. 2, p. 309.
23 Ibid., vol. 1, p. 113.
24 *Selected Letters*, p. 103.
25 Ibid., p. 104.
26 Ibid., p. 221.
27 'The Answer', p. 536.
28 'Animula', *Collected Poetry*, vol. 3, p. 420.

29 'Metamorphosis', ibid., p. 417.
30 'Carmel Point', ibid., p. 399.
31 'An Extinct Vertebrate', ibid., p. 438.
32 'The Broken Balance', ibid., vol. 1, p. 375.

See also in this book

Emerson, Ruskin, Thoreau

Jeffers' major writings

Tamar and Other Poems, New York: Peter Boyle, 1924.
Roan Stallion, Tamar and Other Poems, New York: Boni & Liveright, 1925.
The Women at Point Sur, New York: Liveright, 1927.
Cawdor and Other Poems, New York: Liveright, 1928.
Dear Judas and Other Poems, New York: Liveright, 1929.
Thurso's Landing and Other Poems, New York: Liveright, 1932.
Give Your Heart to the Hawks and Other Poems, New York: Random, 1933.
Solstice and Other Poems, New York: Random, 1935.
Such Counsels You Gave Me and Other Poems, New York: Random, 1937.
The Selected Poetry of Robinson Jeffers, New York: Random, 1938.
The Double Axe and Other Poems, New York: Random, 1948; repr. New York: Liveright, 1977.
Robinson Jeffers: Themes in My Poems, San Francisco, CA: Book Club of California, 1956; repr. in M.B. Bennett, 1966.
Robinson Jeffers: Selected Poems, New York: Vintage-Random, 1965.
The Selected Letters of Robinson Jeffers, 1897–1962, ed. Ann N. Ridgeway, Baltimore, MD: Johns Hopkins University Press, 1968.
Brides of the South Wind, ed. William Everson, Cayucos, CA: Cayucos Books, 1974.
What Odd Expedients and Other Poems of Robinson Jeffers, ed. Robert Scott, Hamden, CT: Shoe String, 1981.
The Collected Poetry of Robinson Jeffers, ed. Tim Hunt, 4 vols, Stanford, CA: Stanford University Press, 1988–.

Further reading

Bennett, M.B., *The Stone Mason of Tor House: The Life and Work of Robinson Jeffers*, Los Angeles: Ward Ritchie, 1966.
Brophy, R., *Robinson Jeffers: Myth, Ritual, and Symbol in His Narrative Poems*, rev. edn, Hamden, CT: Archon, 1976.
—— (ed.), *Robinson Jeffers: Dimensions of a Poet*, New York: Fordham University Press, 1995.
Brower, D. (ed.), *Not Man Apart: Lines From Robinson Jeffers*, San Francisco, CA: Sierra Club/Ballantine Books, 1969.
Carpenter, F.I., *Robinson Jeffers*, New York: Twayne, 1962.
Coffin, A.B., *Robinson Jeffers: Poet of Inhumanism*, Madison, WI: University of Wisconsin Press, 1971.
Zaller, R. (ed.), *Centennial Essays for Robinson Jeffers*, Newark: University of Delaware Press, 1991.

—— *The Cliffs of Solitude: A Reading of Robinson Jeffers*, Cambridge Studies in Literature and Culture, New York: Cambridge University Press, 1983.

MICHAEL McDOWELL

MARTIN HEIDEGGER 1889–1976

> Man is not the lord of beings. Man is the shepherd of Being.[1]

Martin Heidegger was born on 26 September 1889 in the village of Messkirch in southern Germany. After an abortive training for the Roman Catholic priesthood, he studied philosophy at Freiburg University – from 1919 as assistant to the renowned philosopher, Edmund Husserl. His reputation as an incisive and radical thinker was sealed in 1927 with the publication of his magnum opus, *Being and Time*. His reputation as a man, on the other hand, was later sullied by his fervent support of Nazism during the early 1930s. In these early years of the Reich, Heidegger saw in Nazism a means to combat the rise of technologism and globalization and to thereby recover the rootedness of the German people in their homeland. After the mid-1930s, however, he became both increasingly disillusioned with Nazism[2] and increasingly dissatisfied with his earlier philosophical project in *Being and Time*. Now the 'later' Heidegger came to see his earlier work as being infused with the anthropocentrism or 'humanism' of the Western philosophical tradition, a tradition which, he contended, lay at the root of our modern 'technological' estrangement from nature. Accordingly, in his later years Heidegger concerned himself with the possibility of 'recovering' an authentic non-technological 'dwelling' in harmony with nature. Heidegger died on 26 May 1976, and was buried in the churchyard of his beloved Messkirch.

Heidegger's thought has had repercussions throughout the intellectual world, having influenced fields as diverse as literary theory, theology (both Catholic and Protestant), psychology, political theory and aesthetics. In the 'Continental' philosophical tradition, the movements of existentialism, hermeneutics and deconstruction all take their cue from his work, philosophers of the prominence of Sartre, Habermas, Foucault and Derrida all admitting their indebtedness to him. Although Anglo-American philosophers have traditionally dismissed Heidegger's work, in recent years many have come to recognize his status as a (post-) modern thinker of a stature comparable perhaps only to Wittgenstein.

Before considering Heidegger's relevance to environmental thought, one must first come to grips with some basic features of his analysis of the human condition. Heidegger maintains that, at the most profound level, to be a human is not to be a particular type of *thing*, but to be a space or 'clearing' in which things show up *as things* in the first place. In *Being and Time* Heidegger articulates this point by claiming that Being (the process whereby things 'reveal' themselves as things) occurs only within the space provided by human being (referred to in these earlier works as *Dasein*). Later, after the so-called 'Turn' (*die Kehre*) in the direction of his thought, he came to reject this 'existentialist' position in favour of a less anthropocentric conception of humans as humble participants in a wider clearing of Being. For the later Heidegger then, to be truly human is not to determine Being but to keep watch over the revealing of things, to act as a humble 'shepherd of Being'.[3]

Heidegger claims that Being is essentially historical in the sense that different things reveal themselves in different historical epochs (and to different cultures). A witch, for instance, might reveal herself to a medieval but not to a post-Enlightenment European; an individual citizen might reveal themselves to a modern-day American but not to a fourth-century Chinese. (Note the language here: for Heidegger, these are not changes in perspective or worldview but the results of Being 'granting' different things in different epochs.) Heidegger contends that in the modern world we are increasingly finding that things come to reveal themselves 'technologically'.[4] Technological revealing Heidegger associates with a 'setting upon' or 'challenging' of nature. He tells us that it makes the 'unreasonable demand' that all nature submit to human ends, that all things reveal themselves as 'standing reserve' (*Bestand*), as resources for our use.

Heidegger's account no doubt jars with common sense: surely technology consists of various man-made artifacts, food blenders, calculators, dynamos, and so on. And surely, while these technological things may be put to good or bad *uses*, they are themselves neither good nor bad but merely *neutral*. Heidegger, however, would contend that this is precisely what any pragmatically minded individual who was fully inculcated into the technological way of revealing *would* say. For to maintain this is to offer a technological or instrumental explanation of technology and hence to remain blind to the essence of technology as a mode of revealing.

Heidegger claims of the 'technological' mode of revealing the peculiar and 'dangerous' power to 'drive out' all other modes of revealing. As it encroaches into all areas of life, non-technological understandings find themselves levelled down and destroyed: poetry, for instance, becomes nothing more than clever wordplay; great artworks, divested of their intrinsic power, become mere decorations, or perhaps worse, investments. Nowadays, Heidegger would no doubt complain that the authentic appreciation of wild nature has become levelled down to a pitiable concern with the proper *management* of natural *resources*. Heidegger sees the greatest danger in the possibility that technology might eventually come to extinguish all other modes of revealing. In such a nightmarish future all would have been sacrificed to the modern technological idols of efficiency and management. The world would have become a featureless expanse of standing reserve, a domesticated world shorn of 'otherness' and mystery, and impoverished as a result.

How then can we resist this insidious spread of technologism? For Heidegger, the question is inappropriate: we will not, he tells us, be able to halt the encroachment of the technological understanding through an act of *will*, for our will counts for nothing compared to the remorseless 'destining' of history. We and our technological world are the powerless products of the blind dictates of the history of Being. To contend otherwise, he points out, is to exhibit a characteristically technological arrogance. Yet our situation is not entirely without hope: Heidegger affirms the possibility of salvation, not indeed through stubborn resistance to technological developments, but through 'questioning' or meditating on the essence of technology itself. For deep questioning reveals that technology is a mode of revealing itself, and this realization invites us to discover our essential nature as 'clearings' wherein things reveal themselves in the first place. Accordingly, Heidegger calls for us to recognize the flip side of our historical destiny, namely, the *contingency* of the technological mode of revealing. We can, for instance, contemplate the fact that other peoples in other eras – the Ancient Greeks, for instance – were free from the urge to 'technologise' the world. Calm contemplation on technology reveals further that the world is not entirely technological, that other modes of revealing still persist (at least for the moment). Thus Heidegger calls for us to remain open to those facets of life which have so far resisted being subsumed in the technological understanding and have been marginalized as a result.

We must cherish art and beauty, for instance, as well as simple pleasures such as hiking, fishing or laughing and chatting with friends.

Elsewhere, Heidegger offers a series of 'poetic' meditations on the nature of a wholesome, non-technological way of life he terms 'dwelling'.[5] In describing this way of life, he develops a quasi-mythic account of a world consisting of a 'fourfold' of 'earth, sky, mortals and gods'. Dwelling, he writes, involves a way of being which allows things to reveal themselves in such a way that they come to unite or 'gather' these four dimensions. In this manner, even a lowly and unremarkable thing such as an earthenware jug can become resplendent with world, coming to gather the 'dark slumber' of the earth, the cool radiance of the sky, the nobility of authentic mortal life and the promise of divine deliverance. In these meditations, Heidegger seems to be articulating what we might refer to as a 'deep ecological' holistic vision of nature. However, it must be noted that Heidegger's holism does not involve the dissolution of things into some idealized whole – the Environment, Nature, or whatever. Rather, the experience of Heidegger's dweller combines a realization of wholeness with an appreciation of the inherent worth of individual things. Accordingly, dwelling involves, not the reverence of some nebulous idealization of nature, but a 'poetic' sensitivity to particular things, a sensitivity of the sort one might associate with a Zen *haiku* poet, for instance.[6] Heidegger maintains that dwelling 'poetically' in this way enables the dweller to come home to the environment as the milieu in which they live as a worldly being. (In this respect, the deep affection Heidegger retained throughout his life for the land of his birth is surely significant.)

Not all writers however are happy with the possibility of a Heideggerian environmental philosophy. Some critics have expressed concerns that Heidegger's thought cannot be abstracted from his disturbing commitment to what he once referred to as the 'inner truth and greatness of National Socialism'.[7] Indeed, Heidegger himself told Karl Löwith, a former student of his, that his 'political engagement' was based on his philosophical concept of historicity.[8] Considerations of this sort have led some critics to see in Heidegger's appropriation by deep ecologists a cause to fear the rise of totalitarian or so-called 'eco-fascistic' elements in radical ecological thought.[9] Nevertheless, although such concerns are undoubtedly justified, they ought to induce scholars, not to reject Heidegger altogether, but to determine more precisely the connection between his politics and his thought.[10] Such efforts would be worthy indeed, for it would certainly be a great and unnecessary loss if Heidegger's profound insights were lost to contemporary environmental thinkers. Indeed, it seems possible that modern thinkers may find in Heidegger a solid philosophical foundation on which to build a robust conception of an environmentally virtuous way of life.

Notes

1 *Basic Writings*, p. 245.
2 Nevertheless, the fact that he never offered a satisfactory apology for his involvement has struck many as appalling. See, for instance, Emmanuel Levinas, 'As If Consenting to Horror', reprinted in *Critical Inquiry*, 15, pp. 485–8, 1989.
3 Heidegger explains this change of tack in his 'Letter on Humanism', *Basic Writings*, pp. 217–65.
4 See especially 'The Question Concerning Technology' in either *The Question Concerning Technology and Other Essays*, trans. William Lovitt, New York: Harper & Row, 1977, or *Basic Writings*.
5 See Heidegger's essays 'The Thing' in *Poetry, Language, Thought* and 'Building Dwelling Thinking' in *Basic Writings*.
6 Similarities with Zen may not be accidental: Heidegger took a keen interest in East Asian philosophies, Taoism and Buddhism in particular. See Graham Parkes (ed.), *Heidegger and Asian Thought*, Honolulu: University of Hawaii Press, 1987.
7 *An Introduction to Metaphysics*, trans. Ralph Manheim, New Haven, CT: Yale University Press, p. 199, 1959.
8 Cited in Richard Wolin, *The Heidegger Controversy: A Critical Reader*, Cambridge, MA: MIT Press, p. 142, 1993.
9 For an appraisal of this line of criticism see Michael E. Zimmerman, 'Martin Heidegger: Antinaturalist Critic of Technological Modernity', in David Macauley (ed.), *Minding Nature: The Philosophers of Ecology*, New York: Guilford Press, 1996.
10 For a good introduction to the question of Heidegger's politics (and much else in Heidegger's philosophy) see Richard Polt, *Heidegger: An Introduction*, London: UCL Press, pp. 152–64, 1999.

See also in this book

Chuang Tzu, Marx

Heidegger's major writings

Basic Writings, ed. David Farrell Krell, New York: Harper & Row, 1996.
Being and Time, trans. John Macquarrie and Edward Robinson, Oxford: Blackwell, 1996.
Poetry, Language, Thought, trans. Albert Hofstadter, New York: Harper & Row, 1971.

Further reading

Dreyfus, Hubert L., 'Heidegger on the Connection Between Nihilism, Technology and Politics', in Charles Guignon (ed.), *The Cambridge Companion to Heidegger*, New York: Cambridge University Press, 1993.

Taylor, Charles, 'Heidegger, Language and Ecology', in Hubert L. Dreyfus and Harrison Hall (eds), *Heidegger: A Critical Reader*, Cambridge, MA: Blackwell, 1992.

Zimmerman, Michael E., 'Heidegger, Buddhism and Deep Ecology', in Charles Guignon (ed.), *The Cambridge Companion to Heidegger*.

SIMON P. JAMES

RACHEL CARSON 1907–64

> Only within the moment of time represented by the present century has one species – man – acquired significant power to alter the nature of this world ...
>
> The most alarming of all man's assaults upon the environment is the contamination of air, earth, rivers, and sea with dangerous and even lethal materials. This pollution is for the most part irrecoverable; the chain of evil it initiates not only in the world that must support life but in living tissues is for the most part irreversible. In this now universal contamination of the environment, chemicals are the sinister and little-recognized partners of radiation in changing the very nature of the world – the very nature of its life.[1]

Silent Spring (1962) embodies a connectedness with nature, a kinship with other species, a feeling of the responsibility to take personal action. From what wellsprings had such a mature environmental philosophy flowed? What influences led to such profound insights, personal courage and ultimate heroism in face of vilification by the American corporate and scientific establishment? What experiences led Carson to such deep understanding of nature awareness and how it might be conveyed, to her surpassing understanding of human ecology, and to her realization of the end of nature as humans had known it throughout time?

Carson was born in 1907 in Springdale, Pennsylvania, and had positive formative experiences in nature and in literature under the tutelage of her mother, Maria. She was strongly influenced in a romantic view of nature by several children's magazines. She published her first story in *St Nicholas* at age 11. She attended the private Pennsylvania College for Women, now Chatham College, in

Pittsburgh just a few miles from home. Carson was educated at great financial sacrifice to her family and with the aid of several scholarships. She continued her close relationship with her mother throughout college. She had faculty mentors, in writing, Grace Croff, and in biology, Mary Scott Skinker. She followed the latter into science as her academic major, to advanced study at Johns Hopkins University, and into work as a government scientist.

Rachel Carson's writing about the sea fulfilled a childhood inland dream. Her favourite line of poetry while studying English and science in Pennsylvania, proved prescient: 'For the mighty wind arises, roaring seaward, and I go.'[2] She wrote, 'I can still remember my intense emotional response as that line spoke to something within me, seeming to tell me that my own path led to the sea – which then I had never seen – and that my own destiny was somehow linked with the sea ...'[3]

She became a biologist at the US Fish and Wildlife Service, eventually working her way to the position of editor-in-chief. She ultimately became, in writing a trilogy of popular books, *Under the Sea-Wind* (1941), *The Sea Around Us* (1951) and *The Edge of the Sea* (1955), what she called a biographer of the sea. She was captivated by the eternal mysteriousness of the sea. She wrote in such a way as to express the facts as well as the beauty of nature – the knowledge as well as the poetry. According to Paul Brooks, her editor and biographer:

> Though she had the broad view of the ecologist who studies the infinitely complex web of relationships between living things and their environment, she did not concern herself exclusively with the great impersonal forces of nature. She felt a spiritual as well as physical closeness to the individual creatures about whom she wrote: a sense of identification that is an essential element in her literary style.[4]

This writing was in the spirit of John Muir, William Burroughs, Anna Botsford Comstock and others in the nature-study era of North American environmental writing. Through her nature-study writing, voice was given both to nature and to the unexpressed sensibilities of readers. The depth and power of her insights and the authority of her research educated readers to a world they did not know. Carson shared a subject she believed was vital. Through effective, powerful writing, she vivified the sea. Carson contextualized her scientific writing within this vastness of time and space. Her writing conveys a sense of proportion – a soul aware of sitting at the edge of the continent, at not only the edge of the sea, but the edge of the sea of stars. These vast expanses of time and the cycles of recurring natural events situate her insights of the human place in science. She understood the limits of science – even when she enriched its definition to include humans as feeling and socially responsible participants in its study. She saw the power of science to reveal knowledge of natural processes and to raise questions of the human relationship to such processes and to human knowledge of them. Finally, Carson saw science as needing to evoke the sister of identification and knowledge – personal responsibility.

In addition to scientific knowledge of the 'nearly eternal', she would have us feel the poetic essence of our response to nature – and of reverence for it. Rachel Carson's combined knowledge and love of nature has been compared in a feminist critique of her writing to Barbara McClintock's 'feeling for the organism'.[5]

The scholar Vera L. Norwood has plumbed the subtext of Carson's work and its epistemology. Carson's thinking and feeling lead her to question how we know what we know. She has no God-like perspective apart from nature and human nature, rather she struggles to locate herself. In 'The Nature of Knowing: Rachel Carson and the American Environment', Norwood writes:

> The occasions when the economic metaphor shatters against the unwillingness of the natural world to 'produce' meaning provide her most telling critiques of human limitations and lead her to doubt all context, [sic] Carson becomes more than a nature writer; she raises fundamental questions about how human knowledge is constructed, questions that reveal the epistemological hubris underlying much human understanding. These questions prompt her later normative work in *Silent Spring* and *The Sense of Wonder*.[6]

Carson's nature writing has been celebrated for sensitivity, complexity and depth. She taught about life in the sea but also to stand in reverence of how little is known. She educated towards another way to know – to *feel* nature. And finally she raised questions not only about nature, but about the nature of the knowledge by which we know nature.

A qualitatively different stage of Carson's writing began in 1956 when she wrote an article for *Woman's Home Companion* entitled 'Help Your Child to Wonder'. This was the first time in seven years she had not had a book in production. She wanted to leave the sea for a time; she wrote to her editor 'like that old scorpionlike thin in the Silurian, I have come out on land'.[7] She had hopes to develop the article as a book, but soon she was to start her research on pesticides, and she never did. The article was published post-humously in 1965 as a book, *The Sense of Wonder.*

It is in this work that Carson is explicitly an environmental educator and can be best critiqued for her philosophical and pedagogical contributions to the field. She asks:

> What is the value of preserving and strengthening this sense of awe and wonder, this recognition of something beyond the boundaries of human existence? Is the exploration of the natural world just a pleasant way to pass the golden hours of childhood or is there something deeper?
>
> I am sure there is something much deeper, something lasting and significant. Those who dwell, as scientists or laymen, among the beauties and mysteries of the earth are never alone or weary of life.[8]

In the most direct statement of Carson's rationale for her kind of environmental education, she assures us of a deeper meaning, a hidden soul, that lies just beyond our experience in the natural world. Much of education teaches not to trust wonder, intuition and the ineffable sources of human strength. Yet these are part of our knowledge of nature, Carson says. She offers a validation of the power and authority of childhood experience and an invitation to reconsider its depths. The reader is enticed to wonderment in the sensual experience of nature. This work, with its explicit inclusion of affect and questions of value, foreshadows the raising of these questions by educators in the 1970s. She gives permission to explore the actual and perceived landscapes of childhood. Rachel Carson's philosophy of environmental education speaks of sensory creatures in a sensory world, humble citizens of a mysterious universe, and people free to place themselves 'under the influences of earth, sea, and sky and their amazing life'.[9]

In 1958, Carson decided to write a brief article on the impact of DDT spraying upon bird life – her next four and a half years were spent researching and writing one of the most influential books of the age. She told her long-time friend Dorothy Freeman she had proposed an article about it in 1945. The 1945 article became her magnum opus in 1962. *Silent Spring* has demonstrated remarkable vitality. It has been translated into 'nearly every language on the planet'.[10] Thirty-eight years after US publication, it has never been out of print and continues to sell. It is given great credit for changing the way we see our world. According to H. Patricia Hynes, '*Silent Spring* crystallized an "ethic of the environment" which inspired grassroots environmentalism, the "deep ecology" movement and the creation of the Environmental Protection Agency (EPA) and its state counterparts; it influenced the ecofeminist movement and feminist scientists.'[11]

Through her research on pesticides, Rachel Carson saw the vast destruction of which humans are capable. Hynes, in a chapter entitled '*Silent Spring*: A Feminist Reading', writes:

> Rachel Carson told students of Scripps College in 1962 that 'in the days before Hiroshima,' she thought that there were powerful and inviolate realms of nature, like the sea and vast water cycles, which were beyond man's destructive power. 'But I was wrong,' she continued. 'Even these things, that seemed to belong to the eternal verities, are not only threatened but have already felt the destroying hand of man.'[12]

Her dedication of *Silent Spring* is instructive of her environmental world-view. She quoted Albert Schweitzer, 'man has lost the capacity to foresee and to forestall. He will end by destroying the earth.'[13] She was among the very first to appreciate the gravity of the human impact on nature and her writing in this period precedes the concern to follow in the years leading up to Earth Day 1970 and the popular recognition of the seriousness of the environmental crisis. Rebecca Raglon gives Carson a new place in the context of the tradition of women's writing in nature-study and of nature writing by both women and men. She writes:

> *Silent Spring* marks the origin of a new kind of nature writing: a dark new genre that deals with the horrific consequences of human actions upon the earth ... Carson's legacy has insured that such innocent nature appreciation will now have to occur within a much darker context.[14]

Destined to be considered a seminal work in environmentalism, and perhaps one of the most important books of the twentieth century, its writing is meticulously chronicled by Carson's editor Paul Brooks in his biography *The House of Life: Rachel Carson at Work*:

> The storm aroused in certain quarters by the publication of *Silent Spring*, the attempts to brand the author as a 'hysterical woman,' cannot be explained by the concern of special interest groups for their power or profits. The reasons lie deeper than that. Rachel Carson's detractors were well aware of the real danger to themselves in the stance she had taken. She was not only questioning the indiscriminate use of poisons but declaring the basic responsibility of an industrialized, technological society toward the natural world. This was her heresy. In eloquent and specific terms she set forth the philosophy of life that has given rise to today's environmental movement.[15]

Carson's environmental philosophy raises questions about the nature of nature and human knowledge of it; it invites the reader to stand in wonder at the depth of nature's influence upon values and attitudes; and it calls a people to their responsibility to halt its destruction. Indeed, the recent intellectual history of environmental thought owes much to the wisdom of this remarkable scientist, writer, educator, elder and lover of nature.

Notes

1 *Silent Spring*, pp. 5–6.
2 Alfred Lord Tennyson, 'Locksley Hall'.
3 Paul Brooks, *The House of Life: Rachel Carson at Work*, Boston, MA: Houghton-Mifflin Company, p. 18, 1972.
4 Ibid., pp. 7–8.
5 H. Patricia Hynes, *The Recurring Silent Spring*, New York: Pergamon Press, p. 57, 1989.
6 Vera L. Norwood, 'The Nature of Knowing: Rachel Carson and the American Environment', *Journal of Women in Culture and Society*, 12, pp. 747–52, 1987.
7 Brooks, op. cit., p. 201.
8 *The Sense of Wonder*, p. 88.
9 Ibid., p. 95.
10 Michael Brosnan quoted in Caskie Stinnett, 'The Legacy of Rachel Carson', *Down East*, June, p. 43, 1992.
11 Hynes, op. cit., p. 9.
12 Ibid., p. 181.
13 *Silent Spring*, p. v. This is, fascinatingly, a slight misquoting of

Schweitzer, who said, 'Modern man no longer knows how to foresee or
to forestall. He will end by destroying the earth from which he and other
living creatures draw their food. Poor bees, poor birds, poor men ...'

14 Rebecca Raglon, 'Rachel Carson and her Legacy', in Barbara T. Gates
and Ann B. Shteir, *Natural Eloquence: Women Reinscribe Science*,
Madison, WI: University of Wisconsin Press, pp. 198, 207, 1997.

15 Brooks, op. cit., p. 284.

See also in this book

Comstock, Griffin, Muir, Schweitzer

Carson's major writings

Under the Sea-Wind, New York: Oxford University Press, 1941.
The Sea Around Us, New York: Oxford University Press, 1951.
The Edge of the Sea, Boston, MA: Houghton-Mifflin Company, 1955.
Silent Spring, Boston, MA: Houghton-Mifflin Company, 1962.
The Sense of Wonder, New York: Harper & Row, 1965; this edition is now
out of print. The Nature Company has published a new version (1994) as
has HarperCollins (1998).
Lost Woods: The Discovered Writing of Rachel Carson, ed. with an
Introduction by Linda Lear, Boston, MA: Beacon Press, 1998.

Further Reading

Freeman, Martha (ed.), *Always, Rachel: The Letters of Rachel Carson and
Dorothy Freeman, 1952–1965*, Boston, MA: Beacon Press, 1995.
Graham, Frank, Jr, *Since Silent Spring*, Boston, MA: Houghton-Mifflin
Company, 1970.
Lear, Linda, *Rachel Carson: Witness for Nature*, New York: Henry Holt &
Company, 1997.

Further readings and information may be found at the following websites:

http://www.rachelcarson.org
http://www.chatham.edu.RCI/rcinew.htm
http://www.dep.state.pa.us/dep/Rachel_Carson/Rachel_Carson.htm

PETER BLAZE CORCORAN

LYNN WHITE, JR 1907–87

What people do about their ecology depends on what
they think about themselves in relation to things around
them.[1]

Lynn Townsend White, Jr was born in San Francisco, California, on 29 April 1907. After his academic training at the finest schools in the USA,[2] his first academic post was at Princeton University from 1933 to 1937. In 1938 he joined the faculty of his alma mater, Stanford, and remained there until 1943. From 1944 until 1957 Lynn White served as President of Mills College, a women's college in Oakland, California. In the midst of his stint at Mills, White penned a provocative book entitled *Educating Our Daughters*, which spoke to the problems women faced in higher education in the USA at the time. White clearly made his mark at Mills College: a residence hall and an endowed chair still carry his name. In 1958 White joined the History faculty at the University of California – Los Angeles where he remained until retiring from academic life in 1974. Lynn White, Jr is widely and most notably recognized as the 'founder of all serious modern study' of the history of technology in medieval Europe. His most famous and still classic, *Medieval Technology and Social Change*, was once declared by Joseph Needham to be 'the most stimulating book of the century on the history of technology'. On 30 March 1987 Lynn White, Jr died of heart failure; he was 79. In his lifetime White was known for both his scholarly and his more popular writings, for his timely and controversial intellectual boldness and for his insistence that scholarly parochialism was antithetical to the life of the mind. It is said that throughout his life White remained a Christian, a fact which might seem curious to some given the nature of his most obvious contribution to environmental thought.

In the cold of the Washington, DC winter of 1966, Lynn White, Jr presented a ground-breaking and controversial paper at the annual meeting of the American Association for the Advancement of Science. In the paper, which was published the following year in the journal *Science*,[3] White laid much of the blame for our current environmental predicament upon the doorstep of Christianity. It is, therefore, deeply ironic that 'The Historical Roots of Our Ecologic Crisis', a paper considered to be so critical of the Christian tradition, was presented by a Christian thinker on the day after Christmas.

Within just a few short years the published article had already been dubbed a 'classic'. The essay provoked both immediate and long-term reactions: literally dozens of responses to White's essay have since been published,[4] the essay remains a staple for university 'environmental' courses, and it continues to be reprinted in a wide variety of anthologies and textbooks. White seemed surprised by the response to his essay. However, 'as the tide of protest from churchmen flowed across his desk in a growing stream of letters and articles', White apparently kept his sense of humour, joking that he 'should have blamed the scientists'.[5] In an essay White wrote in 1973, responding to his critics, he comments that as the criticisms poured forth he 'was denounced, not only in print but also on scraps of brown paper thrust anonymously into envelopes, as a junior Anti-Christ, probably in the Kremlin's pay, bent on destroying the true faith'.[6]

It is, of course, not ordinary to consider someone a key environmental thinker on the basis of essentially a single essay. But this is no ordinary essay. Seldom has the splash of a single work created such enduring ripples. There are two noteworthy contributions made by White's paper.

Understandably the point most people immediately fixated on was White's attack on Christianity. White begins by pointing out that although 'all forms of life modify their context' current anthropogenic environmental impact 'has so increased in force that it has changed in essence'. Whereas past environmental impact was local and point-source impact, currently we are witness to not just a difference in the degree of environmental impact but a different kind of impact all together. We now possess and exercise an ability to affect the globe as a whole. As White put it: 'the impact of our race upon the environment has so increased in force that it has changed in essence'. In fact, the bulk of his scholarly work was an attempt to show how even quite minute alterations in technology – such as the use of horse power and the resulting heavy plough – can and did eventuate in a radical escalation in the ability of humans to exploit nature.[7] Hence, as an historian, White provides us with an explanation of how it is that humans have impacted and altered the environment so extensively. However, White denies that our current rate of environmental change, resulting in our environmental crisis, is merely a result of an increase in our ability to manipulate our context with the tools of modern science and technology. Instead, White asserts that because the fusion of science and technology during the seventeenth-century Scientific Revolution occurred within a Christian conceptual framework, and because Christianity had been interpreted as dictating an essentially despotic relationship between humans and the rest of nature – a relationship where 'technological advance was seen as superlatively virtuous' – Christianity is ultimately responsible for our contemporary environmental crisis. As White puts it:

The artifacts of society, including its political, social and economic patterns, are shaped primarily by what the mass of individuals in that society believe, at the sub-verbal level, about who they are, about their relation to other people and to the natural environment, and about their destiny. Every culture, whether it is overly religious or not, is shaped primarily by its religion.[8]

According to White, the message we have gleaned from Christianity is that we humans are uniquely created in the image of God, a quality which cuts us out from the rest of creation, making us not only separate but special, and that our role on this earth with regard to the rest of God's creation is to dominate and subdue. Hence, because of these background assumptions, humans feel 'we are superior to nature, contemptuous of it, willing to use it for our slightest whim'. Where 'formerly man has been part of nature; now he was the exploiter of nature'. Hence, Christianity not only allows for the anthropogenic exploitation of nature that has resulted in our environmental crisis, it also sanctions and enforces it.

So, what is the solution to our environmental crisis? White dismisses a focus on an increase in science and technology since 'our science and technology have grown out of Christian attitudes toward man's relation to nature'. According to him, since

what we do about our ecology depends on our ideas of the man–nature relationship ... more science and more technology are not going to get us out of the present ecologic crisis until we find a new religion, *or rethink our old one.*[9]

Since 'Christianity in absolute contrast to ancient paganism and Asia's religions ..., not only established a dualism of man and nature but also insisted that it is God's will that man exploit nature for his proper ends' – and since it is, at least in part, the message taken from the establishment of this dualism – the goal, for White, is to not jettison Christianity but, rather, it 'is to find a viable equivalent to animism'.[10]

White's proposal, then, is to rethink Christianity, to focus on the possibility of an alternative message about the human–nature relationship, a message of stewardship. White ends his essay with a tribute to St Francis of Assisi who, White claims, was not only 'the greatest radical in Christian history since Christ' but who delivered the required nature-sympathetic message of stewardship. In fact, White even goes so far as to propose St Francis as the environmental 'patron saint'. Since the publication of 'Historical Roots' in 1967, Christians seem to have taken up the task White lays out. In fact, the advent of a Christian environmental stewardship is arguably the most powerful thing to happen to the environmental movement since Rachel Carson's *Silent Spring*.

However, apart from attributing the West's brazenly opprobrious environmental behaviour to our narrowly focused and anthropocentric interpretation of Christianity, there is a subtle yet powerful subtext which flows throughout White's essay. White asserts that to solve our environmental crisis we must 'clarify our thinking', 'think about fundamentals', and 'rethink our axioms'. In other words, we must philosophize. White's entire essay, then, is a call for, and stamp of approval on, the new field of environmental ethics, a subdiscipline of philosophy in its infancy when White's challenge broke. In fact, because of this essentially philosophical subtext, environmental philosopher J. Baird Callicott has even gone so far as to dub White's essay 'the seminal paper in environmental ethics'.[11]

Although often misunderstood, Lynn White, Jr's contribution to environmental thought was both important and profound. He boldly challenged us to think deeply about the roots of our environmental problems and to be brave enough to reconsider those fundamental anthropocentric axioms asserting our human superiority. White lays before us a formidable task: we must learn humility. We must learn to care for ourselves as well as for God's creation.

Notes

1 'The Historical Roots of Our Ecologic Crisis', *Science*, 155, p. 1205, 1967.
2 BA, Stanford, 1928; MA, Union Theological Seminary, 1929; MA, Harvard, 1930; PhD, Harvard, 1934.
3 'The Historical Roots of Our Ecologic Crisis', pp. 1203–7.
4 One of the best is James Barr, 'Man and Nature: The Ecological Controversy and the Old Testament', *Bulletin of the John Rylands Library*, 55, pp. 9–32, 1972.
5 Bert S. Hall, 'Obituary of Lynn White, Jr', *ISIS*, 79, p. 480, 1988.
6 'Continuing the Conversation', in Ian Barbour (ed.), *Western Man and Environmental Ethics*, p. 60.
7 See especially *Medieval Technology and Social Change*, p. 1203.
8 White in Barbour, op. cit., pp. 59, 57.
9 Ibid., pp. 1205–6, emphasis added.
10 White in Barbour, op. cit., p. 63.

11 J. Baird Callicott, 'Environmental Philosophy is Environmental Activism: The Most Radical and Effective Kind', in D.E. Marietta and L. Embree (eds), *Environmental Philosophy and Environmental Activism*, Lanham, MD: Rowman & Littlefield, p. 30, 1995.

See also in this book

Callicott, Carson, Saint Francis of Assisi

White's major writings

Latin Monasticism in Norman Sicily, Cambridge, MA: Medieval Academy of America, 1938.

Educating Our Daughters: A Challenge to the Colleges, New York: Harper & Brothers, 1950.

Medieval Technology and Social Change. London: Oxford University Press, 1962.

Machina ex deo: Essays in the Dynamism of Western Culture, Cambridge, MA: MIT Press, 1968.

Medieval Religion and Technology: Collected Essays, Berkeley, CA: University of California Press, 1978.

Further reading

Barbour, Ian G. (ed.), *Western Man and Environmental Ethics: Attitudes Toward Nature and Technology*, Reading, MA: Addison-Wesley, 1973.

Callicott, J. Baird, *Earth's Insights: A Multicultural Survey of Ecological Ethics from the Mediterranean Basin to the Australian Outback*, Berkeley, CA: University of California Press, 1994.

—— 'Genesis and John Muir', in *Beyond the Land Ethic: More Essays in Environmental Philosophy*, Albany, NY: State University of New York Press, 1999.

Gimpel, Jean, *The Medieval Machine: The Industrial Revolution of the Middle Ages*, New York: Holmes & Meier, 1974.

Gottlieb, Roger (ed.), *This Sacred Earth: Religion, Nature, Environment*, London: Routledge, 1996.

Hill, Brennan R., *Christian Faith and the Environment: Making Vital Connections*, Maryknoll, NY: Orbis, 1998.

Kinsley, David, *Ecology and Religion: Ecological Spirituality in Cross-cultural Perspective*, Englewood Cliffs, NJ: Prentice-Hall, 1995.

MICHAEL P. NELSON

E.F. SCHUMACHER 1911–77

The fight against pollution [cannot] be successful if the patterns of production and consumption continue to be of a scale, a complexity, and a degree of violence which, as is

> becoming more and more apparent, do not fit into the
> laws of the universe, to which man is just as much subject
> as the rest of creation.[1]

Schumacher was working on his ideas at a time when the dominant ideology was 'the bigger the better'. Large institutions, multinational corporations, industrial mergers, unlimited economic growth and ever increasing consumption, were considered symbols of progress. Schumacher said, 'We suffer from an almost universal idolatry of giantism.'[2]

In response to such idolatry, Schumacher encapsulated an alternative world-view in his seminal collection of essays, *Small is Beautiful* (1973), which became one of the most popular books amongst members of the British Parliament. The suggestion that many of the environmental and social problems facing the world were the result of idolatry to giantism intrigued Jimmy Carter, President of the USA, and consequently Schumacher was invited to the White House to advise the president in 1977. The Governor of California at that time, Jerry Brown, became so convinced by Schumacher's analysis that he initiated a number of measures embodying the 'small is beautiful' approach.

Ernst Fritz Schumacher was born in Bonn, Germany, in 1911. He came to England in 1930 as a Rhodes scholar to read Economics at New College, Oxford. After a short spell of teaching Economics at Columbia University, New York, followed by dabbling in business, farming and journalism, he became an economic advisor to the British Control Commission in Germany (1946–50), followed by a long career in the National Coal Board in Britain.

It was Schumacher's involvement in the economics of developing countries that challenged and changed his economic philosophy. He realized that the Western pursuit of unlimited economic growth on a gigantic scale is neither desirable nor practicable for the rest of the world. If anything, the West itself needs to learn the simplicity, spirituality and good sense of other cultures which are not yet in the grip of technological imperatives. 'In the excitement of the unfolding of his scientific and technological powers, modern man has built a system of production that ravishes nature, and a type of society that mutilates man.'[3]

The turning point came in 1955 when he was sent as Economic Development Advisor to the government of Burma. He was supposed to introduce the Western model of economic growth in order to raise the living standards of the Burmese people. But he discovered that the Burmese needed no economic development along Western lines, as they themselves had an indigenous economic system well suited to their conditions, culture and climate. As a result of his encounter with this profound and practical Buddhist civilization, he wrote his well-known essay, 'Buddhist Economics' (1966). Schumacher was perhaps the only Western economist to dare to put these two words, Buddhism and economics, together. The essay was printed and reprinted in numerous journals and anthologies.

Recalling his time in Burma he told me that the Burmese needed little advice from him. In fact Western economists could learn a thing or two from the Burmese. They had a perfectly good economic system, which supported a highly developed religion and culture and produced not only enough rice for their own people but also a surplus for the markets of India. He further commented that when he had published his findings under the title of 'Buddhist Economics', a number of his economist colleagues had asked, 'Mr Schumacher, what does economics have to do with Buddhism?' His answer was simply that 'Economics without Buddhism is like sex without love'. Economics without spiritual values can only give temporary and physical gratification; it cannot provide lasting fulfilment. Buddhist economics includes service to fellow human beings and compassion for all life as well as making a profit and working efficiently. We need both economics and spirituality and we need them simultaneously.

During his time in Burma Schumacher encountered the Buddhist concept of the Middle Way. He wanted to apply it to technology. He saw that people are either stuck with the sickle or they seek a combine harvester, thus he developed the Schumacher Principle of the Disappearing Middle, referring to the way that when a new, advanced technology is developed it displaces its immediate predecessor. Consequently what is left is either expensive, sophisticated, state-of-the-art technology or very simple hand tools. Whereas what small farmers and manual workers require is a technology between these two extremes.

In 1970, after many years of gestation, Schumacher founded the Intermediate Technology Development Group (ITDG) with an article in the pages of the *Observer* newspaper. He received an overwhelming response from the general public. ITDG became the practical expression of respect for cultural diversity. It pursued economic development within people's cultural context, rather than looking at the non-industrialized world as 'under-developed'. Intermediate Technology was envisioned to be environment-friendly, non-polluting and non-exploitative of people or nature. Therefore it also became known as 'appropriate technology' or 'alternative technology'. The concept was initially applied to non-industrialized countries, but technologies of renewable energy, of recycling and of ecological restoration in the West, became part of the same movement of a technology for a sustainable future.

Complementary to Intermediate Technology was his involvement with sustainable agriculture; he spent much time on his organic garden and became president of the Soil Association. He believed that 'in the simple question of how we treat the land ... our entire way of life is involved'.[4] He had no doubt that ' a callous attitude to the land and to the animals thereon is connected with, and symptomatic of, a great many other attitudes, such as those producing heedless urbanization, needless industrialization, and a kind of fanaticism which insists on playing about with novelties – technical, chemical, biological and so forth – long before their long term consequences are even remotely understood'.[5]

For Schumacher, care for the land and for the soil was fundamental to caring for the whole of the natural world, as well as a way of creating a just and equitable society. In the 1960s and 1970s attention to 'Mind, Body, Spirit' was becoming popular amongst alternative circles. Schumacher found this too narrow, human-centred and individualistic. It was all about the human mind, human body and human spirit. It left out the issues of social justice and caring for the earth. The spiritual dimension for Schumacher was paramount: individual development and personal growth were necessary, but only in the context of social wellbeing and the wellbeing of the Earth. Therefore Schumacher's philosophy led away from the personal focus of 'mind, body and spirit' to the broader and more inclusive concerns of what I have called, 'soil, soul and society'.

Schumacher was a holistic and ecological economist. Modern economics looks at the world as a resource for ever-increasing profit, and at human beings as units of labour for the profitability and continuity of the economic system. Schumacher saw it the other way round. That is why he subtitled *Small is Beautiful* 'a study of economics as if people mattered'. Furthermore economics must be a way of sustaining, restoring and maintaining the immense diversity and complexity of the biosphere in addition to nourishing, nurturing and fulfilling appropriate human needs. Economics is to serve people *and* planet. In order to achieve this kind of economic system, it must remain under local control and within a human scale, not becoming subservient to the so-called 'economy of scale'.

The importance of small-scale and local production became crystal clear to him when Schumacher saw a lorry full of biscuits being brought from Manchester to London, and minutes later another lorry full of biscuits being taken from London to Manchester. Schumacher gasped: What could be the economic rationale of this activity? Having failed to see any good reason for this transportation which caused air pollution and wasted fossil fuels and human labour, Schumacher said in frustration: 'As I am not a nutritionist, I wonder if the nutritional value of the biscuits is increased by this transaction?! Otherwise, if Manchester has a special kind of biscuit it could simply send the recipe to London on a postcard, and vice-versa.'

To Schumacher it was logical and natural to produce, consume and organize as locally as possible, which inevitably meant on a smaller scale. Therefore to him the question of size was an over-riding and over-arching principle. He refused to accept that largeness was necessary for prosperity: 'Small units are highly prosperous and provide society with most of the really fruitful new developments'.[6] Again, he wrote: 'The question of scale is extremely crucial today, in political, social and economic affairs just as in almost everything else'.[7]

Beyond a certain scale the people involved are disempowered and a bureaucratic machine takes over. For example, in a school of 1,000 children, parents do not know the teachers, teachers cannot know all the children, the children cannot know each other, and the surrounding community is overwhelmed by the influx of pupils who do not belong to that community. In this situation children become numbers, and the aim of education becomes meeting the requirements of the system and the league tables rather than the development of the whole child.

Similarly, large hospitals, large factories and large businesses lose the purpose of enriching human wellbeing and become obsessed with maintaining and perpetuating the organization for its own sake. Therefore it could be said almost invariably that if there is something wrong, there is something too big. Also, big organizations will have big problems, and small organizations will have small problems, which can be solved more easily.

As in economics, so in politics. Schumacher was greatly influenced by the Austrian philosopher Leopold Kohr, whom he considered his mentor. In the book, *Breakdown of Nations*, Kohr outlines the case against giantism and against big nations. In countries such as Sweden or Switzerland, there is much more political participation and flexibility. When people in these countries want to bring about change they can do so with greater ease than in countries like China, India or the USA. So Schumacher believed in small nations, small communities and small organizations. Small, simple, and non-violent were his three philosophical precepts.[8] These were to determine all relationships – economic, political and cultural – within human societies, as well as between humans and the natural world.

Schumacher died in September 1977, in Switzerland. He wrote only two books, *Small is Beautiful* and *A Guide for the Perplexed*, the latter published posthumously. A collection of his speeches was later published under the title of *Good Work*. Yet his influence was vastly greater than the volume of his published work might suggest. He was more than an economist, he was also a very practical man. He inspired many people through his busy schedule of lectures, private meetings and through his support of grassroots projects. 'Pollution must be brought under control and mankind's population and consumption of resources must be steered towards a permanent and sustainable equilibrium'[9] was his advice to the groups with whom he worked.

His legacy continues to be felt. Immediately after his death the Schumacher Society was established in Britain, which continues to promote the ideas of ecological economics. The Society holds annual lectures in Bristol, Liverpool and Manchester. Some of these lectures have now been published. Schumacher Societies have also sprung up in the USA, Germany and India.

His writings have inspired people in different disciplines. In education a number of Small Schools have been established, where the emphasis is on 'education as if children matter'. A College named after him has also been established at Dartington, Devon, exploring an ecological world-view from many different perspectives, while students practise a lifestyle built around the precepts of small, simple, local and non-violent. In economics, the New Economics Foundation encourages ideas of local economies, local currencies and local trading. In the field of development, ITDG continues to promote indigenous and small-scale projects. In the field of energy, the National Centre of Alternative Technology, Wales, attracts thousands of visitors keen to see methods of renewable energy.

Resurgence magazine, for which Schumacher wrote regularly, continues to examine and expound the 'small is beautiful' ethos.

Notes

1   *Small is Beautiful*, p. 247.
2   Ibid., p. 54.
3   Ibid., p. 246.
4   *This I Believe*, p. 181.
5   Ibid.
6   *Small is Beautiful*, p. 53.
7   Ibid., p. 55.
8   *Resurgence*, January/February 1974.
9   Ibid., p. 248.

E.F. Schumacher's major writings

Small is Beautiful; A Study of Economics as if People Mattered, 1973, London: Abacus, Sphere Books Ltd, 1988. Includes the essay 'Buddhist Economics'.
A Guide for the Perplexed, 1977, London: Abacus, Sphere Books Ltd, 1989.
Good Work, London: Jonathan Cape, 1979. A collection of speeches.
This I Believe, Totnes, Devon: Green Books, 1997. A collection of twenty-one articles published in *Resurgence*; includes 'Buddhist Economics'.

Further reading

Button, John, *The Green Fuse*, London and New York: Quartet Books, 1990. A collection of Schumacher Lectures.
Kohr, Leopold, *The Breakdown of Nations*, 1957, London and New York: Routledge & Kegan Paul, 1986.
Kumar, Satish (ed.), *The Schumacher Lectures Vol. I and II*, London: Blond & Briggs, 1980 and 1984 respectively.
McRobie, George, *Small is Possible*, London: Jonathan Cape, 1981.
Wood, Barbara, *Alias Papa: A Biography*, London: Jonathan Cape, 1984.

SATISH KUMAR

ARNE NAESS 1912–

> If you hear a phrase like 'all life is fundamentally one!', you must be open to *tasting* this, before asking immediately 'what does this mean?'. Being more precise does not necessarily create something that is more inspiring'.[1]

Since the early 1970s, when he introduced the expression 'deep ecology', Arne Naess has been the most influential of living environmental philosophers, his voice heard well beyond the confines of academic discussion. Born in Norway in 1912, Naess was Professor of Philosophy at the University of Oslo from 1939 to 1970. During the war he was active in the Norwegian resistance, and after it became recognized as his country's leading philosopher. He was founding editor of the journal *Inquiry* and the central figure in the Oslo School of philosophy. In 1970 he resigned from his Chair in order to play a more active role in the environmental movement. He has spent much of his time in mountain retreats where he writes, skis and enjoys renown as a climber whose prowess belies his years. Lean, fit and with flowing white beard, Naess remains, in his eighties, an instantly recognizable and admired figure in Norwegian intellectual life and in radical environmentalist circles all over the world.

The younger Naess's writings on the philosophy of science and 'empirical semantics' provided little indication of the interest in environmental philosophy which was later to dominate his work. Indeed, his earlier enthusiasm for the natural sciences and logical positivism appears to be at odds with his later metaphysical views, influenced more by Spinoza, Romanticism and Eastern thought than by any empiricist tradition. By 1960, however, Naess's attention was turning more towards the history of philosophy and the comparative study of 'total views' of the world and humankind. The difficulty or impossibility of deciding among such views encouraged a respect, recorded in the 1968 book *Scepticism*, for the undogmatic sceptical stance associated with the Hellenistic philosopher, Pyrrho. The Pyrrhonians had drawn from their sceptical premises certain lessons for the conduct of life, including the adoption of a non-aggressive and tolerant attitude not dissimilar from the one later recommended by Gandhi, on whom Naess had written a book, *Gandhi and the Nuclear Age* (1960). By the early 1970s, Naess was already reflecting on the relevance to environmental issues of the views of Gandhi and other thinkers outside the orthodox Western traditions of science and philosophy.

The initial result of those reflections was a short, staccato, and seminal paper, 'The Shallow and the Deep, Long-range Ecological Movement', published in 1973. Naess characterized the shallow movement as primarily engaged in a 'fight against pollution and resource depletion', its 'central objective' being 'the health and affluence of people in the developed countries'.[2] As subsequent writings show, what he means, more widely, by 'shallow ecology' is an 'anthropocentric' position which argues for responsible treatment of the environment solely on the basis of the broadly material benefits which will accrue to human beings – a position, as Naess perceives, which is adopted in the 1980 *World Conservation Strategy* and is apparent in the goal of 'sustainable development'. Deep ecology is not given a similarly concise characterization. To understand what Naess intends by this label, we might begin with the two words which comprise it. Deep ecology is *deep* because it explores the 'fundamental presuppositions' of our values and experience of the world. It is deep *ecology*, not because it is the empirical science of ecosystems, but because the attitudes it endorses, though inspired by several sources, receive 'rational justification' from the ecologists' demonstration of 'the intimate dependency of humanity upon decent behaviour toward the natural environment'.[3]

Deep ecology is best represented, perhaps, as a set of practical environmental policies underpinned by a set of normative principles which in turn are supported by a scientifically informed, but ultimately philosophical, view of reality and humankind. Among the policies advocated by Naess are radical reduction of the world's population, abandonment of the goal of economic growth in the developed world, conservation of biotic diversity, living in small, simple and self-reliant communities, and – less specifically – a commitment 'to touch the Earth lightly'. The immediate justification for these policies is to be found in normative principles such as 'Natural diversity has its own intrinsic value' and that of 'bio-spherical egalitarianism', which enjoins respect for 'the equal right' of life forms to 'live and blossom'. The failure to recognize these principles reveals 'racial prejudice' against non-human life.[4] (Egalitarianism has its limits, however. Parents have a right, says Naess, to rid the playground of cobras – though he adds that they should have taken care, for the snakes' sake, over the siting of the playground.)

The deep ecologist's case, however, cannot rest with these moral principles. For one thing, 'ethics *follow from* how we experience the world', so that an adequate set of moral principles must be grounded in a proper articulation of experience of the kind that only a philosophy or religion can provide.[5] Second, while a principle like 'Human beings must respect the rights of non-human life!' is fine as a rallying-call, it can also reinforce an assumption to which Naess is resolutely opposed – one which he thinks, moreover, has been largely responsible for our appalling treatment of the natural environment. This is the assumption that humans and non-humans – indeed, beings of any kind – exist independently of one another. Naess is a

213

'holist', arguing that, at a fundamental level, all organisms are 'intrinsically related' in a 'biospherical net or field'. To distinguish man *from* his environment is to think, therefore, at a 'superficial' and artificial level.[6]

It is this holistic vision which, for Naess, grounds the normative principles and policies of deep ecology – or, as he prefers to call it in later writings, 'Ecosophy T', in order to distinguish his particular position from neighbouring ones. An increasingly central component in this vision is Naess's conception of Self, inspiration for which comes partly from Spinoza's thesis of a single substance, describable as God *or* Nature, but more especially from the Hindu notion of *Atman* (Self). Naess approvingly cites VI.29 of the *Bhagavad Gita*: 'He whose self is disciplined by yoga sees the Self abiding in all beings and all beings in Self'. He does not think, however, that we require 'mystical union' to conclude that individual selves are, so to speak, artificial abstractions from a 'comprehensive Self' in which all beings are integrally bound. It is sufficient to reflect on how the identity of each of us is utterly dependent on relations with others and with the world at large, and properly to attend to natural feelings of empathy and sympathy which presuppose that 'one experiences something [as] similar or identical with oneself'.[7]

This conception of a 'comprehensive Self' supports the moral imperatives of deep ecology in two ways. First, someone who genuinely internalizes it will be naturally drawn to a universal 'altruism', since he or she no longer recognizes what is presupposed by 'egoism' – the existence, at a basic level, of independent individual selves. Second, it follows from this conception that 'self-realization' requires sympathetic identification with the good of the whole. 'We seek what is best for ourselves, but through the extension of the self, our "own" best is also that of others', and 'when we harm others, we also harm ourselves'.[8] Deep ecologists are sometimes criticized for elevating the good of environment over human interests: but, for Naess, appreciation of the 'comprehensive Self' implies that this contrast is illusory.

Naess's critics come from several directions. For the most radical Greens and spokespersons for 'animal rights', he does not go far enough, since he accepts that human beings, in virtue of their 'nearness' to one another, are sometimes justified in lending greater moral weight to human wellbeing than to that of non-human life. For more traditional thinkers, his principle of 'biospherical egalitarianism' goes too far, and indeed is belied by his demanding of human beings a degree of self-sacrifice and altruism which it would be absurd to demand of animals.[9] For yet others, Naess's notion of 'self-realization', with its Indian roots, is far too romantic and 'mystical' to provide a foundation for hard-headed environmental policy. This is a charge which Naess, in my opening citation, is rejecting: that a notion cannot be made precise does not mean that we are unable to 'taste' it and be inspired to action by it.

Despite his many critics, Naess's influence has been immense. As a successor to his Chair at the University of Oslo states: 'philosophy's place in Norwegian academic life, as in the society at large, is due in large measure to Naess'.[10] Not the least of his contributions to society at large has been to environmental education. The Norwegian 'core curriculum' and the Norwegian–Latvian Project in Environmental Education, with their emphasis on, for example, self-awareness and the environment, bear the unmistakable stamp of Naess's ideas.[11] On a broader front, Naess's legacy to the deep ecological tendency in contemporary environmental thought and activism is not simply the name of that tendency. As its most distinguished spokesman among professional philosophers, Naess has provided it with a theoretical foundation at which earlier writers of similar sympathies, such as Aldo Leopold, only hinted.

Notes

1 *Ecology, Community and Lifestyle*, p. 8.
2 Naess, 'The Shallow and the Deep, Long-range Ecological Movement', in L. Pojman (ed.), *Environmental Ethics: Readings in Theory and Application*, Boston, MA: Jones & Bartlett, p. 102, 1994.
3 Naess, 'Ecosophy T: Deep *versus* Shallow Ecology', in Pojman, op. cit., pp. 105–10.
4 Ibid., p. 106.
5 *Ecology, Community and Lifestyle*, p. 20, original emphasis.
6 'The Shallow and the Deep, Long-range Ecological Movement', p. 103.
7 'Ecosophy T: Deep *versus* Shallow Ecology', p. 108.
8 *Ecology, Community and Lifestyle*, pp. 174–5.
9 See R. Watson, 'A Critique of Anti-anthropocentric Biocentrism', in Pojman, op. cit., pp. 117–22.
10 A. Hannay, 'Norwegian Philosophy', in Ted Honderich (ed.), *The Oxford Companion to Philosophy*, Oxford: Oxford University Press, p. 627, 1995.
11 See J.A. Palmer, *Environmental Education in the 21st Century*, London: Routledge, pp. 159–63, 244–8, 1998.

See also in this book

Gandhi, Leopold, Spinoza

Naess's major writings

Interpretation and Preciseness: A Contribution to the Theory of Communication, Oslo: Det Norske Videnskapsakademi, 1953.
Scepticism, London: Routledge & Kegan Paul, 1968.
'The Shallow and the Deep, Long-range Ecological Movement', *Inquiry*, 16, pp. 95–100, 1973.
Ecology, Community and Lifestyle, 1976, trans. and ed. D. Rothenberg, Cambridge: Cambridge University Press, 1989.
'The Deep Ecological Movement: Some Philosophical Aspects', in S. Armstrong and R. Botzler (eds), *Environmental Ethics: Divergence and Convergence*, New York: McGraw-Hill, 1993.

Further reading

Devall, B. and Sessions, G., *Deep Ecology: Living as if Nature Mattered*, Salt Lake City, UT: Peregrine Smith Books, 1985.
Mathews, F., *The Ecological Self*, London: Routledge, 1991.
Witoszak, N. and Brennan, A. (eds), *Philosophical Dialogues: Arne Naess and the Progress of Ecophilosophy*, Savage, MD: Rowman & Littlefield, 1999.

DAVID E. COOPER

JOHN PASSMORE 1914–

> [T]he title of this book [*Man's Responsibility for Nature*] is often misquoted, as man's responsibility *to*, rather than *for*, nature. The difference is fundamental. 'Nature' is not a pseudo-person, to whom human beings are responsible ... Human beings are responsible [only] *for* nature.
>
> (xii)[1]

The Australian philosopher and historian of ideas, John Passmore, published the pioneering book referred to in the above passage in 1974. A decade later it could still be described as 'the one authoritative treatment of environmental ethics so far produced',[2] and nearly three decades after its appearance its arguments and conclusions remain ones with which all serious environmental philosophers feel obliged to engage.

Passmore was born in 1914, in Manley, New South Wales. A graduate of Sydney University, he taught philosophy at his alma mater until 1949, when he went to teach in New Zealand. After returning to his own country in 1955, he became Professor of Philosophy at the Australian National University, where he remained until retirement. He is now Emeritus Professor at the ANU.

Passmore has written widely in most fields of philosophy, but it was with his magisterial history of the subject since the mid-nineteenth century, *A Hundred Years of Philosophy* (1957), that he first attracted a wide readership. (Thirty years later, a supplement appeared, *Recent Philosophers*, in which Passmore brought the story up to date, and expressed his well-known antipathy to fashionable trends in 'Continental' philosophy, such as deconstruction.) Passmore's skill as a historian was again displayed in *The Perfectibility of Man* (1970). It was to be *Man's Responsibility for Nature* (MRN) and some associated papers, however, which made his name known well beyond the confines of philosophy. A second edition of MRN appeared in 1980, now boosted by a useful new Preface and an Appendix based on a lecture entitled 'Attitudes to Nature', the best succinct introduction to Passmore's position. In recent years, his interests have not focused on environmental questions, but he has continued to write the occasional piece in this area, including an incisive article on the political aspects of environmentalism in an edited volume on contemporary political philosophy. His last book, *Serious Art* (1991), reflects a knowledge of and feeling for a dimension of human life which readers of MRN will already have discerned.

MRN was inspired by two convictions: first, that 'men cannot go on living as they have been living, as predators on the biosphere', but second, that irrationalist tendencies in the burgeoning environmentalist movement are threatening to make matters worse (xiii). The book has three fairly specific aims: to examine historically the religious and other ideas which have shaped current attitudes and behaviour towards the natural world; to argue for a number of solutions to our most pressing environmental problems; and to 'remove the rubbish' of fashionable, obfuscating ecological views which hinder solutions to these problems. These aims are connected by an 'over-arching intention: to consider whether the solution of ecological problems demands a moral or metaphysical revolution' (xiv). Passmore's conclusion is that we require neither. A balanced history of ideas will show that at least the 'seeds' of appropriate environmental action are to be found in the Western tradition. The best way to tackle the pressing problems is to call upon a tradition of scientific reason and upon moral convictions with a long pedigree. Finally, in 'removing the rubbish', one demonstrates the bankruptcy, dangers and sometimes hypocrisy in the calls – by 'deep ecologists', 'nature mystics', 'eco-feminists' and others – for a 'new' morality and metaphysics. What we need, writes Passmore, is 'not so much a "new ethic" as a more general adherence to a perfectly familiar ethic' (187).

A main purpose in the historical chapters is to counter the familiar accusation that the Judaeo-Christian legacy is responsible for our 'predatory' treatment of nature. Passmore concedes that there is 'a strong Western tradition that man is free to deal with nature as he pleases' (27). First, however, the roots of this idea are not Jewish, but Greek, for it was the Stoics who bequeathed to Christianity the teaching that the world was created for the sake of human beings. Second, that teaching cannot, by itself, inspire pernicious treatment of the environment, since it is more likely to encourage the 'quietist' belief that God's world is fine as it is, without our intervention. In order for 'anthropocentrism' to become pernicious, it required the much more recent idea – which emerges in Francis Bacon, and reaches its zenith in Marxist images of nature as wax in man's hands – that the proper life for human beings is one of active transformation of the world about them. Finally, although there indeed exists this 'predatory' tradition, there have also been countervailing ones, emphasizing people's prudent responsibilities towards nature and duties to 'perfect' the world in which they live (39).

It is those 'minority traditions' to which we should turn in addressing the most pressing modern problems – those of pollution, conservation of resources, preservation of relatively untouched areas and overpopulation. In each case, Passmore strives to instil a sense of realism and to strike a balance between extremes. It is a waste of time to propose solutions which for political reasons, say, are totally unworkable. He is especially critical of bland calls to reduce the human population which would require gross violation of the democratic process. In this instance, as in others, environmentalists too readily ignore the question 'How are we to get from here to there?'[3] Workable solutions, he argues, must steer between 'primitivism' and 'despotism' (39): between wholesale rejection of a concern for economic progress or material welfare and the unconstrained, short-sighted pursuit of such goals. Such solutions require the application of scientifically and technologically informed cost–benefit analysis of our present practices and the alternatives to them, together with judgement on the political viability and moral acceptability of these alternatives (71). In keeping with his 'overarching intention', Passmore argues that there is no need to introduce 'new' moral considerations, such as the 'absurd' idea that nature has 'rights'. Instead, we may justify conserving resources for future generations as an extension of a natural, 'loving' concern for children and grandchildren, just as we can condemn the destruction of wildernesses as a 'vandalism' of the kind always censured by Western morality (125).

The 'rubbish' which Passmore wants removed from recent environmental debate is a mixed pile. To begin with, there is ' "mystical rubbish", the view that mysticism can save us, where technology cannot', and the related view that 'nature is sacred' (173–5). Such views, Passmore argues, not only rest on an implausible metaphysics but, unless supplemented by other considerations, have no 'environmentally friendly' implications. The idea that nature is sacred, for example, can also encourage Emerson's confidence that, as 'part and parcel of God', nature cannot really be harmed *whatever* we do to it (176). Second, he is critical of any 'primitivist' rejection of modernity in favour of forms of human life which leave nature untouched. Aside from belonging to the realm of fantasy, such proposals often smack of hypocrisy, since the few who might 'return to nature' will be parasitic on the many who do not. 'The Jain priest can walk abroad only because there are other, less spiritual, men ... to sweep the paths for him' (126). Relatedly, Passmore is dismissive of those who regard man as a 'planetary disease' or 'obscene defiler' of 'flower-sweet Earth', purveyors of 'masochistic nonsense' who are blind both to the achievements of civilization and to the legitimate interests of human beings (181).[4]

In the Preface to the second edition of MRN, Passmore wryly observes that 'it is more than a little disconcerting to be cited both as one of the more virulent critics of economic growth and as an uncritical defender of the status quo' (vii). Certainly his critics have come from opposite directions. However, although early on he was attacked by economists and planners who resented the intrusion of environmental considerations into the pursuit of economic growth, the bulk of the critics have been fellow environmental thinkers. The most common charge is one of excessive 'conservatism'. At its mildest, the complaint has been that Passmore's cost–benefit approach to the solution of environmental problems, such as pollution, allows for insufficiently radical revaluation of the policies whose costs and benefits are to be assessed.[5] Less mildly, it has been argued that Passmore scores a hollow victory in showing that traditional values suffice for moral appraisal of proposed solutions to our problems, since he wrongly refuses to recognize any moral problems except those which concern the interests of human beings.[6]

Passmore's most hostile critics, unsurprisingly, come from the ranks of those writers whom he has accused of purveying the 'rubbish' discussed above. Deep ecologists, eco-feminists and others convict Passmore of a complacent and speciesist 'human chauvinism'. His way with such critics is, on the one hand, to charge them with misunderstanding or distorting his position. He points out, for example, that in denying that non-human life can enjoy 'rights', he is in no way denying that we can and do act in morally wrong ways towards animals and environments. Moreover, he might add, his hostile critics overlook the genuinely radical shifts in human attitudes for which he is calling. In some of the most interesting passages of MRN, Passmore argues that it is not just economic greed which has been responsible for our ecological problems. So, ironically, have a 'puritanism' and 'asceticism' which make it difficult for people simply to *enjoy* the world around them 'as itself an object of absorbing interest, not ... a resource' (126). A more 'sensuous society' than our own, in which people are 'ready to enjoy the present moment for itself', would never have endured 'the desolate towns ... the slag-heaps [and] the filthy rivers' which now surround them (188–9). Passmore's other response to his radical critics is simply and unapologetically to accept their labelling him a 'chauvinist', 'speciesist' and 'shallow'. If the pejorative point of such labels is to condemn anyone who treats human interests as paramount, then Passmore is content to stand condemned (187).

John Passmore's book *Man's Responsibility for Nature* remains the most authoritative statement of a main tendency in environmental ethics, constantly cited by both adherents and opponents of that tendency. Within philosophical circles, it may be the 'deeper' ecological tendency represented by Arne Naess which has attracted more attention in recent years. But it is surely the 'shallower' approach of Passmore which has done more to inform the environmental policies of governments and other organizations for which, ultimately, the interests of human beings must be of paramount concern.

Notes

1   All page references in the text are to the second edition of *Man's Responsibility for Nature*.
2   R. Attfield, *The Ethics of Environmental Concern*, p. ix.
3   Passmore, 'Environmentalism', in R. Goodin and P. Pettit (eds), *A Companion to Contemporary Political Philosophy*, Oxford: Blackwell, p. 479, 1993.

4   Passmore's targets here are Ian McHarg and W.S. Blunt ('the ecologist's poet-laureate' (p. 180)).
5   See, e.g., C.A. Hooker, 'On Deep versus Shallow Theories of Environmental Pollution', in R. Eliot and A. Gere (eds), *Environmental Philosophy*, Milton Keynes: Open University, pp. 58–84, 1983.
6   See R. Attfield, *The Ethics of Environmental Concern*, pp. 4ff, and Val Routley, 'Critical Notice of John Passmore's *Man's Responsibility for Nature*', pp. 171–85.

See also in this book

Bacon, Emerson, Marx, McHarg, Naess, Plumwood

Passmore's major writings

A Hundred Years of Philosophy, London: Duckworth, 1957; republished by Penguin, Harmondsworth, 1968.
The Perfectibility of Man, London: Duckworth, 1970.
Man's Responsibility for Nature: Ecological Problems and Western Traditions, London: Duckworth, 1974; second and enlarged edition, 1980.

Further reading

Attfield, R., *The Ethics of Environmental Concern*, New York: Columbia University Press, 1983.
Hooker, C.A., 'Responsibility, Ethics and Nature', in D.E. Cooper and J.A. Palmer (eds), *The Environment in Question*, London: Routledge, pp. 147–64, 1992.
Routley, V. (now V. Plumwood), 'Critical Notice of John Passmore's *Man's Responsibility for Nature*', *Australasian Journal of Philosophy*, 53, pp.171–85, 1975.

DAVID E. COOPER

JAMES LOVELOCK 1919–

> The idea that the Earth is alive is at the outer bounds of scientific credibility. I started to think and then to write about it in my early fifties. I was just old enough to be radical without the taint of senile deliquency.[1]

It is an idea that has absorbed James Lovelock for more than thirty years, the idea that is encapsulated in the name 'Gaia'. The name itself was suggested by the novelist William Golding, a friend and at one time a neighbour, in the course of one of the long walks the two men used to take together in the Wiltshire countryside. In Greek

221

mythology Gaia, or Ge, was the Earth. She sprang from Chaos and gave birth to Uranos, the Heavens, and Pontus, the Sea. She was not a goddess. She preceded the gods and goddesses and provided the context, the environment if you will, in which the gods could exist. Her name lives on in those words in our language that begin with 'ge-' – ge-ography, ge-ology, ge-odesy, ge-ometry, and all the rest. The image is powerful and Lovelock's Gaia hypothesis is conceived on an appropriately grand scale.

James Ephraim Lovelock was born on 26 July 1919 in Letchworth Garden City, Hertfordshire. His father was a keen gardener with a highly developed moral awareness that owed little to formal religious belief, but appears to have been based on a mixture of folk Christianity and traditions and superstitions that used to be widespread in rural Britain. He communicated his love of the countryside to his son and, with it, an enthusiasm for walking. Later, James became a keen hill walker, and in 1999 he and his wife celebrated his eightieth birthday by walking along the Cornwall Coast Path. This runs along the tops of high sea cliffs, then plunges down narrow, steep-sided valleys only to climb again on the far side. It is strenuous walking for anyone.

When he left school, James Lovelock worked in a laboratory, studying chemistry in the evenings but by day learning the practical laboratory techniques that were to serve him well in later years. Eventually he left to become a full-time student at Manchester University, graduating in chemistry in 1941.

It was wartime and the young graduate was absorbed into the national effort, going to work for the Medical Research Council at the National Institute for Medical Research, in London. At the end of the war, in 1946, he went to work at the Common Cold Research Unit, in Wiltshire. There, he and his colleagues found that the search for a cure for colds was fruitless, but they were able to design ways to prevent their transmission. He remained at the Unit until 1951.

He received his Ph.D. in 1948, from the London School of Hygiene and Tropical Medicine. This degree was in medicine. He received the degree of D.Sc. in biophysics in 1959, from the University of London.

His path lay ahead of him, clearly defined. He could have remained a scientist employed by the Civil Service. Year by year his salary would have increased, his standard of living would have been fairly high, and eventually he would have retired with an index-linked pension that would have kept him in reasonable comfort. It was not enough. James wanted more. He once told me that his enthusiasm for science arose from the opportunities it provides for finding answers to questions. It can deliver intellectual freedom, however, only to those who are able to frame the questions for themselves. As a Government scientist his researches would necessarily have been directed towards the resolution of matters of public interest. His imagination would have been constrained.

Still an employee of the National Institute for Medical Research, in 1954 he was awarded a Rockefeller Travelling Fellowship in Medicine. He spent it at Harvard University Medical School, in Boston, and in 1958 spent a year working at Yale as a visiting scientist. He resigned from the National Institute in 1961 in order to take up an appointment as Professor of Chemistry at Baylor University College of Medicine, in Houston, Texas.

His particular skill had always been intensely practical. He had a talent for constructing instruments that would measure with a fine sensitivity whatever anyone wished to measure. A time came when it seemed that this skill might provide him with a means of earning a living while leaving him sufficient free time to pursue his own interests – to find answers to his own questions. So, in 1964, he became a freelance research scientist. He holds more than 50 patents, most of them for instruments used in chemical analysis.

Many of these instruments have developed and refined the technique of gas chromatography. In this, the substance to be analysed is vaporized, and the vapour is mixed with a gas and then introduced into a stationary column filled with a finely powdered solid or liquid. Different components of the specimen react with the column at different rates. This separates them in a way that allows them to be identified – originally by their colour, hence the name of the technique.

In 1957 Lovelock invented the electron capture detector. This is still one of the most sensitive of all detectors. It revealed the presence of residues of organochlorine insecticides such as DDT throughout the natural environment, a discovery that contributed to the emergence of the popular environmental movement in the late 1960s. Later it registered the presence of minute concentrations of CFCs (chlorofluorocarbon compounds) in the atmosphere.

In the early 1960s, while living in Texas, James was a consultant at the Jet Propulsion Laboratory (JPL) of the California Institute of Technology, in Pasadena. At the invitation of NASA he had already helped with some of the instruments used to analyse lunar soil and he was then asked to advise on various aspects of instrument design for the team of scientists planning the two Viking expeditions to

Mars. Instruments can help in finding answers to questions, but before a new instrument can be devised the question must be framed clearly. Asking the right question is often more difficult than finding the answer to it.

James was not directly involved with the question that has always been central to all Martian exploration: Is there, or has there ever been, life on Mars? Nevertheless, as he contemplated the ways in which the team proposed to seek answers, he found himself driven to ask more fundamental questions. The Viking experiments were based on the assumption that Martian biology would resemble that of the only living organisms of which we have any knowledge at all – the ones on Earth. But to James Lovelock this seemed to be a huge assumption with nothing to justify it. 'How can we be sure that the Martian way of life ... will reveal itself to tests based on Earth's life style?' 'What is life, and how should it be recognized?'[2]

When his colleagues at JPL asked how he would set about finding answers, the only thought he could offer was that living organisms must, in one way or another, increase the amount of order in the world around them. It was not much help, but the idea seeded itself in his brain.

No matter what its composition or biochemical pathways might be, any living organism must take certain chemical substances from its surroundings and use them to build and repair its own tissues. This will generate waste products that the organism will dispose of into its surroundings. Eventually this metabolic process will alter the composition of its surroundings, increasing the abundance of certain substances and depleting it of others. In this way the organism will modify its own environment, giving it a chemical composition markedly different from the one it would have if it were allowed to reach a state of chemical equilibrium.

That difference, James maintained, should be detectable, and he and Dian Hitchcock, a philosopher employed to assess the logical consistency of NASA experiments, decided the place to seek it was in the atmosphere. The atmosphere has a much smaller mass than the solid or liquid components of a planet, and so perturbations to its composition would be more easily detectable. Also, the atmosphere is more easily accessible to investigators on another planet or in space.

When the atmosphere of Earth is compared with those of Mars and Venus the chemical disequilibrium of our atmosphere becomes immediately evident. It contains both methane and oxygen, for example. These react naturally to yield carbon dioxide and water, so some process must constantly replenish the methane, and at the pressure and temperature prevailing on Earth, it is only biological reactions that are capable of releasing methane. Were some process not releasing gaseous nitrogen, Earth's atmosphere would have lost any it had billions of years ago, as it was oxidized by lightning to stable oxides that are soluble in water and were washed to the surface by rain. In fact, nitrogen is released by denitrifying bacteria. Both other planets, in contrast, have atmospheres that are in chemical equilibrium.

The atmospheres of Mars and Venus consist mainly of carbon dioxide. Our atmosphere contains very little of this gas (about 3.5 per cent by volume). This has not always been the case. At one time our atmosphere contained much more carbon dioxide. It has reacted to produce carbonates and now forms the chalks and limestones that are among the commonest of sedimentary rocks. James calculated that carbon has been removed from the atmosphere by this means at a rate far faster than could have been achieved by simple, inorganic chemical reactions. Living organisms played a major part, principally by building seashells from calcium carbonate, and carbonate rocks are predominantly of biological origin.

Since the time when the earliest organisms are believed to have appeared on Earth, the Sun has increased its output of energy by about 30 per cent. Carbon dioxide is a so-called 'greenhouse gas' and James concluded that a consequence of its progressive removal from the atmosphere was that the surface temperature remained fairly constant. It was removed by organisms and its removal maintained the climate most favourable to them. In other words, living organisms were regulating the global climate.

With this realization the basis was established of what was to grow into the Gaia hypothesis. It grew as James became increasingly persuaded of the extent to which biological regulation pervades the environment. In 1979, in his first book on the subject, he defined Gaia as 'a complex entity involving the Earth's biosphere, atmosphere, oceans, and soil; the totality constituting a feedback or cybernetic system which seeks an optimal physical and chemical environment for life on this planet'.[3] From this the concept developed of the Earth itself as a single, discrete, living organism, equipped with biological mechanisms for maintaining its overall homoeostasis.

The search then intensified for evidence to test the thesis. In 1987 it was established that cloud formation over the oceans is initiated by particles released by single-celled marine algae.[4] The suggestion of long-term climate regulation was confirmed in 1989.[5] Other predictions arising from Gaian theory have also been confirmed, and the status of the idea has advanced from hypothesis to theory.

Its reception has been mixed. Environmentalists, especially those of a more mystical bent, have embraced it enthusiastically, cheerfully overlooking some of its implications. Gaia, if it (she?) exists, has no great concern for the fate of organisms more complex than micro-organisms. She would remain unmoved by the extinction of elephants, whales, tigers or humans. This is in keeping with her mythical origin, of course. The Greek Gaia was destroyer as well as creator: she buried people.

At this point the name became a handicap. It had never been meant as more than a metaphor and a preferable alternative to some ungainly acronym, but it was proving too evocative and what James intended as a rigorous scientific proposal began to look like sentimentalism. As one critic expressed it:

> The conflict between accepting what science teaches us and what the human heart would like to believe is well illustrated by James Lovelock's Gaia concept. It is a lovely thought, a tempting one too, because it is a form of religion and the human soul requires the comfort of a guided universe; it needs religion. Alas, it is also unnecessary, because the world as it was, has evolved, and now exists, is not explicable. It is merely very complex, and life plays a role in it, but not the main one.[6]

Some scientists warmed to the idea, however. In its less-extreme form it is hardly novel. The influence of living organisms on the cycling of minerals has been known for many years. Indeed, the cycles are described as biogeochemical cycles. What Gaia added was an over-arching, unifying concept leading to a new way to approach problems relating to the functioning of the planet. New questions could be asked about possible perturbations along the lines of 'How would the totality of living organisms respond?' Where environmental difficulties could be analysed they could be remedied with the help of living organisms. This is now a well-established technique, known as bio-remediation. It was used, for example, to clean up Alaskan beaches following the oil spill from the *Exxon Valdez*.

Such an approach can be described, not too fancifully, in something approximating to medical terminology. Environmental problems can be seen as 'ailments' the nature of which can be 'diagnosed' and to which 'therapies' can be applied. James's first doctorate was in medicine and one of his heroes is James Hutton (1726–97), one of the founders of modern geology, whose training was also in medicine. Hutton told a meeting of the Royal Society of Edinburgh in 1785 that the Earth was a super-organism and that its proper study should be physiology. The 'Gaian' study of the Earth is now called 'geophysiology'.

Extend the concept beyond this, however, and it remains controversial. Most evolutionary biologists reject it. Evolution occurs at the very local level of individuals. Genes spread and mutations are fixed in populations as individuals inheriting genes that confer a reproductive advantage produce more offspring than individuals who do not. It is difficult to see how this Darwinian process can link to planetary regulation. The even stronger idea, that the Earth itself is a single organism, finds little support from biologists.

Nevertheless, Gaia remains one of the most interesting and influential ideas of modern times, and its author has been rewarded for it, as well as for his other contributions to science. In 1974 he was elected a Fellow of the Royal Society and he has received many prizes for his contributions to chromatography, climatology and environmental sciences. In 1997 he was awarded the prestigious Blue Planet Prize for helping in the resolution of global environmental problems. He has received honorary doctorates in science from eight universities. In 1990 he was awarded a CBE.

James Lovelock now lives in a converted mill on the border between Devon and Cornwall. He owns some of the adjacent land and over the years has planted native trees and encouraged wildlife to flourish, so in effect he lives at the centre of his own nature reserve, his private corner of Gaia. He has not retired. There is no end to the questions, no end to the search for answers, and no end to his delight in producing radical ideas.

Notes

1   *The Ages of Gaia*, p. 3.
2   *Gaia: A New Look at Life on Earth*, p. 2.
3   Ibid., p. 11.
4   R.J. Charlson, J.E. Lovelock, M.O. Andreae and S.G. Warren, 'Oceanic Phytoplankton, Atmospheric Sulphur, Cloud Albedo, and Climate', *Nature*, 274, pp. 246–8, 1987.
5   David W. Schwartzman and Tyler Volk, 'Biotic Enhancement of Weathering and the Habitability of Earth', *Nature*, 340, pp. 457–60, 1989.

6   Tjeerd H. Van Andel, *New Views on an Old Planet: A History of Global Change*, Cambridge: Cambridge University Press, 1994.

See also in this book

Darwin, Ehrlich, Schumacher, Spinoza, Wilson

Lovelock's major writings

Gaia: A New Look at Life on Earth, Oxford: Oxford University Press, 1979.
The Ages of Gaia, Oxford: Oxford University Press, 1989.
Gaia: The Practical Science of Planetary Medicine, London: Gaia Books, 1991.

Further reading

Allaby, Michael, *Guide to Gaia*, London: Optima, 1989.
Allaby, Michael and Lovelock, James, *The Great Extinction*, London: Secker & Warburg, 1983.
—— *The Greening of Mars*, London: André Deutsch, 1984.
Joseph, Lawrence E., *Gaia: The Growth of an Idea*, New York: St Martin's Press, 1990.
Volk, Tyler, *Gaia's Body: Toward a Physiology of Earth*, New York: Copernicus, 1998.
Westbroek, Peter, *Life as a Geological Force: Dynamics of the Earth*, New York: W.W. Norton & Co., 1992.

MICHAEL ALLABY

IAN McHARG 1920–

'What are the natural determinants for the location and form of development?' The answers are vital to administrators, regional and city planners, architects and landscape architects. The landscape architects, in fact, work within a profession historically concerned with the relation of man to nature and of the natural sciences to the making of the urban environment.[1]

Ian McHarg is this century's leading planner and designer of ecologically based projects. He has, through his prolific public speaking, become one of the leading critics of the world's consumption of physical resources. He is a leading advocate for preservation and for change in planning and design and is a leading educator of professionals in the visual arts, embodying a new ecological aesthetic.

McHarg began his career as a city planner, critically observing the destruction that modern development was causing to the natural environment. He was influenced in his early education by the noted urbanist, Lewis Mumford, and his book, *The Culture of Cities*. McHarg found Mumford to be the only person who correctly appraised the dangers of modern architecture. 'It has achieved its objectives, and they are hollow.' Modern architecture was 'deficient of technology', yet it used technologies and their explicit materials to derive its aesthetic. McHarg furthered the Mumford criticism, stating that science, itself, was 'resolutely excluded' from the architectural definition of cities. 'The wisest man I have ever known was Lewis Mumford', he concluded.

Accepting the challenge of G. Homes Perkins, the then Dean of the College of Fine Arts, at the University of Pennsylvania, to set up a new programme in landscape architecture, McHarg reasoned that any meaningful change to this profession had to commence first with fundamental changes to the education of the professional landscape architect. He assembled a faculty comprised predominantly of natural scientists rather than other landscape architects trained under the old arcane methods. He took this bold step so that it would be possible to discover the true nature and unity among the varying and separate natural sciences. Each scientist would contribute their knowledge and discipline to the process of understanding and it would be the role of the landscape architect, as a master builder, to establish common ground, synthesis and integration of built form in the environment. McHarg attracted young, eager and talented natural scientists to teach in the new curriculum of landscape architecture and liberally supplemented permanent faculty with distinguished lecturers, and national leaders in the environment.

The University of Pennsylvania degree was proposed as a graduate degree for advanced studies of the profession by professionals who had already earned bachelors' degrees in planning, architecture and landscape architecture. To attract only the best, McHarg placed advertisements of his new curriculum in leading international newspapers. With one of the best faculties and student bodies in place, the Department of Landscape Architecture and Regional Planning at the University of Pennsylvania achieved international prominence.

McHarg proceeded to find common ground and solutions to American urban and suburban developmental problems. He sought to achieve built environments that were more compatible with their natural environment. He began by basing his environmental premise on Charles Darwin's assertion: 'The surviving organism is fit for its environment, lest it not survive'. McHarg re-stated the Darwin theory: 'Survival is the first criterion; extinction measures failure ... and ... all organic life fitted within one of either of two systems: syntropic-fitness-healthy or entropic-misfit-morbidity/death' (Ian McHarg, *A Quest for Life*, pp. 244–5). Receiving foundation funding, McHarg hosted a CBS television show, *The House We Live In*, inviting each week a major environmentalist to discuss issues and solutions. Guests included Lewis Mumford, Paul Ehrlich, Abraham Heschel, Gustave Weigel, Paul Tillich, Margaret Mead and Alan Watts. Many of these same speakers came from his roster of lecturers at the University of Pennsylvania series, *Man and the Environment*. It was through these lectures that McHarg developed an environmental agenda.

Using the university as a vehicle of research, McHarg founded the Center for Environmental Studies and sought both developmental problem types and regions impacted for the employment of a new planning and design methodology – one that would be ecologically based. Then, using the traditional landscape architectural design studio as a process-based laboratory, McHarg was able to bring student research power to the Center's agenda. The Center's studies included the New Jersey Shore; the Route Selection Study I-95 for the Delaware–Raritan Citizens Committee, New Jersey; the Potomac River Basin study; and the Metropolitan Philadelphia Open Space study for the Urban Renewal Agency for the States of Pennsylvania and New Jersey.

With proven results from these studies, McHarg formed a partnership with David Wallace, a University of Pennsylvania Professor of Urban Design, to provide professional services utilizing the newly developed environmental planning and design process. Many of the same types of studies as those undertaken at the university, were now the new firm's professional studies for governmental agencies. They advanced the thinking and development of an ecological planning methodology by providing realistic regulation and planning criterion, inter-agency review and evaluation, and most importantly, methods of implementation via the conclusions as policy of the new ecological planning process. The studies of Wallace, McHarg and Associates (WMA)[2] included the famous Baltimore Inner Harbor master plan, the Richmond Parkway study, the plan for Green Springs and Worthington Valleys, Baltimore, Maryland, and the plan for Staten Island, New York.

With more than twenty studies of various problem types completed and a fairly firm notion of process, Ian McHarg set out to formalize both philosophy and methodology, and in 1967 wrote the book *Design With Nature*; Lewis Mumford wrote the book's Introduction. The book discussed the concepts of a limited planet and, therefore, limited natural resources. It discussed natural processes that were beyond the control of man and concluded therefore man's folly to build and settle in these regions. It showed environments where man and nature could co-habit together, if wise planning and design were undertaken, and demonstrated how new highways, suburbs and metropolitan areas could be restructured and soundly designed for future growth. *Design With Nature* was the first book of its kind to define the problems of modern development and present a methodology or process prescribing compatible solutions. It contained powerful prose that delineated the ecological imperative. It also contained an abundance of maps, charts and graphics that illustrated a step-by-step analysis, synthesis and conclusionary methodology.

> If nature is viewed from the vantage of the man who would intervene with intelligence and even aspire to art, then we can see that nature is process; it has values and opportunities for human use, but it also reveals constraints and even prohibitions. Furthermore, process can be measured in terms of creation and destruction ... We can employ this concept for both diagnosis and prescription, in both planning and design. The application of ecology to human affairs is so recent, however, that there is not yet a formal method. I offer my own rudimentary conception of the ingredients of an ecological method.[3]

McHarg's method was not rudimentary, as he so stated, but quite sophisticated and well reasoned. The initial methodology was subjected to testing, criticism and re-evaluation. It remained constant in principle and only differed throughout the years in specifics related to project type and/or environmental type. His methodology, as expressed in *Design With Nature*, was never so succinct as that delineated in an article for a University of Pennsylvania's scholarly journal, *VIA 1, Ecology and Design*. McHarg's method was a ten-step 'diagnostic and prescription'.[4]

1 *Ecological inventory* An analysis of all natural systems and physical conditions of a region and subregion by environmental

types. The analysis universally included, climate, physiography, geology, soils, hydrology, plant and animal associations, existing land uses and cultural developments. This analysis was predominately inventory mapping of each natural science physical characteristic. The mapping was accomplished by using overlays at the same scale thus facilitating comparative interpretations. This overlay comparative methodology has become synonymous with McHarg's name and has influenced the computer language and methodology developed as part of all current GIS (Geographical Information Systems) programs.

2 *Description of natural processes* An analysis of all major physical and biological processes directed towards defining the interactions of one natural science to another and in turn, all natural system's interactions to human needs and development.

3 *Historical Inventory* An analysis of human adaptation to the environmental system which emphasizes the match of development to technological changes over time.

4 *Expression of the 'given' form* A conclusionary analysis that delineates the natural identity of the region and its sub-parts.

5 *Expression of the 'made' form* A conclusionary analysis that delineates man's response in settlements to the 'given' form. McHarg often termed this analysis the *Genus Loci*, a word derived from Greek meaning the appropriate and skilled relationship of civilization to its locale.

6 *Attribution of relative value* A mathematical or matrix comparative analysis determining the relative degree of appropriateness or conflict of any one land use to a subarea's 'given' form.

7 *Interpretation of intrinsic land use* A mapping analysis that utilizes the relative values of a land use to each regional subarea to determine the relative 'suitability' of one land type over another for each differing land use. Suitability maps were determined, at a minimum, for urban development, agriculture and conservation. In a more expanded format, the analysis included, residential, industrial, recreation and parks, forestry and mineral resource extractions as other land uses studied.

8 *Conclusions as to compatible land use* 'By use of a matrix with all possible land uses on both axes, a selection is made of the maximum number of compatibly concurrent land uses.' In many of McHarg's studies, this analysis also included a very dramatic graphic, where under a typical section through the region showing all differing subregion, a bar chart of suitable land uses are arrayed to correlate their appropriate location.

9 *Formulation of alternative land use plans* An alternative plan is formulated that focuses on the predominance and optimization of one of each of the studied land uses. Within each of the areas, appropriate development is shown and integrated with areas for conservation as determined by the natural systems analysis. Accompanying the plan is a set of guidelines to insure the management of the development to environmental conditions and concerns.

10 *Implications for the new 'made' form* A proposed optimum plan is developed where critical ecological systems are conserved and appropriate urban and/or human developments are compatible. Where more than one type of land use is appropriate, rather than allow conflicting land uses, the concept of 'a highest and best use' for safeguarding the natural environment is chosen and incorporated into the plan. The new 'made' form is described in the new ecologically based ethics and aesthetics. The guidelines, then, become policy recommendations.

Design With Nature illustrates this methodology, drawing from a number of executed master plan and ecological studies across northeast America. Included are:

- The Department of Landscape Architecture and Regional Planning, University of Pennsylvania

  — The Delaware River Basin Study, New York, New Jersey, Pennsylvania and Delaware
  — The Potomac River Basin Study, Virginia, Maryland, and the District of Columbia
  — The New Jersey Shore
  — City of Philadelphia, Pennsylvania: Health and Pathology

- The Institute for Environmental Studies, University of Pennsylvania

  — Metropolitan Open Space from Natural Processes, Urban Renewal Administration of Pennsylvania and New Jersey

- The Center for Ecological Planning, University of Pennsylvania.

  — The Delaware River Basin, New York, New Jersey, Pennsylvania and Delaware

— The New Jersey Pine Barrens
— Tocks Island Region, Pennsylvania and New Jersey
— The Neshaminy Watershed, Pennsylvania

- Wallace, McHarg, Roberts and Todd, Philadelphia, Pennsylvania

  — The Potomac Task Force Study, American Institute of Architects
  — Toward a Comprehensive Landscape Plan for Washington District of Columbia

The years after *Design With Nature* were equally productive in the further development of the McHarg methodology. By McHarg's own statement, three projects in particular contributed new insight and yielded new attributes towards a universal model for solving twentieth-century ecological planning and design problems.

> That book [*Design With Nature*] included a relatively short experience in the United States, from 1954 to 1968, whereas my subsequent experience has been much longer over twenty-five years. In this subsequent period have occurred many of my proudest accomplishments: Woodlands New Town, Pardisan, A Comprehensive Plan for Environmental Quality and the Medford study.
>
> (Ian McHarg, *A Quest for Life*, p. 206)

Medford [5]

This small township in southern New Jersey was under severe impact in 1974 from suburban expansion of the nearby cities of Philadelphia, Pennsylvania, and Camden and Trenton, New Jersey – all less than sixty miles away. Simultaneously, all rural and suburban towns in New Jersey were under Federal court order to de-segregate housing ordinances which by their restrictive zoning ruled out affordable housing. Medford decided to test the use of ecological determinism to foster the continuity of its current rural character and to control the type of future growth. This, they reasoned, should be achieved not through the traditional 'master plan', but through the formulation of new ordinances that were defendable to its citizenry and the courts alike. The Township hired McHarg and the University of Pennsylvania to conduct a study to determine the 'performance requirements for the maintenance of social values represented by the natural environment'. Understanding that any restrictions, new or old, would probably constitute a 'taking' of developmental 'rights' and would require a financial burden to the property owner for which the Township would have to compensate, and that any restriction would additionally be scrutinized by the courts, all aspects of the study were to be 'subject to stringent requirements ... and (it was) essential that all data and interpretation be conducted by competent scientists'.

The McHarg methodology developed a system of cross-evaluation of all ecological data to four social values:

- *'Inherently hazardous to human life and property'* These included, for example, areas within the fifty-year flood plane, and zones of high forest fires (predominantly the Pine Barrens areas).
- *'Hazardous to human life and health by specific human action'* Areas adjacent to streams of good-to-excellent quality, severely high water table, aquifer recharge areas, zones of municipal wells, soils not suitable for septic deployment, etc.
- *'Irreplaceably unique and scarce resources; and'* These included, for example, historic monuments, public lands, locations of mature specimen trees and unique forest habitats (cedar swamps, bogs and lowland successional meadows).
- *'Valuable resources where unregulated utilization will result in social costs'* Areas of rare and unique beneficial wildlife habitat, scenic resources, extractable sand and gravel resources and recreational potential sites.

The above four social environmental factors were then evaluated against four developmental options:

- *Rural Urban Development* This land use assumed detached single family development on larger-than-an-acre lots with on-site sewage disposal.
- *Suburban Development* This assumed lots of an acre with lightweight, slab on soil foundations, some site grading and a predominance of fertilized lawn.
- *Clustered Suburban Development* This assumed various multi-family-type dwelling units, extensive paved surfaces, extensive site grading, extensive loss of tree cover, municipal sewer connection and heavy building foundations.
- *Urban Development* This assumed multi-family residential and

other land use types, all requiring heavy foundations, full site coverage, extensive paved surfaces and municipal sewer connections.

It is not surprising that the highest and best type of new development within Medford Township would predominately be *Rural Urban Development*. *Suburban* and *Clustered Suburban Developments*, although possible, would be limited to a small band of lands. *Urban Development* was the least allowable land use and would be highly regulated by restrictive zoning ordinances. As all lands were to some degree developable, no 'rights' were denied and therefore no financial 'takings' were required.

The McHarg study, for the first time, successfully proved the argument of local conservationists and environmentalist for maintaining the status quo, explicitly illustrating the social, and by consequence, economic loss that the standard 'by right' development was causing to small towns and rural communities across America. But it also illustrated that rich and affluent suburban communities could use ecological determinism to foster restrictive zoning and limit any, and all, development.

Woodlands New Town[6]

In 1971 McHarg was approached by the developer George Mitchell to study and plan a new community for 200,000 acres and 50,000 persons north of Houston, Texas. While there had been many previous attempts at new communities in America, most had met limited success and some had failed completely. Mitchell was a Texas oil billionaire but was a fiscally conservative man. He reasoned that ecological planning and design could be accomplished with the least cost and yield the most attractive community, provide sound sales meeting projected cash flow demands and, therefore, would yield the highest profit. McHarg, for the first time, would have the opportunity to totally plan a new development and control site development. In the process, developmental impacts could be minimized and alternative solutions be investigated.

Woodlands evolved several new methodologies; primarily site density and coverage. Exacting calculations were made to find the proper ratios for all types of housing and land uses while conserving vegetation and water regimens. New concepts of storm water management were developed for both insuring positive drainage in flood-prone areas but additionally for recharging of the region's aquifers. Using these ingenious systems, a significant reduction of needed site infrastructure, i.e. storm drains, was accomplished, yielding substantial savings and insuring economic feasibility of the project. Economic determinism could be subverted. Of all the new American communities attempted, Woodlands today remains the most successful and the most ecologically sustainable of McHarg's work.

Pardisan[7]

The culmination of all of McHarg's theories on physical planning based on the ecological sciences was the development of an environmental park in Tehran, Iran. Called Pardisan, a Farsi derivative of paradise, the park, as proposed in 1975, was like no other environmental park or preserve. It differed entirely, in concept and organization.

McHarg opposed throughout his career the view that each of the natural sciences – biology, geology, hydrology, etc. – were isolated sciences. In this, he struck at the very core of their professions, their professional scientific repositories. McHarg reasoned that arboretums and botanical gardens were very artificial and unscientific collections. Botanical gardens, for example, were in no way suitable for teaching the ecological aspects of vegetation. Their grouping by families or growth characteristics spoke nothing of the climate, soil or the inter-dependencies of species. The same was true for zoos and aquariums. These were collections of fauna in cages and walled environments with, at best, token environmental support – water tanks and rocks fashioned naturalistically. Natural history museums, planetariums and other 'housed' institutions separated man from the real environment. McHarg's proposal for Pardisan was for an environmental demonstration park that displayed man within all of earth's many environments. Each display would show a region and its ecology in full successive stages and sub-communities and illustrate man's harmonious adaptation to these environmental sets.

This institution should embrace the functions of the great museums and zoological and botanical gardens that were built in several countries in previous generations and, indeed, up to the present.

Thus (at Pardisan), the diachronic evolution of man's environment, and the present synchronic state of diversification of species and cultures throughout the world will be presented as a national and international base and center for environmental activities of conservation, research, edification and recreation.[8]

Proposed at a time when Middle East nations were awash in petro dollars and proposed to the pre-Islamic revolutionary government of the Shah Reza Palave, Pardisan was to be the showpiece of Iran. It was to demonstrate to the Iranian people the country's cultural and ecological heritage. To this end, the Wallace, McHarg, Roberts and Todd firm was joined by the Iranian architect, Nader Ardalan, who brought to the collaboration, research and sympathetic understanding of traditional adaptive Iranian architecture. Ardalan had documented the growth and architecture of the great city of Isphahan. He knew the architectural response to climate and native materials would demonstrate Iran's designed compatibility to its harsh desert environment. But, it was McHarg who added the rest of the world to Pardisan's purpose.

On the proposed extensive site, the environmental park was planned: 300 hectares on the south-facing foothills of the Elburz Mountains just to the north of Tehran. Based first on climatic types, then upon continental regions, North and South America, Europe, Africa, Asia and Oceania are represented equal to Iran. Within each region, separate zones representing climatic differences would house a collection of imported native soil and geological formations, plant communities representing differing water regimens and correlated fauna. Man's habitat, agricultural and related native crafts would be reproduced from their original location by skilled native craftsmen. The purpose was clearly educational, but also it was to be a warehouse in the traditional scientific sense. It was to represent the world's gene pool. In reality, McHarg was proposing an ecological theme park, a model not too dissimilar to Disney World. It would be far more adventuresome than the environmental displays of the San Diego Zoo, the only other such existing facility . Since then, we have seen the proliferation of theme parks and their movement to representational habitats as well as the environmental habitat displays in today's zoos and aquariums.[9] But none of these developments had the comprehensive breadth nor the scientific thoroughness of the McHarg Pardisan environmental park.

In addition to Medford, Woodlands and Pardisan, the post-*Design With Nature* years, up to the end of the WMRT firm, also included significant professional projects such as

- Department of Landscape Architecture and Regional Planning, University of Pennsylvania

  — Prototype Database for a National Ecological Inventory
- Wallace, McHarg, Roberts and Todd, Philadelphia, Pennsylvania

  — The Lower Manhattan Plan, New York
  — Amelia Island, Florida
  — Indian River Shores, Vero Beach, Florida
  — Laguna Creek Study, Sacramento, California
  — Easton's Neck Point, Long Island, New York
  — Ponchartrain, New-Town-in-Town, New Orleans, Louisiana
  — Lake Austin Growth Management, Texas
  — The Toronto Central Waterfront, Ontario, Canada
  — San Francisco Metropolitan Regional Impact Study, California
  — The Denver Metropolitan Area-wide Environmental Impact Statement, Colorado
  — The 208 Study for Detroit Metropolitan Area, Michigan
  — The 208 Study for Toledo, Ohio
  — Nigerian National Capital Site Selection, Abuja, Nigeria
  — Kenneth Square Human Ecology Study, Pennsylvania

Pardisan, was the end to a significant chapter in McHarg's life. Almost concurrent with the completion of the project but before final acceptance and fee payment, the Iranian government of the Phalavi Dynasty collapsed and Iran was thrown into revolution. The banks were stripped and fee payments held in escrow pending completion by WMRT. The financial impact almost bankrupted the firm, and in a partnership call Ian McHarg was removed from the firm. Without this base, McHarg was unable to demonstrate and continue development of his ecological methodology in planning and design. He has since taken on the writing of his autobiography, *A Quest for Life*, as well as continuing with public speaking engagements, teaching his ecological planning and designing a gospel of change.

McHarg's impact on environmental planning and policy is legendary. In sheer numbers of graduates alone, he has changed the profession of landscape architecture and has accomplished this on a global scale.[10]

Notes

1 McHarg, *Ecology, For the Evolution of Planning and Design; VIA 1, Ecology and Design*, Philadelphia, PA: Graduate School of Fine Arts, University of Pennsylvania, pp. 44–5, 1968.
2 The firm has changed over the years with the addition and deletion of its principals: Wallace, McHarg and Associates (1962–1967); Wallace, McHarg, Roberts and Todd (1967–78); Wallace, Roberts and Todd (1978–1991); WRT, Inc. (1991–).
3 McHarg, *Ecology*, pp. 44–5.
4 Ibid. Italics indicate directly quoted subjects. A brief explanation of each is by the author.
5 Narenda Junega, *Medford: Performance Requirements for the Maintenance of Social Values Represented by the Natural Environment of Medford Township, New Jersey*, Philadelphia, PA: Center for Ecological Research in Planning and Design, Department of Landscape Architecture and Regional Planning, University of Pennsylvania, 1974. This is a report published by the Center as a summary document.
6 Ian McHarg, *A Quest for Life*, pp. 256–63.
7 *Pardisan: Plan for an Environmental Park in Tehran*, Philadelphia, PA: WMRT, 1975. This is a report privately published by the firm as a summary document.
8 The text appears under the title 'Introduction', by Eskandar Firouz, who was the Iranian Director of the Department of the Environment. They are clearly the words of Ian McHarg as is noted in the Table of Contents: 'This Report was written by Ian McHarg'. It was common practice to feed text to governmental officials for such introductions.
9 This recent development is not surprising for some of the major design firms leading this movement grew out of a Philadelphia base and contained many McHarg graduates on staff.
10 Professor Edmond Waller, a 1971 Penn graduate, has conducted a survey of Australia and Asia and in an IFLA Symposium presentation of 1995 cites hundreds of McHarg protégés throughout the region and at least one in every country. More importantly, he cites the McHarg methodology universally used for planning and design throughout this region.

See also in this book

Darwin, Ehrlich

McHarg's major writings:

Design With Nature, Garden City, NY: Natural History Press, 1969; second edn: New York: John Wiley & Sons, 1994.
Composer avec la nature, Paris, France: Cahiers de l'AURIF, 1980.
Progetto con la natura, Padova, Italy: Franco Muzzio Editore, 1989.
A Quest for Life, New York: John Wiley & Sons, 1996.

There are numerous professional reports published by the Center for Ecological Studies, Department of Landscape Architecture and Regional

Planning, University of Pennsylvania, and by the firm, Wallace, McHarg, Roberts and Todd, Philadelphia, PA. These are privately held.

R. TERRY SCHNADELBACH

MURRAY BOOKCHIN 1921–

> Social ecology advances a message that calls not only for a
> society free of hierarchy and hierarchical sensibilities, but
> for an ethics that places humanity in the natural world as
> an agent for rendering evolution – social and natural –
> fully self-conscious and as free as possible ... We stand at
> a cross-roads of conflicting pathways: either we will
> surrender to a mindless irrationalism that mystifies social
> evolution ... or we will regain the activism, that is
> denigrated today, and turn the world into an ever-broader
> domain of freedom and rationality. This entails a new
> form of rationality, a new technology, a new science, a new
> sensibility and self – and above all, a truly libertarian
> society.[1]

Murray Bookchin is one of the most well-known and influential activist-theorists of radical green politics, and has been for over forty years. He was born in New York City on 14 January 1921, to Russian immigrant parents. In the 1930s he entered the communist youth movement, but by the late 1930s had become disillusioned with its Stalinist, authoritarian character. He was involved in organizing activities around the Spanish Civil War, and the fight against European fascism, and remained with the communists until the Stalin–Hitler pact of September 1939, when he was expelled for 'Trotskyist-anarchist deviations'. He has been active in radical politics (both left-wing, anarchist, and ecological) since the 1930s, has written widely on ecological politics and has been active in the ecological movement in America for over thirty years.

After returning from service in the US Army during the 1940s, he was an autoworker and became deeply involved in the United Auto Workers (UAW). In time, he became a left-libertarian anarchist and in the 1960s he was deeply involved in counter-cultural and New Left movements almost from their inception, and he pioneered the ideas of social ecology in the USA. His first American book, *Our Synthetic Environment* (written under the pseudonym Lewis Herber), was published in 1962, preceding Rachel Carson's *Silent Spring* by nearly half a year, while his first published piece of work was in 1952 on the socio-economic origins of environmental pollution and chemicals in food.

In the late 1960s Bookchin taught at the Alternative University in New York, one of the largest 'free universities' in the USA, then at City University of New York in Staten Island. In 1974, he co-founded and directed the Institute for Social Ecology in Plainfield, Vermont, which went on to acquire an international reputation for its advanced courses in eco-philosophy, social theory and alternative technologies, all subjects which reflect his ideas. In 1974, he also began teaching at Ramapo College of New Jersey, becoming Full Professor of Social Theory, and retiring in 1983 in an emeritus status. Now approaching his eighties, Bookchin lives in semi-retirement in Burlington, Vermont.

Bookchin's main contribution to green politics has been the development of 'social ecology', a radical and revolutionary form of green political theory and action which he has developed and espoused since the 1960s. His earlier thinking laid the basis for this later development, particularly his focus on critical social theory (following Marcuse to some extent), the liberatory potential of technology (in the tradition of Lewis Mumford), and the creation of a 'post-scarcity society'. For Bookchin, echoing Ivan Illich, post-scarcity does not mean the Marxist 'abundance of material affluence' but rather 'a sufficiency of technical development that leaves individuals free to select their needs autonomously and to obtain the means to satisfy them'.[2] The extent to which he renounced his earlier commitment to Marxism can be seen in his well-known, acerbic, blunt and refreshingly irreverent essay 'Listen Marxist!'.

Social ecology can be described as a form of eco-anarchism, in which the cause of the ecological crisis lies in structures of hierarchy and power associated with the modern bureaucratic state and corporate capitalism. Bookchin has summarized social ecology as made up of 'an organic way of thinking ... dialectical naturalism ... a mutualistic social and ecological ethics ... the ethics of complementarity ... a new technics ... eco-technology; and ... new forms of human association ... eco-communities'.[3]

The main principles of social ecology are:

1   that the domination of nature by humans has its roots in the historical emergence of patterns of hierarchy and domination within human society;
2   a dialectical approach to understanding the relationship between human society and the natural world. Underpinning

many of his ideas is a reworking of dialectical thinking which combines Hegel's dialectical system of logic with ecological thinking in order to 'naturalize' the dialectical tradition. His 'dialectical naturalism' contrasts with Hegel's dialectical idealism and Marx's dialectical materialism.

3   a rejection of eco-centrism and the idea that humans are 'simply one species amongst others', and anthropocentric views which pit humans over, above or against nature. This is expressed by the notion of how 'first nature' (non-human world) 'grades into' 'second nature' (human culture), and how the latter is derived from the former;

4   a philosophy of nature in which values and practices such as freedom, subjectivity and mutualism are present in germinal form within nature and constitutive of its evolutionary *telos*;

5   a rejection of both the modern nation-state and corporate capitalism and a revolutionary-utopian vision of decentralized, ecologically sustainable, participatory democratic communities in which the economy is run on mutualist and co-operative lines.

Bookchin's work on social ecology has recently developed into what he calls 'libertarian municipalism'. In the words of his partner and fellow theorist of social ecology, Janet Biehl, libertarian municipalism is 'the revolutionary forms of freedom that give organizational substance to the idea of freedom. In brief, libertarian municipalism seeks to revive the democratic possibilities latent in existing local governments and transform them into direct democracies.'[4]

For Bookchin, libertarian municipalism is defined as, 'a confederal society based on the co-ordination of municipalities in a bottom-up system of administration as distinguished from the top-down rule of the nation-state'.[5] It differs from bioregionalism in its concern with the issue of interaction between communities and the rejection of the bioregional model of small-scale, self-sufficient communities, promoted by other environmental thinkers such as Rudolf Bahro. The confederal nature of the arrangement means it is a voluntary political association of autonomous communities with sovereignty retained at the local level. Yet, the relativism that typifies some anarchist political arrangements is explicitly ruled out. As he puts it, 'Parochialism can ... be checked not only by the compelling realities of economic interdependence but by the commitment of municipal minorities to defer to the majority wishes of participating communities.'[6] Here economic-ecological interdependence goes hand in hand with political autonomy and self-determination.

Autarky is not a central principle of social ecology, as it is for other radical green decentralist approaches such as bioregionalism.

The distinction between libertarian municipalism and other forms of anarchism/eco-anarchism (including bioregionalism) can be seen in their different understandings of community. Whereas for bioregionalists and 'pure' anarchists, community is understood as some version of *Gemeinschaft*, libertarian municipalism is presaged on the idea of a 'democratic community'. *The community is defined politically not ecologically.* The aim of libertarian municipalism is to recapture the classical political values of the *polis*, and 'authentic' politics of the community, in opposition to the 'inauthentic', modern politics which Bookchin views as 'statecraft'.

Bookchin is well known as a polemical writer and has spent much time and energy criticizing those aspects of the ecological movement which he sees as based on flawed and dangerous political and moral principles, aims and analyses of the ecological crisis. His most vehement critiques have been levelled at deep ecology. In 1988 he stated that deep ecology was 'the same kind of ecobrutalism [that] led Hitler to fashion theories of blood and soil that led to the transport of millions of people to murder camps like Auschwitz', and at other times he has called deep ecology 'eco-lala'.[7]

Bookchin's polemical and uncompromising stance had led him to vehemently disagree and disown those within the broad 'social ecology' school with whom he disagrees. According to Clark:

> Although Bookchin develops and expands the tradition of social ecology in important ways, he has at the same time also narrowed it through dogmatic and non-dialectical attempts at philosophical systems-building, through an increasingly sectarian politics, and through intemperate and divisive articles on 'competing' ecophilosophies and on diverse expressions of his own tradition. *To the extent that social ecology has been identified with Bookchinist sectarianism, its potential as an ecophilosophy has not been widely accepted.*[8]

Bookchin's combative style often obscures the originality of his thought, and while one could say that Bookchin's style is a classic example of how 'exaggeration is when the truth loses its temper', the dogmatism of his presentation for many also betrays a dogmatism in the content of his work, which of course stands at odds with his libertarian, anarchist thrust.

Bookchin leaves a mixed legacy: a combination of ground-breaking and impressive scholarly analysis, critique and prescriptions for explaining and combating the ecological crisis, combined with an obsession with combating any perceived threat to his position from deep ecology and any other non-Bookchin forms of ecological thought and action. For many, even those broadly sympathetic to social ecology, this polemical and dogmatic propensity is damaging, to his legacy, to social ecology and to the wider cause of finding political and economic solutions to socio-ecological problems. Andrew Light suggests that 'the question is whether the approach to political ecology that Bookchin champions, including a tendency to make judgements about interlocutors based on a few extreme examples, is what we need today'.[9]

Bookchin has been enormously influential within the ecological movement in North America and Europe, and within academic theorizing about green moral and political values, principles and aims, and has been an invigorating and original thinker about ecological politics for over four decades.

According to Peter Marshall, Bookchin's main achievement is to have

combined traditional anarchist insights with modern ecological thinking ... In this way he has helped develop the powerful libertarian tendencies within the Green movement. Just as Kropotkin renewed anarchism at the end of the last century by giving it an evolutionary dimension, so Bookchin has gone further to give it an ecological perspective. In his view, the creation of an anarchist society is now the only way to solve the threat of ecological disaster confronting humanity.[10]

Notes

1 *Remaking Society*, p. 204.
2 *Towards an Ecological Society*, p. 251.
3 Murray Bookchin, in Steve Chase (ed.), *Defending the Earth: A Dialogue between Murray Bookchin and Dave Foreman*, Boston, MA: South End Press, p. 131, 1991.
4 Janet Biehl, *The Politics of Social Ecology: Libertarian Municipalism*, p.viii.
5 Bookchin, 'Libertarian Municipalism', *Society and Nature*, 1:1, pp. 94–5, 1992.
6 Ibid., p. 97.
7 Bookchin, 'Social Ecology vs Deep Ecology', *Socialist Register*, 18 (3), p. 13, 1988.

8 John Clark, 'A Social Ecology', *Capitalism, Nature, Socialism*, 8 (3), p. 9, 1997, emphasis added.
9 Andrew Light, 'Introduction', in Andrew Light (ed.), *Social Ecology after Bookchin*, p. 4.
10 Peter Marshall, *Demanding the Impossible: A History of Anarchism*, London: Fontana Press, p. 602, 1993.

See also in this book

Bahro, Marx

Bookchin's major writings

Our Synthetic Environment, pseud. Lewis Herber, New York: Harper & Row, 1962.
Post-scarcity Anarchism, London: Wildwood House, 1971.
The Spanish Anarchists, New York: Harper & Row, 1977.
Towards an Ecological Society, Montreal/Buffalo: Black Rose Books, 1980.
The Modern Crisis, Philadelphia, PA: New Society Publishers, 1986.
Remaking Society, Montreal and New York: Black Rose Books, 1990.
The Ecology of Freedom: The Emergence and Dissolution of Hierarchy, rev. edn, Montreal and New York: Black Rose Books, 1991.
Urbanization without Cities: The Rise and Decline of Citizenship, Montreal: Black Rose Books, 1992.
The Philosophy of Social Ecology, rev. edn, Montreal/Buffalo: Black Rose Books, 1994.
Re-Enchanting Humanity: A Defence of the Human Spirit against Antihumanism, Misanthropy, Mysticism and Primitivism, London: Cassell, 1995.

Further reading

Biehl, J. (ed.), *The Murray Bookchin Reader*, Montreal: Black Rose Books, 1997.
—— *The Politics of Social Ecology: Libertarian Municipalism*, Montreal: Black Rose Books, 1998.
Clark, J. (ed.), *Renewing the Earth: The Promise of Social Ecology*, Basingstoke: Green Print, 1990.
Light, A. (ed.), *Social Ecology after Bookchin*, London: The Guilford Press, 1998.
Watson, D., *Beyond Bookchin: Preface for a Future Social Ecology*, New York and Detroit: Autonomedia, 1996.

JOHN BARRY

EDWARD OSBORNE WILSON 1929–

When the century began, people could still think of

themselves as transcendent beings, dark angels confined to Earth awaiting redemption by either soul or intellect. Now most or all of the relevant evidence from science points in the opposite direction: that having been born into the natural world and evolved there step by step across millions of years, we are bound to the rest of life in our ecology, physiology, and even our spirit. In this sense, the way in which we view the natural world, Nature has changed fundamentally.[1]

Edward Osborne Wilson was born in Birmingham, Alabama, in 1929, the son of a travelling government accountant. His scientific career, which began with the study of ants and ultimately generated theories that were to influence profoundly concepts of biodiversity, sociobiology and, most recently, the unification of all knowledge, have earned him many of the highest academic honours. In 1996 he was described by *Time* magazine as one of America's twenty-five most influential people. By the time of his retirement in 1997 he had become recognized as one of the greatest evolutionary biologists of the twentieth century.

Wilson describes his early life as 'blessed', although he was often beset by difficult emotional and physical circumstances. These included the divorce of his parents and an itinerant schooling where he attended fourteen different schools in eleven years. The loss of an eye in a fishing accident denied him access to a military career but left him with eyesight characteristics that he turned to his advantage in science. Gradual, partial loss of hearing during his adolescence influenced his choice of studies in scientific research, deflecting him away from ornithology towards the study of ants. He considers that three formative experiences during his youth influenced his later career and personal philosophy: an intimate knowledge of natural history that first developed in his childhood; an induction into military discipline and the virtues of hard work at the Gulf Coast Military Academy; and a Southern Baptist upbringing that left him with the conviction that religion and science might be reconciled by the understanding of the former by means of the latter.

After gaining bachelor's and master's degrees at the University of Alabama, Edward Wilson studied for his Ph.D. at Harvard University, where he taught from 1953 until his retirement in 1997. He was successively a Harvard Professor of Zoology, Curator of Entomology at the Museum of Comparative Zoology, Baird Professor of Science, Mellon Professor of the Sciences and Pellegrino University Professor. He is currently Pellegrino Professor Emeritus.

His scientific awards include the US National Medal for Science, the Swedish Academy of Sciences Crafoord Prize, Germany's Terrestrial Ecology Prize, Japan's International Prize for Biology and the French Prix du Institut de la Vie.

Edward Wilson's influence is in no small part attributable to his skill as a writer, whose elegant prose has confirmed his status as one of the finest communicators of science in the twentieth century. Several of his fluent, beautifully written books are at once important academic sources and accessible, engrossing works of popular scientific literature. They have also earned him many literary honours, including two Pulitzer Prizes, the Los Angeles Times Book Prize, the Publishers' Marketing Association Benjamin Franklin Award, the Sir Peter Kent Conservation Book Prize and the John Hay Award from the Orion Society.

Even if his scientific career had been confined to mymecology, Wilson's reputation as an outstanding biologist would be indisputable. His taxonomic and behavioural studies on ants have made him a leading international expert on these insects. *The Ants*, published in 1990 with Bert Hölldobler, was not only an authoritative study of their anatomy, taxonomy, ecology and social behaviour, but also a winner of the Pulitzer Prize, acclaimed as much for its detailed information and taxonomic keys for specialists as for its engaging accounts of ant social behaviour for the interested layman.

Inevitably, close field-based study of such a complex, diverse and widely distributed group of insects brought Wilson in close contact with the biodiversity of numerous temperate, sub-tropical and tropical ecosystems. In 1967 he and Robert MacArthur published *The Theory of Island Biogeography*, describing how the number of species in an isolated patch of habitat – whether a true oceanic island or an island of surviving natural vegetation in a once continuous tropical forest – could be determined with reference to a simple mathematical expression and distance to the nearest source of immigrant species. The theory showed that a balance between new species immigration and extinction of established species was eventually reached, and that the extent of biodiversity in such islands was determined by their size. The theory was successfully validated by denuding a small island in the Florida Keys of all animals and then following in detail the pattern of re-colonization.

Subsequently MacArthur and Wilson's theory of island biogeography has been criticized and modified, but remains immensely influential in the design of nature reserves, emphasizing the importance of conserving the largest possible patches of natural, undisturbed habitat. More controversially, the theory has been used to calculate probable rates of extinction, since it also provides a means of calculating species loss as habitats become fragmented, isolated and reduced in size. The development of this theory coincided with novel methods for measuring biodiversity, such as those of Terry Erwin,[2] who proposed vast increases in estimates of species diversity based on extrapolation from sub-samples of beetle biodiversity measured on a single tree species in the Panamanian rain forest. New estimates of total biodiversity were pitched at 10, 30 or even as many as 100 million species, when only about 1.5 million species have been scientifically classified. Wilson's work indicated that extinction rates due to habitat degradation and destruction were far higher than anyone had hitherto imagined. His writings have tirelessly warned of the disastrous consequences of the likely rapid loss of a large proportion of Earth's biodiversity which, he warns, is 'the folly our descendants are least likely to forgive us'.

Edward Wilson's behavioural studies of ant societies were the foundation for a second great theme of his scientific career, the study of sociobiology. His proposal that there were genetically determined elements in human behaviour – that evolution has generated certain patterns of neural connections that predispose human behaviour towards certain courses of action – was instantly controversial. His book *Sociobiology* brought him into conflict with Richard Lewontin[3] and Steven Jay Gould, whose ideological predispositions abhorred any suggestion that nature rather than nurture could be a guiding force of human behaviour. In retrospect, Wilson's admission that 'at my core I am a social conservative, a loyalist. I cherish traditional institutions, the more venerable and ritual-laden the better' made it probable that there might be no easy accord with those of a more Marxist disposition in American society. The possibility that characteristics such as altruism or aggression in humans might be even partially governed by instinctive, genetically determined algorhythms had profound consequences for sociology, civil rights and justice. The reception for *Sociobiology* was at times abusive and even violent, as when Wilson was doused with water by protestors at a sociobiology symposium in Washington in 1978. Subsequently, accumulated circumstantial evidence and data from molecular biological studies has reinforced the notion that there are genetic components in human behaviour. Wilson's Pulitzer Prize-winning *On Human Nature* was to some degree a rebuttal of his detractors' politically motivated criticisms of sociobiology.

Wilson's rapid rise in academic status and public recognition coincided with developing tensions between Harvard's traditionalists in biology, whose work was based on the study of whole organisms, and the growing power of the reductionist molecular geneticists who sought to explain the complexity of nature through exploring its constituent molecules. Wilson, a whole-organism traditionalist through and through, with an upbringing that had instilled Old World courtesies, civility and good manners in academic debate, magnanimously describes himself as 'being blessed with brilliant enemies' but admits to despising 'the arrogance and self-regard so frequently found amongst the very bright'. He has made no secret of his personal dislike and professional admiration for Nobel Laureate James Dewey Watson, co-discoverer of the structure of DNA, who he describes in his autobiography *Naturalist* as 'the Caligula of biology'.

Some, then, perceive a certain irony in Wilson's most recent work, described in *Consilience*, which seeks to unify all knowledge – including religion, economics and aesthetics – in terms of reductionist physical and biological principles. The term 'consilience' was originally coined by the nineteenth-century philosopher William Whewell, to describe the solving of problems by the combined use of inferences drawn from disparate sources, a process which is common practice in science. Harking back to the controversial concepts first outlined in *Sociobiology*, *Consilience* proposes that an understanding of the biological mechanisms underlying human behavioural characteristics, assembled during the evolution of the brain, will ultimately provide the framework for understanding the decisions that we make about our interactions with our environment and with each other. Predictably, this attempt to reduce the arts and social sciences to an understanding of genetic programming has not received a warm welcome amongst most practitioners in those disciplines, but perhaps it might prompt the re-examination of their intellectual legitimacy, in much the same way that whole-organism biologists were compelled to reconsider their future in the face of the molecular biological revolution. In the somewhat safer home territory of conservation of biodiversity, Wilson has proposed the concept of biophilia, which he defines as 'the innately emotional affiliation of human beings for other organisms' and believes may be resident in our genes. He has argued that biophilia governs our aesthetic response to the living world and acts as a powerful driving force in environmental ethics.[4]

Ernst Mayr, another of the twentieth century's outstanding

evolutionary biologists, considers the most memorable lesson he learned from Darwin is that 'the most important thing in scientific research is not to add to the accumulation of facts, but to ask challenging questions and to try to answer them'.[5] Edward Wilson is one of a small cadre of contemplative evolutionary biologists, imbued with a deep knowledge of field natural history from an early age, who, in a career that has combined meticulous observational and experimental study with scholarship, has asked challenging questions, providing answers that have consistently generated controversy, and which by doing so have stimulated whole fields of scientific endeavour.

Notes

1   From the author's *Prelude*, in E.O. Wilson, *Naturalist*, p. xii.
2   Terry Erwin, 'Tropical Forests: Their Richness in Coleoptera and Other Arthropod Species', *Coleopterists' Bulletin*, 36 (1), pp. 74–5, 1982.
3   R.C. Lewontin, *The Doctrine of DNA*, pp. 87–104.
4   'Biophilia and the Environmental Ethic', in *In Search of Nature*.
5   Ernst Mayr, 'Understanding Evolution', *Trends in Ecology and Evolution*, 14 (9), pp. 372–3, 1999.

See also in this book

Darwin, Ehrlich, Malthus

Wilson's major writings

MacArthur, Robert H. and Wilson, E.O., *The Theory of Island Biogeography*, Princeton, NJ: Princeton University Press, 1967.
The Insect Societies, Cambridge, MA: Belknap Press, 1971.
Sociobiology: The New Synthesis, Cambridge, MA: Harvard University Press, 1975.
On Human Nature, Cambridge, MA: Belknap Press, 1978.
Lumsden, Charles J. and Wilson, E.O., *Genes, Mind and Culture*, Cambridge, MA: Harvard University Press, 1981.
—— *Promethean Fire*, Cambridge, MA: Harvard University Press, 1983.
Biophilia, Cambridge, MA: Harvard University Press, 1984.
Wilson, E.O. and Peter, Frances M., *Biodiversity*, Washington, DC: National Academy Press, 1988.
Hölldobler, Bert and Wilson, E.O., *The Ants*, Cambridge, MA: Belknap Press, 1990.
The Diversity of Life, Cambridge, MA: Belknap Press, 1992.
Kellert, S.R. and Wilson, E.O. (eds), *The Biophilia Hypothesis*, Washington, DC: Island Press, 1993.
Hölldobler, Bert and Wilson, E.O., *Journey to the Ants: A Story of Scientific Exploration*, Cambridge, MA: Belknap Press, 1994.

Naturalist, Washington, DC: Island Press, 1994.
In Search of Nature, Washington, DC: Island Press, 1996.
Consilience, London: Little, Brown & Co., 1998.

Further reading

Ehrlich, Paul R. and Wilson, Edward O., 'Biodiversity Studies: Science and Policy', *Science*, 253, pp. 758–62, 1991.
Futuyma, Douglas J., *Evolutionary Biology*, 3rd edn, Sunderland, MA: Sinauer, 1997.
Leopold, Aldo, *A Sand County Almanac and Sketches Here and There*, New York: Oxford University Press, 1947.
Lewontin, R.C., *The Doctrine of DNA*, Harmondsworth: Penguin, 1995.
MacArthur, Robert H. and Wilson, Edward O., 'An Equilibrium Theory of Insular Zoogeography', *Evolution*, 17 (4), pp. 373–87, 1963.
Simberloff, Daniel S. and Wilson, Edward O., 'Experimental Zoogeography of Islands: Defaunation and Monitoring Techniques', *Ecology*, 50 (2), pp. 267–78, 1969.
Williamson, Mark, *Island Populations*, New York: Oxford University Press, 1981.

PHILLIP J. GATES

PAUL EHRLICH 1932–

'Nothing less is at stake than the fate of human civilization'[1] is Paul Ehrlich's motto both now and for much of his academic career. Of all the fields of the natural sciences, it might be expected that biology might produce the most thinkers on environmental matters, and the entry on Aldo Leopold is another example of this. But of all the recent (post-1960) contributors to the provision of information and to participation in public debate, Ehrlich is the most prominent. Born in 1932, he took his first degree at the University of Pennsylvania and his Ph.D. at the University of Kansas (1957); an appointment as Professor of Biology at Stanford University in 1966 was the first of a series of posts in that institution. From this secure base he has published a series of books and papers, travelled widely and engaged in numerous debates and acts of public service. His contributions to environmental thought and action have brought him honours such as medals from the World Wildlife Foundation, the MacArthur prize and the Heinz Award as well as Membership of the National Academy of Sciences and Fellowship of the AAAS.

Although Ehrlich has had a high public profile in the USA and in certain world forums, most people are influenced by his published work. There are perhaps four strands to this: (1) basic research in the natural sciences, and in particular on the population ecology of birds and butterflies; (2) advocacy on the subject of human population growth, with a strong neo-Malthusian outlook which suggests that many, if not most, problems of the human species are the result, immediately or indirectly, of rapid population growth; (3) human ecology: the connection of human activities to the biophysical systems of the planet in areas such as biodiversity and agriculture; and (4) widely read popular works and student texts on population–resource–environment linkages.

Category (1) is perhaps of least obvious interest to us here. It includes work on birds, butterflies and coral reefs in the classic team mode of the natural sciences,[2] but it is worth noting that a 1965 paper on the co-evolution of butterflies and plants has become a Citation Classic in the ISI *Current Contents* series.[3] The ways in which the central concern with basic biology (which has acted as a grounding for all the other work throughout) include a concern for the extinction of species, the conservation of both tropical and temperate forests, and even the effect of scientific study upon butterfly populations.[4] The key point is that although Ehrlich became mostly known for his advocacy – and indeed polemic – on environmental concerns, his attention to basic science has been constant.

As a result of rapid immigration and industrial anabasis, coupled with an affluent and well-educated population, California in the 1960s became a centre of 'alternative' thinking about population–resource–environment relations. The 'hippy' movement with its attention to communal lifestyles and illegal substances was one strand, but another was a more intellectual and factually well-informed questioning of the gospels of growth and development as they appeared in that state, in the USA, in the industrial nations, and finally in the world as a whole. One pointer was the volume of essays edited by S. von Ciriacy-Wantrup and J.J. Parsons,[5] which brought many of the issues into focus, another the radical questioning of 'growth' by the geographer D.B. Luten,[6] yet another the expansion of the influence of the Sierra Club (which is based in San Francisco) as an environmental campaigning body rather than a mountaineers' organization. In this *Zeitgeist* the strongly expressed views of Ehrlich on population, for example, were not seen as extreme, and indeed the outlooks developed in category (2) fitted well into the relatively radical sets of ideas being developed at the time.

Thus it was that the publication by the Sierra Club of *The*

Population Bomb in 1968 which propelled Ehrlich from a base in which notions of population control in the affluent countries was not seen as necessarily controversial to a wider public discussion in which it certainly was. The USA was described as the world's largest consumer and so strong was its effects, example and influence that, '[W]e must have population control at home ... by compulsion if voluntary methods fail. We must use our political power to push other countries into programs which combine agricultural development and population control.'[7] The book created considerable interest world-wide and has been reprinted and translated into several languages. Hardback and paperback reprints were still on offer in on-line bookshops in 1999. The uncompromising neo-Malthusian message, combined with some startling prophecies (the Prologue's second sentence starts, '[I]n the 1970's the world will undergo famines – hundreds of millions of people are going to starve to death ...'[8]), not only presented a series of challenges to development-minded agencies in the USA and internationally, but was sufficiently well expressed to propel Ehrlich into the status of a media-figure and global guru. In particular it confronted the orthodox position of the Roman Catholic Church on chemical and physical methods of contraception (mathematics was however allowed), although these were not particularly strongly obeyed in most developed countries: growth rates in, for example, Latin America were then very high. The term 'Vatican roulette' inspired the inclusion in the book of the text of letters to the then Pope and the local Archbishop suggesting that the Church modify its position: the letter to Paul VI seems not to have been passed to his successor. Famines did occur in the 1970s, though mostly in zones of civil strife rather than in areas with especially rapid population growth (of course competition for resources of any kind may have a demographic component), and there have been some notable downturns in population growth rates though the highest in Africa, for example, are not associated with an especially Catholic culture, and AIDS has rather transformed the demographics of several African nations.

The bulk of *The Population Bomb* was however devoted to extending the ideas of Malthus in the sense that it was not the absolute size of the population that mattered, but its relation to its resource base. So the foundations were laid in that book for explorations of the linkages of population growth to the new world of intensive agriculture, of high rates of per capita mineral and energy use, of the production of environmental contaminants and even of the crowding of recreation space. Small wonder therefore that such ideas were contested: by those whose 'boosterist' heritage came under attack, and by those whose stance was fundamentally in favour of population growth as producing a responsive innovation in technological development and who in the end saw each extra human as the producer of a resource rather than a consumer. The refinement of the neo-Malthusian argument has however been a continuing theme of subsequent years, with more and more attention being paid to the social context of population growth and the contexts in which policy decisions are made about, for example, US aid to family planning programmes overseas. These more developed ideas were brought together in *The Stork and the Plow: The Equity Answer to the Human Dilemma*,[9] though the use of the definite article in the subtitle perhaps suggests that there is still held to be a central relationship which determines most if not all of the others. The forcing function of population in all those linkages was underlined by a paper that used energy consumption as a surrogate for human impact on the environment to calculate the optimum population size.[10] This came out at 1.5×109 people (1.5 billion) using 4.5 TW of energy. The population in 1999 was 6 billion and the energy consumption in the order of 15 TW, so the difference is large.

The more detailed exploration of the relationships between human populations, resource use and environmental impact has been explored by Ehrlich (usually with co-authors and most frequently with Anne Ehrlich) in a number of papers in relatively specialized journals, as well as in sources with a wider circulation. These comprise category (3) of his output. The topics include, but are not confined to, food security and production[11] and the nuclear winter debate.[12] Inevitably, during the 1990s the term 'sustainability' enters the discussion and an integrated attempt to bring together several aspects of the relations of population, technology and environment can be found in the 1992 paper where the social dimensions of the perceived problems are linked to those provided by more mainstream ecological science: '[S]ound science ... can give minimal guidance at best regarding the issues surrounding the question of the kinds of lives people would choose to live.'[13] Their bottom line, not one popular with either democratic governments, large corporations or dictatorships, is that technology cannot make biophysical carrying capacity infinite, though there is presumably a stage somewhere when a vastly increased world population is one half of a food–humans monoculture. The working-out of detail in the topics of food, energy, wildlife, toxicology, water and minerals is at the heart of a number of books which are aimed at college students in the USA and which convey the Ehrlich world-view as well as a great deal of factual material,[14] as well as popular books designed to raise awareness among lay people.[15]

As a result of the study of these connections, Ehrlich was often ready to make predictions. The putative famines of the 1970s were accompanied by suggestions that smog in Los Angeles and New York might kill 200,000 people (predicted in 1969), that England would not exist in the year 2000 (said in 1969) and that accessible minerals would be facing depletion before 1985 (dated 1976). If we read 'United Kingdom' for the common misuse of 'England' by North Americans, then we can see that it is still physically in existence in spite of the excesses of the last days of 1999. The minerals issue was the focus of a public dispute with the optimistic economist and business advocate Julian Simon, whose work consistently argued that all measures of human welfare were tending upwards. The dispute came to a bet in which the change in price of metals between 1980 and 1990 was held by Ehrlich to be upwards and by Simon to be downwards. Copper, chrome, nickel, tin and tungsten were Ehrlich's choice. Adjusted for inflation, copper fell by -18.5%, chrome by -40%, nickel by -3.5%, tin by -72% and tungsten by -57%. Ehrlich paid up. Most of the prices fell because of technological improvements and substitutions, though tin crashed because of the break-up of a price-fixing cartel. In other words, social factors plus the robustness of neo-classical economic thought outran the purely Malthusian calculations of the ecologist. Simon then offered to bet on any trends relating to 'basic human material welfare'. Ehrlich and Schneider offered a list of 15 items for consideration of a mostly biological nature but Simon declined these parameters and died (in 1977) before any new betting conditions could be drawn up. In general, however, Ehrlich made fewer large-scale prophetic statements after about 1970, in common with many environmentalists who had adopted especially pessimistic outlooks in the 1960s. In the long run (whatever that is), they may of course turn out to be right or indeed their anxieties might be averted by the very act of prediction.

Category (4), widely read student texts and semi-popular works, does not need extended discussion here except to note that throughout this period Ehrlich has been concerned to disseminate his work to as many people as possible. In part, this seems like the action of any advocate who is convinced of his or her case, but it also seems to stem from the fundamental and laudable trait of scientists to expose their work to sceptical audiences. There is no lack of audience for the latter in Ehrlich's case, of course, and the anti-Ehrlich viewpoints have had no shortage of outlets, both in academia and especially in business publications. Books with titles like *The End of Affluence*[16] (published in the US bi-centenary year of 1976) strike at the vitals of the American way of enthusiasm. So there has been, and continues to be, a 'brownlash' of anti-environmentalist rhetoric designed to show that everything is getting better and better: one commentator's summary in 1997 was 'technology has thwarted Ehrlich's projections, and you needn't be Nostradamus to know it always will'.[17] The response to much of this polemic is in the book of that year which takes up the theme of some previous publications, namely the reception of the findings of the human ecology-environmentalist strand in US thought during the post-1960 period.[18]

A few considerations might strike the non-American and guardedly sympathetic commentator. One is the persistently North American tone of the debate, both pro and contra. In early works, a global set of scenarios was often discussed, but the emerging tone from a period of intensive reading is one of a rootedness in the discourses of the world's richest nation. The particular diversities of the many poorer countries seem to be elided. In part at the beginning, this appears to be the consequence of a biologist's view of humans as behaviourally homogenous diversivores, and although there are some strenuous efforts to encompass the social context of change, the cultural context is often given a rather minimal position as if the whole debate over the social construction of 'environment' in the post-structuralist sense had not happened (some movement in the direction of the management of cultural change, largely in North America, is given in a book which stresses that the human mind and its features are mismatched with the world as it now is[19]). It is perhaps then surprising that there seems to be a consistent underestimation of the role of technology in the human--environment relationship. While the consumption of commercial energy may be a good broad-scale indicator of the penetration of technology, it undervalues agents of change such as the microelectronics that make possible vast and immediate transfers of capital. If it is accepted that technology and its associated cultural metaphysics of acceptability have been at the heart of the great changes in ecology and economy such as the spread of agriculture from its hearths, and the dissemination of industrialism based on fossil fuels, then to deny it a central and high-profile role in current and near-future metamorphoses tends to the eccentric. True, it may not in the very long run allow humanity to escape certain biophysical constraints, but it may buy time (as did the Green Revolution), and in a world that is often held to be best described by versions of chaos theory rather than linear equations, there is no telling what synergisms may emerge. Not many commentators, after all, forecast the 'soft' revolutions of the late 1980s in Eastern Europe and the place of 'green' thinking that was one of the factors in those mass convulsions.

The greatest contribution of Ehrlich to environmental thought since the later 1960s has been his energetic lack of fear. To engage for over thirty years in continued controversy with a powerful opposition, while still producing basic science, is an example of stamina which deserves every plaudit. Even those not convinced by all the arguments and for whom a high-profile role is not part of their personality have to engage with the central question of what numbers of the human species the Earth could support at what quality of life for them and for other species as well. Ehrlich's role can perhaps be measured by the fact that this question is now always part of the schedule in any serious environmental debate or research programme.

Notes

1 P.R. Ehrlich, 'Recent Developments in Environmental Sciences', address at presentation of the H.P. Heineken Prize for Sciences, 25 September 1998, accessed at http://dieoff.com/page157.htm on 29 July 1999.

2 For example, P.R. Ehrlich, A.E. Launer and D.D. Murphy, 'Can Sex Ratio Be Defined or Determined? The Case of a Population of Checkerspot Butterflies', *American Naturalist*, 124, pp. 527–39, 1984; P.R. Ehrlich, 'Population Biology of Checkerspot Butterflies and the Preservation of Global Biodiversity', *Oikos*, 63, pp. 6–12, 1992.

3 P.R. Ehrlich and P.H. Raven, 'Butterflies and Plants: A Study in Coevolution'; see ISI, 'Citation Classics', *Current Contents*, 37, p. 16, 1984.

4 S. Harrison, J.F. Quinn, J.F. Bauman, D.D. Murphy and P.R. Ehrlich, 'Estimating the Effects of Scientific Study on Two Butterfly Populations', *American Naturalist*, 137, pp. 227–34, 1991.

5 S. von Ciriacy-Wantrup and J.J. Parsons (eds), *Natural Resources: Quantity and Quality*, Berkeley and Los Angeles, CA: University of California Press, 1967.

6 T.R. Vale, *Progress Against Growth: Daniel B. Luten on the American Landscape*, New York and London: Guilford Press, 1986.

7   *The Population Bomb*, prologue; rev. and updated as P.R. Ehrlich and A.H. Ehrlich, *The Population Explosion*.
8   *The Population Bomb*, prologue.
9   P.R. Ehrlich, A. Ehrlich and G.C. Daily, *The Stork and the Plow: The Equity Answer to the Human Dilemma*, New York: Putnam, 1995.
10 G.C. Daily, P.R. Ehrlich and A. Ehrlich, 'Optimum Human Population Size', *Population and Environment*, 15, pp. 469–75, 1994.
11 G.C. Daily and P.R. Ehrlich, 'Population, Sustainability, and Earth's Carrying Capacity'; P.R. Ehrlich, A.H. Ehrlich and G.C. Daily, 'Food Security, Population and Environment'; G.C. Daily and P.R. Ehrlich, 'Socioeconomic Equity, Sustainability, and Earth's Carrying Capacity'.
12 For example, P.R. Ehrlich, A.H. Ehrlich and H.C. Mooney (eds), *The Cold and the Dark: The World After Nuclear War*, New York: Norton, 1984.
13 G.C. Daily and P.R. Ehrlich, 'Population, Sustainability and Earth's Carrying Capacity', p. 770.
14 P.R. Ehrlich and A.H. Ehrlich, *Population Resources Environment: Issues In Human Ecology*, San Francisco, CA: Freeman, 1970, and subsequent editions; P.R. Ehrlich, A.H. Ehrlich and J.P. Holdren, *Ecoscience: Population, Resources, Environment*, San Francisco, CA: Freeman, 1977.
15 P.R. Ehrlich and R.L. Harriman, *How To Be a Survivor*, New York: Ballantine Books, 1971; D.C. Pirages and P.R. Ehrlich, *Ark II. Social Response to Environmental Imperatives*, San Francisco, CA: Freeman, 1974; P.R. Ehrlich and A.H. Ehrlich, *Healing the Planet*.
16 P.R. Ehrlich and A.H. Ehrlich, *The End of Affluence: A Blueprint For Your Future*, Rivercity, MA: Rivercity Press, 1976.
17 S. Milloy, 'Doomsayer Paul Ehrlich Strikes Out Again', accessed at www.junkscience.com/news/fumento.htm on 23 July 1999.
18 P.R. Ehrlich and A.H. Ehrlich, *Betrayal of Science and Reason*.
19 R. Ornstein and P.R. Ehrlich, *New World, New Mind. Changing the Way We Think to Save Our Future*, London: Methuen, 1989.

See also in this book

Darwin, Goethe, Leopold, Malthus, Wilson

Ehrlich's major writings

Professor Ehrlich maintains a web page at www.stanford.edu/dept/biology/oldsite/fac/ehrlich.html

Ehrlich, P.R. and Raven, P.H., 'Butterflies and Plants: A Study in Coevolution', *Evolution*, 18, pp. 586–608, 1965.
The Population Bomb, New York: Sierra Club/Ballantine Books, 1968.
Ehrlich, P.R. and Roughgarden, J., *The Science of Ecology*, New York: Macmillan, 1987.
Ehrlich, P.R. and Ehrlich, A.H., *The Population Explosion*, New York: Simon & Schuster, 1990.
—— *Healing the Planet*, New York: Addison-Wesley, 1991.

Daily, G.C. and Ehrlich, P.R., 'Population, Sustainability, and Earth's Carrying Capacity', *BioScience*, 42, pp. 761–71, 1992.

Ehrlich, P.R., Ehrlich, A.H. and Daily, G.C., 'Food Security, Population and Environment', *Population and Development Review*, 19, pp. 1–32, 1993.

Daily, G.C. and Ehrlich, P.R., 'Socioeconomic Equity, Sustainability, and Earth's Carrying Capacity', *Ecological Applications*, 6, pp. 991–1001, 1996.

Ehrlich, P.R. and Ehrlich, A.H., *Betrayal of Science and Reason: How Anti-environmental Rhetoric Threatens Our Future*, New York: Putnam, 1996.

Hughes, J., Daily, G.C. and Ehrlich, P.R., 'Population Diversity: Its Extent and Extinction', *Science*, 278, pp. 689–92, 1997.

Further reading

Johnson, S., *The Politics of Population: The International Conference on Population and Development, Cairo 1994*, London: Earthscan, 1995.

Lutz, W. (ed.), *The Future Population of the World: What Can We Assume Today?*, London: Earthscan for IIASA, 1995.

Myers, N. and Simon, J.L., *Scarcity or Abundance? A Debate on the Environment*, New York and London: Norton, 1995.

Simon, J.L. (ed.), *The State of Humanity*, Oxford and Cambridge, MA: Blackwell, 1995.

—— *The Ultimate Resource 2*, Princeton, NJ: Princeton University Press, 1996.

United Nations Environmental Program, *Global Outlook 2000*, London: Earthscan, 1999.

Vitousek, P. *et al.*, 'Human Domination of Earth's Ecosystems', *Science*, 277, pp. 494–9, 1997.

IAN G. SIMMONS

HOLMES ROLSTON III 1932–

Holmes Rolston III is widely recognized as the 'father' of environmental ethics as an academic discipline. Although others planted seeds before Rolston, theirs were mainly inspirational. More so than any other, he has shaped the essential nature, scope and issues of the discipline.

Throughout Rolston's many books and articles, he holds that intrinsic value entails duties. In *Environmental Ethics*, he states:

> Duties arise to the individual animals and plants that are produced as loci of intrinsic value within the system ... These duties to individuals and species, so far from being in conflict with duties to ecosystems, are duties toward its

products and headings. The levels differ, but, seen at depth, they integrate. Perhaps on some occasions duties to the products will override duties to the system that produced them, but – apart from humans who live in culture as well as in nature – this will seldom be true.[1]

Especially influential were Rolston's early, ground-breaking article in the journal *Ethics* (1975), and his mature, comprehensive formulation of his ethical theory in the book *Environmental Ethics* (1988). In 1997, he gave the prestigious Gifford Lectures at the University of Edinburgh in Scotland, published under the title *Genes, Genesis and God* (1999).

Holmes Rolston III was born 19 November 1932, the son and grandson of Presbyterian ministers, whose names he shares. Except for summers spent in Alabama on his mother's parents' farm, Rolston spent his childhood in the Shenandoah Valley in the state of Virginia, where his father was a Presbyterian minister and respected theologian. In these rural places, Rolston grew to love nature and to value simplicity. The Maury River flowed in front of the family home, which was nestled in the woods, and the Blue Ridge Mountains shaped the horizon. The house lacked electricity, and water came from cisterns.

As an undergraduate at Davidson College, Rolston wanted to study nature and so completed his degree in physics (BS, 1953), with occasional excursions into biology. Planning to be a Presbyterian minister like his father and grandfather, Rolston next obtained a divinity degree from Union Theological Seminary in Richmond, Virginia (BD, 1956), and then a PhD in theology and religious studies at the University of Edinburgh in Scotland (1958). For the next decade, he was a minister in the Appalachian Mountains of western Virginia near the Tennessee and North Carolina borders. He and his wife, Jane, have two children, a daughter and son.

In his spare moments while serving as minister, Rolston attended classes at East Tennessee State University, explored the biology, mineralogy and geology of the southern Appalachian Mountains, becoming a recognized naturalist and bryologist. He also worked as an activist to conserve wildlife, to preserve Mount Rogers and Roan Mountain, and to maintain and relocate the Appalachian Trail.

While studying the natural world, Rolston felt a need to study philosophy in an attempt to explain the values he found in nature and to resolve the intellectual conflicts between his religious faith and the non-theistic naturalism of the biological sciences. Leaving his beloved Virginia, he studied philosophy of science at the University of Pittsburgh. There he began to formulate his theory of the intrinsic value of nature and his objections to the naturalistic fallacy. After finishing a master's degree in 1968, Rolston was appointed Professor of Philosophy and Religion at Colorado State University, Fort Collins, where during the ensuing decades he achieved international academic recognition and currently holds the prestigious position of University Distinguished Professor. In addition to his many academic achievements, he has continued his ordained status in the local Presbytery.

Five concepts frequently recur throughout Rolston's writings: (1) the intrinsic value of nature, which value is non-anthropocentric and even anti-anthropocentric since it is independent of and apart from humankind; (2) ecological-systemic holism; (3) the derivation of duties to nature from the intrinsic value of nature, which logically entails the denial of the naturalistic/is-ought fallacy; (4) the intrinsic value of species as forms, or groupings, of life; and (5) biocentrism, that is, the intrinsic value of and derivative duty to respect every individual living organism.

Central to Rolston's theory of environmental ethics are the concepts 'intrinsic value' and 'holism'. Aldo Leopold proposed holism under the rubrics 'community' and 'land ethic'. Holism is an essential concept in ecology, and has become a key component in every contemporary theory of environmental ethics. In Rolston's theory, ecological wholes are intrinsically valuable. His ethic is explicitly an ethic of duties, duties he derives from intrinsic value.

Rolston clearly names and identifies two 'rules' or 'principles': the Homologous Principle and the Principle of Value Capture.[2] He also uses at least four other principles, for a total of at least six. Others may need to be added. These six principles are:

1 The Homologous Principle: Follow Nature
2 The Value-Capture Principle
3 The Organic Principle: Respect for Life
4 The Species Principle: Preserve 'Forms' of Life
5 The Ecosystemic Principle
6 The Three 'Environments' Principle: Urban, Rural and Wilderness (or, the Nature–Culture Principle)

By 'nature', Rolston generally means *non-human* nature. He carefully distinguishes 'nature' and 'culture'. Culture is an artifact made possible by human self-awareness and thoughtfulness, which are found to such an advanced degree in no other species, and which make possible the acquisition and transfer of knowledge, information, science, technology, art, and a host of other human achievements. In contrast to 'deliberative' culture, nature is 'spontaneous' and 'non-reflective'.[3] Natural processes are law-like, orderly though also probabilistic, and open to historical novelty, as evidenced in the creativity in evolving ecosystems. Natural selection, combining with genetics, results in the genesis of value.

Rolston acknowledges that humans are in nature and part of nature in many important respects. The biology of our bodies, for instance, is fully natural. He often says that humans (and human culture) 'emerged' out of nature. For Rolston, 'wilderness' is a synonym for the environment of nature wherever it is free of human interventions. Wilderness, rural culture and urban culture make up the present world's three 'environments', each having its own particular intrinsic goods.[4]

Understanding Rolston's metaphysical commitments is essential to understanding his ethic. His explicit commitments are deeply biological and evolutionary. Yet, he parts company with contemporary theoretical evolution when he denies that nature operates by 'nothing but chance'.[5] Rolston's philosophy, in addition to being deeply biological, is also deeply theistic. The ultimate explanation for the origin, order and historical novelty in nature is God.[6]

Rolston's denial of chance is consistent with his Organic Principle, which is the assertion that every individual organism, from the simplest cell to the most complex multi-cellular organism, is intrinsically valuable and, therefore, worthy of appropriate respect. Unlike inorganic things, living organisms have 'vitality'. In contrast to inorganic things, every living organism has four features: (1) each individual has an identity; (2) it defends itself; (3) it functions for an end (*telos*); and (4) it has within itself, in its DNA, information that is passed on, or communicated, to others via reproduction. By virtue of these traits, organisms are centres of valuing; even when unconscious, what happens to them matters. In addition, natural organic evolution is projective in value in the sense that the values are captured and carried forward in time, producing increases both (a) in numbers (quantity) of individuals and species, and (b) in complexity (quality) of the forms of life.[7]

Denying the is-ought fallacy, Rolston argues for a naturalistic ethic in which morality – including both values and duties – is derivative from the holistic character of the ecosystem. 'Substantive values', Rolston contends, 'emerge only as something empirical is specified as the locus of value.'[8] Like it or not, all values are objectively grounded and supported by the possibilities and limitations within the Earth's ecosystem.

Rolston concedes that the *concepts* of value essential to holism, namely, the Leopoldian concepts of beauty, stability and integrity, are human and perhaps non-natural. Nevertheless, the *values* are a product of the inter-relationship and interaction of human persons with an objective environment. What counts as beauty, stability and integrity emerges from the interaction of world and concept. Rather than being located solely in human persons, values are collectively relocated in human persons in the environment. The value of the ecosystem is not imposed on it but is discovered already to be there: 'we find that the character, the empirical content, of order, harmony, stability is drawn from, no less than brought to, nature'. Because the substantive, empirical content is in nature, and in nature independent of human and other valuing beings, the value is appropriately and most clearly called 'intrinsic value'. Rolston asserts that ' ... here an "ought" is not so much *derived* from an "is" as discovered simultaneously with it'.[9]

As a theory of value, ecological holism claims that everything, whether an individual thing or a collective ecosystem, is in some sense morally relevant and valuable. Rolston argues that value is both in the thing and in the system directly and intrinsically, not just indirectly – or instrumentally – as the thing or system is related to humans or other beings that are rational, sentient, conative or alive.

To use a term favoured by Rolston, the value that emerges at the evolutionary ecosystem level is 'systemic'.[10] Rolston asserts that systemic value is intrinsic. In addition, he seems to hold that systemic intrinsic value is qualitatively richer than – greater than – the intrinsic value of the component parts and sub-systems, whether these components are considered as discrete things or sub-systems, or whether their discrete intrinsic values are totalled. In other words, the value of the whole is greater than the sum of the parts; the systemic intrinsic value of the whole exceeds the net sum of the intrinsic values of the individuals, things and sub-systems making up the whole system. Moreover, when the system is compared to any component part or sub-system, the qualitatively richer intrinsic value of the whole system seems to entail that, whenever the health or integrity of the system is threatened, the parts are expendable. The system as a whole captures lower intrinsic values and qualitatively enhances them, thereby exceeding the net sum of their individual intrinsic values.

In support of his notion of natural systemic intrinsic value, Rolston cites research in evolutionary history. He argues that the explanation for the accumulated diversity of species in nature is systemic: nature is organized in such a manner as to produce greater diversity and complexity of life forms. This generalization seems to be true, despite the four or five catastrophic extinctions in the fossil record. The natural tendency of the Earth's ecosystem is to increase species diversity – and to do so without any evident limit. It is this natural value that Rolston calls 'systemic'. Natural systemic values are also intrinsic values, and as such they entail duties and obligations, Rolston argues.[11]

Systemic value does not prohibit instrumental use of the component parts, provided the health and integrity of the system are not threatened. According to Rolston's Principle of Value-Capture, any human action should not destroy anything of intrinsic value unless the action produces something else of equal or greater intrinsic value.

Conflicts of intrinsic value occur only rarely in nature, Rolston contends, and conflicts between individuals and ecosystems are a problem for culture, not nature. In other words, Rolston claims that a feature of evolution is the generation of increasingly greater kinds and amounts of intrinsic value. When bacteria infect and kill a mammal, for instance, they contribute to greater emergent value. Evolution is producing greater diversity of life forms, greater complexity of life forms, and greater populations of individuals. Except for human intrusions that shut down evolutionary progress, values are enhanced and increased in nature.

Rolston argues that because humans are only members – one of many members – of the biotic community, holism is non-anthropocentric, if not anti-anthropocentric. Moral value is attributed to the natural environment considered as an ecological-systemic whole, independent of humans and human interests, except insofar as humans are naturally part of the whole. In contrast, anthropocentric-humanistic approaches treat ecosystems as resource values to be exploited for human ends. A scientifically enlightened humanist would have no reason not to use the planet as a mere resource according to long-term ecological science and the highest humanistic values.

Rolston rejects the anthropocentric view that ecology is merely enlightened and expanded human self-interest. We preserve the environment, not merely because it is in our best long-term economic, aesthetic and spiritual self-interest, but because there is no firm boundary between what is essentially human and what is essentially ecosystem. Human and environmental interest merge; egoism becomes 'ecoism'. Since the boundary between the individual and the ecosystem is diffuse, 'we cannot say whether value in the system or in the individual is logically prior'. The individual is not suppressed but enriched.[12]

A scientific ecological fact is that complex life forms evolve and survive only in complex and diversified ecosystems. If 'human' as we know it is to survive, we must maintain the oceans, forests and grasslands. To convert the planet entirely into cultivated fields and cities would impoverish human life. We also ought to preserve the ecosystem to enable the further evolution of the planet, including that of human mental and cultural life.[13]

Echoing Leopold, Rolston maintains that normatively right actions – our duties – are those actions that preserve ecosystemic beauty, stability and integrity. Preserving the ecosystemic status quo, however, may not be entailed because humans can improve and transform the environment. Borrowing a metaphor from contemporary physics, Rolston holds that integrity is a function of a 'field' interlocking species and individuals, predation and symbiosis, construction and destruction, aggradation and degradation. Since human life-support is part of the ecosystem, domestication is enjoined in order maximally to utilize the ecosystem. Biosystemic welfare allows alteration, management and use. 'What ought to be does not invariably coincide with what is.'[14]

Regarding species, Rolston contends that our duties are to the species as forms of life rather than to the individual members of the species. The species is the form; whereas, the individual member re-presents the form. 'The dignity resides in the dynamic form; the individual inherits this, instantiates it, and passes it on.' Biologically and ecologically, the individual is subordinate to the species.[15]

Although extinctions do occur in nature, natural ones are open-ended, usually producing diversification, new ecological niches and opportunities, new species and ecological trade-offs. In contrast, extinctions caused by humans are dead ends destroying diversity, producing monocultures and shutting down evolution. Species diversity is essential to continuing evolution. Consequently, duties towards species begin whenever human conduct endangers any species. Our duties include preserving not only species but entire ecosystems. This is because, unless preserved *in situ* in their ecosystems, species will not be preserved and evolution will halt.

Scholarly objections to Rolston's thought have taken mainly five directions. First, ecofeminists and social ecologists contend that Rolston is too hierarchical in his notions of intrinsic value, value-capture and the emergent complexity in evolutionary nature. Second, pragmatists, especially Bryan Norton, have rejected the meaningfulness of the concept of intrinsic value, preferring instead the rubric 'non-instrumental' value. Others, notably J. Baird Callicott and Eugene C. Hargrove, contend that value necessarily has a subjective component, namely, unless someone – a mind or subject – does the valuing, there is no value. Third, most philosophers continue to regard the naturalistic fallacy as legitimate. The fallacy takes a variety of logical forms, and Rolston needs a more detailed analysis of the precise form to which he is objecting. Fourth, Rolston concedes that his philosophy is merely the beginnings of a full theory and casuistry of environmental ethics. Many conflicts, usually involving particular cases as well as broader practical and theoretical issues, still need to be resolved. Finally, the present author has argued that Rolston's theory of ethics produces at most a very weak *prima facie* duty of beneficence that is easily overridden in practice. Strict duties cannot be derived directly from values, including intrinsic values, because an intermediate premise is needed in which the duty is asserted as an obligation to promote the good or prevent the harm. Instead of being a theory about non-consequential duties, Rolston's theory seems to be a consequential-ism in which the general obligation is the obligation to produce good.

Notes

1 *Environmental Ethics*, p. 188.
2 Ibid., pp. 61, 79, *passim*.
3 *Conserving Natural Value*, p. 4.
4 *Philosophy Gone Wild*, pp. 40–6.
5 *Environmental Ethics*, p. 207.
6 See *Genes, Genesis, and God*.
7 *Environmental Ethics*, chap. 6.
8 *Philosophy Gone Wild*, p. 19.
9 Ibid., pp. 19–20.
10 *Environmental Ethics*, pp. 186–9; *Conserving Natural Value*, pp. 68–100.
11 Ibid., pp. 155–7. Rolston cites D.W. Raup and J.J. Sepkoski, *Science*, 215, pp. 1501–3, 1982.
12 *Philosophy Gone Wild*, p. 25.
13 Ibid., pp. 22–4.
14 Ibid., p. 25.
15 Ibid., p. 212.

See also in this book

Callicott, Leopold

Rolston's major writings

A full bibliography may be found at: http://lamar.colostate.edu/~rolston

'Is There an Ecological Ethic?', *Ethics*, 85, pp. 93–109, 1975.
Philosophy Gone Wild: Essays in Environmental Ethics, Buffalo, NY: Prometheus, 1986.
Science and Religion: A Critical Survey, Philadelphia, PA: Temple University Press, 1987.
Environmental Ethics: Duties to and Values in the Natural World, Philadelphia, PA: Temple University Press, 1988.
Conserving Natural Value, New York: Columbia University Press, 1994.
Genes, Genesis and God: Values and Their Origins in Natural and Human History, The Gifford Lectures, University of Edinburgh, 1997–8; New York: Cambridge University Press, 1999.

Further reading

Callicott, J. Baird, *In Defense of the Land Ethic*, Albany, NY: State University of New York Press, 1989.
Katz, Eric, 'Searching for Intrinsic Value: Pragmatism and Despair in Environmental Ethics', in Andrew Light and Eric Katz (eds), pp. 307–18.
Kheel, Marti, 'From Heroic to Holistic Ethics: The Ecofeminist Challenge', in Greta Gaard (ed.), *Ecofeminism: Women, Animals, and Nature*, Philadelphia, PA: Temple University Press, pp. 243–71, 1993.
Light, Andrew and Katz, Eric (eds), *Environmental Pragmatism*, London: Routledge, part 4, 1996.
The Monist, 75, 2 (April), 1992; topical issue on 'The Intrinsic Value of Nature'. Articles include: Eugene C. Hargrove, 'Weak Anthropocentric Intrinsic Value', pp. 183–207; Bryan Norton, 'Epistemology and Environmental Values', pp. 208–26; Jim Cheney, 'Intrinsic Value in Environmental Ethics', pp. 227–35; Holmes Rolston III, 'Disvalues in Nature', pp. 250–78; and others.
Partridge, Ernest, 'Values in Nature: Is Anybody There?', *Philosophical Inquiry*, 8, pp. 1–2, 1986; reprinted with responses by Holmes Rolston III in Louis J. Pojman (ed.), *Environmental Ethics: Readings in Theory and Application*, 2nd edn, Belmont, CA: Wadsworth, pp. 81–92, 1998.
Weston, Anthony, 'Beyond Intrinsic Value: Pragmatism in Environmental Ethics', in Andrew Light and Eric Katz (eds), pp. 285–306.

JACK WEIR

RUDOLF BAHRO 1935–97

> To bring it down to the basic concept, we must build up
> areas liberated from the industrial system. That means,
> liberated from nuclear weapons and from supermarkets.
> What we are talking about is a new social formation and a
> new civilisation.[1]

Rudolf Bahro was a communist dissident, an early member of the
German Greens and a leading proponent of spiritual green political
thought and action. Bahro originally became well known as the
author of *The Alternative in Eastern Europe*, which he wrote during
the 1970s while he was a dissident Marxist and party member in the
former East Germany. This work was described by Herbert Marcuse
as 'The most important contribution to Marxist theory and practice
to appear in several decades'. In it Bahro argued that Eastern
Europe's non-capitalist, communist development path has been
shaped and corrupted by the same growth and materialist aims as
Western capitalism. In 1977, the ruling communist government
sentenced him to prison for his dissident activities and writings, and
in 1979 he was deported to what was then West Germany, in part
due to international protests at his imprisonment.

Soon after he arrived in West Germany, Bahro became involved
with the nascent German Greens (Die Gruen), affirming that 'red
and green go well together',[2] and urged communist groups to
dissolve themselves and work within the Die Gruen. As such he was
strongly identified with the 'eco-socialist' wing of the Green
movement, arguing for a synthesis of green and socialist ideals
and aims. He was clear that such a rapprochement required the
critical reconstruction of socialist politics, a central aspect of which
was a rejection of the productivist and 'materialist abundance'
dimensions of Marxist socialism, and the emergence of what Bahro
called a 'historical compromise' between the labour movement and
new social movements (environmental, feminist, peace), and a
rejection of Marxist 'class politics' and proletarian revolution. While
a resolute critique of capitalism and consumerism, Bahro's view
(which had much in common with Antonio Gramsci's 'anti-
hegemony' political strategy) was that what was required to defeat
capitalism and create a more sustainable, just, democratic and
peaceful social order, was a 'rainbow coalition' of all anti-capitalist
social forces, and not just the labour movement and the industrial
working class. Thus at this stage, Bahro's politics shared much with
that of Andre Gorz's 'red-green' position.

An example of where Bahro differed from Marxists was in relation to the creation of a more egalitarian and just world order in terms of the present inequalities between the developed 'North' and un/underdeveloped 'South'. The classic Marxist view would be that what is needed is 'communism on a world scale', a constitutive aspect of which would be the raising of the living standards and lifestyles of the 'Third world' to 'First world' levels. Against this 'cornucopian' view, Bahro, expressing a common Green view that this Marxist myth is both ecologically unsustainable (i.e. physically impossible) and spiritually undesirable/unworthy, proposed that, 'With a pinch of salt one might say ... the path of reconciliation with the Third World might consist in our becoming Third World ourselves'.[3]

Throughout the early 1980s, Bahro became an increasingly vocal and public critic of the 'realo' wing of the German Greens (those who became generally committed to competing for, winning and exercising parliamentary power) and became a leading spokesperson for the 'fundi' or fundamentalist wing of the party. The 'fundi–realo' split within the German Greens, a division which also emerged in other Green parties and green political thinking, owes much to the passion and conviction with which Bahro railed against what he saw as the corruption and co-option of the radical and emancipatory potential of Die Gruen by 'the system'. He ultimately left the party in 1985. He and his companion Christine Schröter called for an end to all animal experimentation. The party agreed, but decided to make exceptions in the case of medical research, which was unacceptable to Bahro's uncompromising position.

In the mid 1980s, in keeping with his disillusionment with Die Gruen and 'normal' democratic politics, he began to speak less in political terms and more in religious terms, asking that 'the emphasis [be] shifted from politics and the question of power towards the cultural level ... to the prophetic level ... Our aim has to be the "reconstruction of God"'.[4] Bahro had come to the view that if the Greens were to address the ecological crisis by radically changing society, they had to focus their efforts on psychological, cultural and spiritual levels. As he put it: 'If we take a look in history at the foundation on which new cultures were based or existing ones essentially changed, we always come up against the fact that in such times people returned to those strata of consciousness which are traditionally described as religious.'[5] For Bahro, personal inner transformation was a necessary and desirable part of the wider social and cultural transformation of Western civilization away from its ecologically destructive path.

Green politics must be based on spiritualistic values, in Bahro's view, because, as Eckersley points out, for Bahro 'the challenge of ecological degradation is primarily a cultural and spiritual one and only secondarily an economic one'.[6] Echoing the redemptive character of deep ecology, Bahro's vision of Green politics is of a life- and earth-affirming spirituality and the primary aim of Green politics is to be uncompromising in bringing about radical cultural change from which political and economic change will follow. A central part of Bahro's analysis focuses on the failures of Western civilization and the Enlightenment as a whole, and his argument is that nothing less than change at the level of civilization will prevent what he calls the 'logic of exterminism' within the 'mega-machine' (a term he borrows from Lewis Mumford) from destroying the planet and humanity along with it.

Bahro's 'social ecology' is a combination of spirituality, deep ecology, 'post-industrial utopianism',[7] and what Eckersley calls 'eco-monasticism',[8] a form of Green politics in which the strategy is of 'withdrawal and renewal' or 'opting-out' from the life-denying logic of the industrial 'mega-machine', and the creation of 'Liberated Zones'. These Liberated Zones provide protection for alternative ecological practices and values, places within which experiments in sustainable living can take place, and finally bases from which ecological, cultural and spiritual renewal and change will come. This eco-monastic perspective he shares with Theodore Roszak. As Ferris notes, Bahro's view is that 'Greens should opt out of industrial society by adopting a new lifestyle and living in small self-sufficient communes. Eventually, the communes would demonstrate a qualitatively better way of life that others would wish to adopt.'[9] His ultimate vision of the 'sustainable society' is of an 'eco-anarchist' federation of communes, comprised of small-scale, face-to-face communities which produce and consume the vast majority of what they require, a way of 'living lightly' on the planet.

In 1989, Bahro co-founded a combination educational centre and commune near Trier, the Lernwerkstatt (an 'ecological academy for one world'), whose purpose is to synthesize spirituality and politics, to put the eco-monastic ideal into practice. It presents lectures, cultural events and weekend workshops on various New Age themes, including deep ecology, ecofeminism, Zen Buddhism, holistic nutrition, Sufism, and other alternative theories, therapies and practices. Bahro also held a professorship at Humboldt University in Berlin in 'social ecology', but Bahro's work is not to be confused with the social ecology conceived and developed by Murray Bookchin.

Critics, both within and outside the Green movement, were concerned at the nationalistic tone of his position which was in contradiction with the Green slogan of 'act locally, think globally'. For example, in the early 1980s peace movement, he alarmed many by enunciating nationalistic arguments against the deployment of Pershing missiles. Some accused him of flirting with fascism, authoritarian spirituality and linking ecological politics to right-wing/conservative/nationalistic values and principles. He has been portrayed as believing that the ecological crisis is resolvable only through authoritarian, non-democratic means. He calls for a spiritually based and hierarchically elitist 'salvation government' or a 'god-state' (Gottesstaat) 'that will be run by a "new political authority at the highest level": a "prince of the ecological turn"'.[10] Bahro's apocalyptic analysis leads him to suggest that what is required is a 'rescue government', which would be an emergency or crisis government which while possessing absolute power, and thus a non-democratic political order, would be a transitional rather than permanent political arrangement. Standing above Bahro's later analysis of and political prescriptions for the ecological crisis seem to be modern, Green descendants of Plato's Guardians – dedicated, knowledgeable and wise elite stewards who will guide society in the right direction away from ecological, spiritual and cultural disaster, who govern without any democratic input from the people. His thought also echoes aspects of the early eco-authoritarian diagnosis of the ecological crisis put forward by William Ophuls,[11] and his call for an 'ecological Leviathan', as well as some deep ecological arguments that 'What is required is a new type of warrior – a person who is intense, centred, persistent, gentle, sincere, attentive and alert.'[12]

A rather startling example of the distance he had travelled since his early 'red-green' position is his statement that

> The most important thing is that ... [people] take the path 'back' and align themselves with the Great Equilibrium, in the harmony between the human order and the Tao of life. I think the 'esoteric'-political theme of 'king and queen of the world' is basically the question of how men and women are to comprehend and interact with each other in a spiritually comprehensive way. *Whoever does not bring themselves to*

cooperate with the world government [Weltregierung] will get their due.[13]

While the mystical/spiritual-cum-cultural focus of Bahro's analysis here would find some support in the wider Green movement, the anti-democratic and indeed anti-political sentiment would not. His call for a return to the something that was lost is close to other 'green mavericks', such as the British ecological writer and activist (and founder of *The Ecologist* magazine) Ted Goldsmith's prescription of a return to 'the way',[14] and Paul Shepard's 'back to the neolithic' suggestion.[15] However, it is the increasingly authoritarian dimension of Bahro's thought and strategy for dealing with the ecological crisis that many Greens find most worrying.

Ultimately, Bahro's legacy is a mixed one: a formative influence within the largest and most successful Green party in the world, the German Greens; a central figure in the fundi/realo split within the latter, and at the same time 'exporting' this 'radical/reformist' dichotomy to the theory and practice of the global Green movement as a whole; a defender of the position that Green politics and action is resolutely 'beyond left and right', and committed to a utopian, total transformation of the current socio-political order. However, for some, Bahro's thought and action could be said to be a study in someone who starts on the left and moves progressively to the right. At the same time, and consistent with the evolution of his thought, Bahro put the cultural, psychological and spiritual aims of the Green movement on the map as central substantive and strategic issues that had to be dealt with as essential to the resolution of the associated crises that together make up the ecological crisis.

Notes

1 *Building the Green Movement*, p. 29.
2 Bahro, quoted in Werner Hülsberg, *The German Greens: A Social and Political Profile*, London: Verso, p. 93, 1988.
3 *Building the Green Movement*, p. 88.
4 *From Red to Green*, pp. 220–2.
5 *Building the Green Movement*, p. 90.
6 Robyn Eckersley, *Environmentalism and Political Theory*, p. 164.
7 Boris Frankel, *The Post-industrial Utopians*.
8 Eckersley, op. cit., p. 164.
9 John Ferris, 'Introduction', in Helmut Wiesenthal, *Realism in Green Politics: Social Movements and Ecological Reform in Germany*, Manchester: Manchester University Press, p. 13, 1993.
10 *The Logic of Salvation*, p. 325.

11   William Ophuls, *Ecology and the Politics of Scarcity*, San Francisco, CA: Freeman, 1977.
12   Bill Devall, *Simple in Means, Rich in Ends: Practising Deep Ecology*, London: Green Print, p. 197, 1990.
13   Bahro, quoted in Jutta Ditfurth, *Feuer in die Herzen: Plädoyer für eine Ökologische Linke Opposition*, Hamburg: Carlsen Verlag, pp. 207–8, 1992, emphasis added.
14   Edward Goldsmith, *The Way: 87 Principles for an Ecological World*, London: Rider, 1991.
15   Paul Shepard, 'A Post-Historic Primitivism', in Max Oelschlaeger (ed.), *The Wilderness Condition*, Washington, DC, and Covelo, CA: Island Press, 1994.

See also in this book

Bookchin, Marx

Bahro's major writings

The Alternative in Eastern Europe: Towards a Critique of Real, Existing Socialism, London: New Left Books, 1979.
Socialism and Survival, London: Heretic Books, 1982.
From Red to Green, London: Verso, 1984.
Building the Green Movement, London: Heretic Books, 1986.
The Logic of Salvation. Logik der Rettung: Wer kann die Apokalypse aufhalten? – Ein Versuch über die Grundlagen ökologischer Politik, Stuttgart and Vienna: Transaction Press, 1987.
Avoiding Social and Ecological Disaster: The Politics of World Transformation, Bath: Gateway Books, 1994.

Further reading

Dobson, A., *Green Political Thought*, 2nd edn, London: Routledge, 1995.
Eckersley, R., *Environmentalism and Political Theory*, London: UCL Press, 1992.
Frankel, B., *The Post-industrial Utopians*, Oxford: Basil Blackwell, 1987.

JOHN BARRY

GRO HARLEM BRUNDTLAND 1939–

Our message is directed towards people, whose well-being is the ultimate goal of all environment and development policies. Unless we are able to translate our words into a language that can reach the minds and hearts of people young and old, we shall not be able to undertake the extensive social changes needed to correct the course of development.[1]

Dr Gro Harlem Brundtland was born in Oslo, Norway, on 20 April 1939. At the age of 10 she moved with her family to the United States, where her father had been awarded a Rockefeller Scholarship. A few years later, the family moved to Egypt, where Gro Harlem's father served as an expert on rehabilitation for the United Nations. By profession, he was a doctor; a specialist in rehabilitation medicine. From childhood, Gro Harlem's career ambition was to follow in her father's footsteps. As a newly qualified doctor herself, and indeed a young mother, she won a scholarship to Harvard School of Public Health. There, her great interest in public health issues and environmental concerns, which were later to bring her fame in the international arena of global environmental thinking, was nurtured and developed as she worked alongside distinguished public health experts.

In 1965 Dr Brundtland returned to Norway, to commence a nine-year period of working in the Ministry of Health and other positions in the medical field in Oslo. At the Ministry she specialized in children's health issues, including breastfeeding, cancer prevention and other diseases. She also worked in the children's department of the National Hospital and Oslo City Hospital and became Director of Health Services for Oslo's school-age children.

Alongside this early career in medicine, Gro Brundtland pursued her other great interest in public life, namely party politics. As a 7-year-old child she had become enrolled in the children's section of the Norwegian Labour Movement and as an adult she worked her way up through the Labour Party hierarchy and represented her country in international political conferences. Her commitment both to the Labour Party and also to a vision of health which extends beyond the medical world to encompass environmental issues and human development were the motivational factors leading to a change of career. In 1974 Dr Brundtland was offered and accepted the post of Minister of the Environment. In 1981, at the age of 41, she was appointed Prime Minister of Norway. She is noted for being the first woman and the youngest person in the country to hold this post. She led her party to election victory three times, and was Head of Government for more than ten years.

Dr Gro Harlem Brundtland, medical doctor and Master of Public Health, thus spent ten years as eminent physician and scientist in the Norwegian public health system and more than 20 years in senior public office. It was during the 1980s, when Prime Minister, she gained international recognition for championing and promoting the principle of sustainable development. In 1983 the then United

Nations Secretary General invited her to establish and chair the World Commission on Environment and Development. This Commission, without doubt best known for developing the broad political concept of sustainable development and for promoting global dialogue on the concept, published its report *Our Common Future*, otherwise known as *The Brundtland Report*, in April 1987.

The recommendations of the World Commission on Environment and Development (WCED) led to the staging of the largest-ever gathering of heads of state and others with a concern for the global environment and sustainable living – the United Nations Conference on Environment and Development (UNCED) in Rio de Janeiro in 1992.

Dr Brundtland stepped down from office as Prime Minister in 1996 and in 1998 she was nominated for and successful in being elected to the position of Director General of the World Health Organization (WHO).

The WCED, or Brundtland Commission, included individuals from twenty-two nations. Membership included six Commissioners from Europe (with three from Eastern Europe), five from Africa, five from Asia (including one from the Middle East), three from North America and three from South America. Its daunting task was to investigate the state of the world, to suggest ways into the twenty-first century that would allow the planet's rapidly growing population to meet its basic needs and to come up with a 'global agenda for change'. The Commission engaged in a great deal of empirical research and debate. The group, composed of ministers, scientists, diplomats and law makers, studied, debated and held public hearings on five continents over almost three years. The final Report, *Our Common Future*, consisting of almost 400 pages, includes a widely quoted definition of sustainable development, and promotes the view, perhaps for the first time in a globally promoted document, that conservation and development can co-exist.

> Humanity has the ability to make development sustainable – *to ensure that it meets the needs of the present without compromising the ability of future generations to meet their own needs.* The concept of sustainable development does imply limits – not absolute limits but limitations imposed by the present state of technology and social organisation on environmental resources and by the ability of the biosphere to absorb the effects of human activities. But technology and social organisation can be both managed and improved to

make way for a new era of economic growth. The Commission believes that widespread poverty is no longer inevitable ... A world in which poverty is endemic will always be prone to ecological and other catastrophes ... Sustainable development is not a fixed state of harmony, but rather a process of change ... We do not pretend that the process is easy or straightforward. Painful choices have to be made. Thus, in the final analysis, sustainable development must rest on political will.[2]

Thus the Report identifies two key concepts that are tied to the process of sustainable management of the Earth's resources:

1   The basic needs of humanity – for food, clothing, shelter, and jobs – must be met. This involves, first of all, paying attention to the largely unmet needs of the world's poor, which should be given over-riding priority.
2   The limits to development are not absolute but are imposed by present states of technology and social organisation and by the impacts upon environmental resources and upon the biosphere's ability to absorb the effect of human activities. But technology and social organisation can be both managed and improved to make way for a new era of economic growth.[3]

It advanced the following list of critical objectives for sustainable development policies:

• Reviving economic growth
• Changing the quality of growth
• Meeting essential needs for jobs, food, energy, water, sanitation
• Ensuring a sustainable level of population
• Conserving and enhancing the resources base
• Reorienting technology and managing risk
• Merging environment and economics in decision-making processes

Our Common Future is subdivided into three main sections:

1   Common Concerns

• A Threatened Future

- Towards Sustainable Development
- The Role of the International Economy

2 Common Challenges

- Population and Human Resources
- Food Security: Sustaining the Potential
- Species and Ecosystems: Resources for Development
- Energy: Choices for Environment and Development
- Industry: Producing More With Less
- The Urban Challenge

3 Common Endeavours

- Managing the Commons
- Peace, Security, Development and the Environment
- Towards Common Action: Proposals for Institutional and Legal Change

The Report contains many specific recommendations for institutional and legal change. By way of summary, the Commission's main proposals fall within six priority areas:

Getting at the Sources   International and regional organizations and national governments must start making bodies directly accountable for the environmental effects of their actions.

Dealing with the Effects   Agencies formed to protect and restore the environment should be reinforced, especially the United Nations Environment Programme.

Assessing Global Risks   The capacity to identify, assess and report on risks to the environment must be improved. This should not only be the responsibility of individual governments; a new independent coordinating body should be set up.

Making Informed Choices   The public, NGOs, scientists and industry should all have the opportunity to participate in decision making.

Providing the Legal Means   National and international law is being outpaced by events. Governments must fill the major gaps.

Investing in our Future   The overall cost effectiveness in halting pollution has been shown over the last decade. But a commitment to sustainable development has large financial implications, and a new priority and focus must be taken up by financial institutions, aid agencies and governments. Developing countries need a strong infusion of financial support from international sources for environmental restoration, protection and improvement. Major lending agencies like the World Bank, the International Monetary Fund, and the regional development banks should upgrade their environmental programmes.

The Brundtland Report concludes with a 'Call for Action' which asks the UN General Assembly to 'transform this report into a UN Programme of Action on Sustainable Development'. Sustainable development is seen not as a fixed state, but rather as a process of change in which each nation achieves its potential for development, whilst also improving the quality of the environmental resources upon which the development is based. Throughout, the Report argues that sustainable development at a global level can be achieved only through major changes in the ways in which our planet is managed. Suggested changes include those in political systems, which allow effective citizen participation in decision-making processes; in economic systems, which lead to the ability to generate surpluses and technical knowledge on a self-reliant and sustained basis; in social systems, which provide solutions to tensions arising from our present form of development; in production systems, which respect the obligation to preserve the ecological base for development; in technological systems, which can search continuously for new solutions; in international systems, which foster sustainable patterns of trade and finance; and in administrative systems, which promote flexibility and have the capacity for self-correction.

The Report of the WCED was ambitious, and based on a vast array of accumulated evidence and wisdom. One criticism made of it is that it set a very broad and complex agenda for change in the direction of achieving sustainable development, without identifying the many and specific barriers that exist to achieving the intended goals. Mechanisms for achieving the end results appear as rather vague statements, lacking in precision or guidelines for translating them into specific actions.

More fundamental criticism attacks the Report's definition of development. The Commission's analysis is based on a certain conception of development, and thus of economic growth. Mishra, an Indian environmentalist, comments:

We should not assume that we can look for solutions to our

279

problems within the framework of the current development pattern. It would be folly to think the Brundtland Commission can find solutions within the 'counter-productive framework' of governments, the United Nations, the World Bank, and so on. Because the present structures have given us the disease, is it then logical that they should also provide the cure? This seems to be the limitation of this Commission, because it itself stemmed from the current framework.[4]

By 'the current framework' is meant that dominant pattern of development today, based on Western culture. This has created a universal order: universal values, universal economics, universal science. Critics of this culture, including Shiva and Bandyopadhyay, emphasize its emphasis on private endeavour, interests and profits, and indeed its non-sustainability:

> The ideology of the dominant pattern of development derives its driving force from a linear theory of progress, from a vision of historical evolution created in eighteenth and nineteenth century Western Europe and universalised throughout the world, especially in the post-war development decades. The linearity of history pre-supposed in this theory of progress, created an ideology of development that equated development with economic growth, economic growth with expansion of the market economy, modernity with consumerism, and non-market economics with backwardness. The diverse traditions of the world, with their distinctive technological, ecological, economic, political and cultural structures, were driven by this new ideology to converge into a homogeneous monolithic order modelled on the particular evolution of the west.[5]

According to this viewpoint, the dominant development paradigm disregards the true complexity and inter-relationships of all processes on Earth, a complexity encompassed in the Gaia Hypothesis, an alternative paradigm articulated and developed by James Lovelock (1979). For Lovelock, the

> entire range of living matter on Earth ... could be regarded as a single living entity, capable of manipulating the Earth's atmosphere to suit its overall needs. This organism, of which

human society is a part, but only one part, regulates her activities in a very complex and subtle way.[6]

This is a radically different view from that embodied in the 'dominant development paradigm', which sees the Earth and its resources merely as a place with raw materials to be used by its inhabitants. Proponents of the Gaia theory accept a concept of development that is based on restoring internal control, creating stability and peaceful co-operation.

> Such a concept does not allow strong external influences. It will maximize 'stocks' (physical, intellectual, ecological) and will minimize the movement and export of things (in the form of goods, natural resources, capital and so on). Essentially, it runs contrary to the open market system. Sustainable development in this sense will demand solving the problem of domination in society élites ... It is in this sense also a critique and action against the dominant paradigm of development. Sustainable development therefore means solving a conflict which is rooted deep in our images of the world and the organisation of our society.
> The Report of the World Commission on Environment and Development does not contain any of this critique ...[7]

Such criticisms fuelled important and far-reaching dialogue. Whatever views may be held on these fundamental issues, the fact remains that the Brundtland Commission successfully steered the world's thinking and debate on the formulation and re-orientation of policies relating to environment and development. Consideration of the environmental consequences of any action had been placed firmly on the agenda of governments, NGOs and international agencies alike.

By late 1989 the Report had been published in seventeen languages and had generated many other publications that provided commentaries on or developed aspects of its policy recommendations. A Centre for Our Common Future was established in Geneva as a focal point for the environmental activities of governments, multilateral institutions, scientific bodies, industry and NGOs.

Within the United Nations systems, the Brundtland Report inspired the planning of a major global Conference on Environment and Development to take place in 1992 in Rio de Janeiro, marking the twentieth anniversary of the UN Conference on the Human Environment that had been held in Stockholm.

Dr Gro Harlem Brundtland, variously described as 'tough and efficient', 'energetic and committed' and 'a master survivor', continues on her life's pathway in a position where her undoubted talents as doctor, politician, activist and manager can come together in the shaping of global policy on health and the environment. In her acceptance speech for the World Health Assembly in 1998, she declared:

> What is our Key mission? I see WHO's role as being the moral voice and technical leader in improving health of the people of the world. Ready and able to give advice on the key issues that can unleash development and alleviate suffering. I see our purpose to be combating disease and ill-health – promoting sustainable and equitable health systems in all countries.

The Director General herself possesses the vital combination of necessary skills, personality and motivation to be both powerful voice and effective leader.

Notes

1   Brundtland in the 'Foreword' to *Our Common Future*, World Commission on Environment and Development, p. xiv.
2   Ibid., p. 8, emphasis added.
3   IIED/Earthscan, 1989.
4   T. de la Court, *Beyond Brundtland: Green Development in the 1990s*, p. 13.
5   Ibid., p. 128.
6   J.E. Lovelock, *Gaia. A New Look at Life on Earth*, Oxford: Oxford University Press, 1979.
7   Court, op. cit., p. 135.

See also in this book

Lovelock, Shiva

Further reading

Court, T. de la, *Beyond Brundtland: Green Development in the 1990s*, London: Zed Books, 1990.
Elliot, J.A., *An Introduction to Sustainable Development: The Developing World*, London: Routledge, 1994.
Goodland, R.J.A., *Environmentally Sustainable Economic Development: Building on Brundtland*, Paris: UNESCO, 1992.

Starke, L., *Signs of Hope: Working Towards Our Common Future*, Oxford: Oxford University Press, 1990.
Timberlake, L. and Holmberg, J., *Defending the Future: A Guide to Sustainable Development*, London: Earthscan, 1991.
World Commission on Environment and Development, *Our Common Future*, Oxford: Oxford University Press, 1987.

JOY A. PALMER

VAL PLUMWOOD 1939–

> The logic of domination and the deep structures of dualism create 'blind spots' in the dominant culture's understanding of its relationship to the biosphere, understandings which deny dependency and community to an even greater degree than in the case of human society. The distorted perceptions and mechanisms of denial which arise from the master rationality are an important reason why the dominant culture which embodies this identity in relation to nature cannot respond adequately to the crisis of the biosphere and the growing degradation of the earth's natural systems.[1]

Val Plumwood began her work on environmental philosophy in collaboration with her then husband Richard Routley (later Sylvan, also an important environmental philosopher) in the early 1970s when the ecological crisis of the modern West was becoming more obvious. Within the framework of analytical philosophy in which they were both trained, the most obvious tools with which to explain the crisis were those of the ethics of value and respect. Routley and Plumwood argued that one reason why the dominant global culture of the West had been able to expand and conquer indigenous cultures as well as nature itself was that it lacked their respect-based constraints on the use of nature – a thought that cast Western triumphalism in a rather different and more dangerous light. This lack of respect had, as one of its main philosophical expressions, the deep-seated Western conviction that only humans could be of any direct ethical significance or value.[2]

The view that value and moral consideration were confined to humans, for which Plumwood and Routley coined the term 'human chauvinism', was supported by the assumption (too deeply embedded to be thought of as needing explicit statement or defence) that the natural world could have only indirect or instrumental value, as a means to human ends, or as mattering to human beings.

283

They exposed the arrogance of this assumption, and tried to clarify concepts like respect and instrumental value so as to make available alternative modes of valuing and respecting nature as an independent other. They met objections that there was no rational alternative to purely instrumental values by showing that this involved an infinite regress. They argued that, since its supposed 'naturalness' and inevitability were ultimately based on fallacies that closely paralleled those of philosophical egoism, human chauvinism had no more legitimacy than human-group chauvinism.[3]

For Plumwood, however, value is too narrow a focus for explanation and activism. While she agrees that we must value the natural world more highly, she holds that value and the failure to discern and accord it provides only a very incomplete explanation of environmental failure. This last, she holds, stems equally from what she calls 'the standpoint of mastery': overconfidence, failure to recognize the other's agency and limits, and other kinds of insensitivity that come from the dominant rationalist and colonizing frameworks by which historically we have understood and created human/nature relations.

After Plumwood's collaboration with Routley ended in 1981, she refined their initial, relatively unanalysed concept of 'human chauvinism' by exploring the implications of seeing human-centredness as a parallel concept to androcentrism and eurocentrism.[4] From this she came to the conclusion that human-centred stances are subject to similar blind spots and distortions of conception and perception; for example, seeing the other as radically separate and inferior, the background to the self as foreground, as one whose existence is secondary, derivative or peripheral to that of the self or centre, and whose agency is denied or minimized.

A major feature of human-centred frameworks is the denial of human dependency on nature. In 'Anthrocentrism and Androcentrism' Plumwood draws on many aspects of women's oppression to theorize 'hegemonic otherness', the condition, in androcentric frameworks, in which women appear as appendages to men and in which their agency is treated as lesser or denied altogether. She sees a strong parallel here with the treatment of nature's agency in human-centred frameworks. When nature as agent and collaborative partner is similarly denied, she claims, we get blind spots about human dependence and vulnerability, which help to make such frameworks dangerous and misleading.[5]

In her book *Feminism and the Mastery of Nature* (1993), in which many of the main themes of her work are articulated, Plumwood argues the ecofeminist thesis that the West's problems with the non-human world (now global problems) must be understood in the context of its larger dualist problematic. Traditions excluding non-humans from the spheres of ethics, mind, culture, and reason are matched on the other side by traditions excluding humans from the realm of nature and animality, to form what she calls a 'hyperseparation' between humans and nature which is entrenched in the dominant traditions of the West. She sees the drive to hyperseparation as part of a colonizing conceptual dynamic which places the human colonizer radically apart from, and above, those he conceives as part of the subordinated realm of nature.

Plumwood identifies human/nature dualism as a key part of the system of gendered dualisms that have helped to shape Western world-views. These include reason/nature, human/animal and mind/body dualisms, which have been historically central to environmental thinking, as well as male/female, reason/emotion and civilized/primitive, which are associated with other forms of colonization. Thus she links the treatment of non-humans to the treatment of women and other groups, such as indigenous people, who have also been considered part of the inferior realm of nature. The influence of dualistic approaches that separate the truly human from, and inferiorize, the ecologically situated body and the perishable order of biological life is traced in the Greeks, and is especially clear in the work of Plato (a major influence in the development of 'flesh-inferiorising dualisms' in Christianity).[6]

For Plumwood, dualism has deeply marked both concepts of nature and concepts of reason. The environmental crisis, she argues, should be seen as a crisis of dualistic reason, a form of rationality expressed especially in the contemporary global market, which conceives rationality as self-interest in opposition both to the emotions (including care for others) and to the ecologically situated body. Challenging this form of rationality challenges current social and political systems and means, bringing economic and ecological rationality together. Plumwood argues from an eco-socialist perspective against the treatment of animals and nature as property under capitalism and for the ecological virtues of more egalitarian and democratic social systems.[7]

Employing this analysis of human-centredness and of human/nature dualism, Plumwood argues that the dominant problematic of modern environmental ethics is set up in an anthropocentric way, focusing on establishing the qualifications of non-humans for moral consideration (usually through establishing some basis of similarity to the human), rather than on the problems of a human-centred ethical system and epistemology which excludes them. In many current forms of academic environmental philosophy, she says, the question of value seems to be taken to be the only issue. Moreover, valuing itself is too often treated as the stance of someone looking on at nature as a separate and passive entity; evaluating, ranking and assessing its existence, somewhat in the manner of a property appraiser. Such a stance, she claims, assumes that it is our human prerogative to order the world in terms of some generalized species-ranking, to assign or withhold 'value' according to whim or to degrees of rationality or consciousness.[8]

Equally problematic for Plumwood is the sort of environmental ethic that extends moral consideration only to some 'higher' animals on the basis of their similarity to the human (especially their similarity in consciousness) in the fashion of some animal rights positions.[9] This type of environmental ethics is rejected as neo-Cartesian and implicitly human-centred, a minimum-change position that relocates, rather than cancels, the radical break between the human and non-human. Plumwood argues that it is based implicitly on sameness, extending a human model of reason or consciousness, and is therefore both human-centred in a damaging sense and unable to acknowledge that difference does not mean inferiority.

In *Feminism and the Mastery of Nature* Plumwood delineates a position in environmental ethics that, on the one hand, is distinct from the conventional neo-Cartesianism that would extend moral consideration just to conscious beings ('like us'), and, on the other hand, from deep ecology. She agrees with deep ecology that it is a major problem that the modern West positions itself as 'outside nature'. But Plumwood's analysis seeks to explain the West's misunderstanding of its ecological embedment not as the outcome of a separation from nature that departs from the deep ecological ideal of unity between humanity and nature, but as an aspect of dualistic hyperseparation, in which normative human identity excludes features traditionally associated with nature and the animal sphere.[10] Plumwood's underlying metaphysics refuses the demand to base ethical consideration on sameness that underlies both moral extensionism and deep ecology, the two conventional choices. Rejecting also the pure appeal to difference as the source of value which appears in some post-modernist positions, Plumwood insists on both continuity with, and difference from, the human as sources of value and consideration.

According to Plumwood, both sameness and difference are

required to counter human/nature dualism. In the Western tradition especially, she admits the need to stress continuity between self and other, human and nature, in response to the gulf created by the hyperseparated and alienated models of nature and of human identity that remain dominant. These models define the truly human as (normatively) outside of nature and in opposition to the body and the material world, and conceive nature itself in alienated and mechanistic terms as lacking elements of mind. But she contends that we also need to stress the difference and divergent agency of the other in order to defeat that further part of the colonizing dynamic that seeks to assimilate and instrumentalize the other, recognizing and valuing it only as a part of or similar to the self, or as means to the self's ends. Vague concepts of unity and identity of the sort stressed by deep ecology, she argues, provide very imprecise and inadequate correctives to our historical denial of continuity with and dependency on nature. She maintains that ethical theories based on unity cannot provide a good model of mutual adjustment, communication and negotiation between different parties and interests, and are unhelpful in the key areas where we need to construct dialogical, ethical relationships.

Plumwood is not opposed to spirituality and the sacred *per se* but thinks that dominant forms of Western spirituality have located the sacred in the wrong place – above and beyond a fallen Earth. An ecological spirituality would need to relocate it in the immediate world around us, as in much indigenous spirituality. In this location the sacred can be experienced as in and of the Earth, but need not, and should not, be overly singularized and centralized as it is in some forms of Gaia theory. In her forthcoming book, *Environmental Culture*, she sees such a 'materialist spirituality' as one component of a strong environmental culture which needs to be developed over a wide range of areas to counter the excesses of the dominant form of rationality.

To defeat human/nature dualism, Plumwood argues that we need to revise conceptions of human virtue which are based on excluding, from the ideal human character, the supposedly oppositional elements to reason, especially emotionality, embodiment and animality. She also emphasizes 'counter-hegemonic virtues', ethical stances which can help to minimize the influence of the oppressive ideologies of domination and self-imposition that have formed our conceptions of both the other and ourselves. She advocates the adoption of philosophical strategies and methodologies that maximize our sensitivity to other members of our ecological communities and openness to them as ethically considerable beings in their own right, rather than strategies that minimize ethical recognition or that adopt a dualistic stance of ethical closure that insists on sharp moral boundaries and denies the continuity of planetary life.

Among such strategies, she stresses the need for communicative virtues of listening and attentiveness to the other to help counter the backgrounding which obscures and denies what the non-human other contributes to our lives and collaborative ventures. Openness and attentiveness involve giving the other's needs and agency more weight, being open to unanticipated possibilities and aspects of the other, and re-conceiving and re-encountering the other as a potentially communicative and agentic being, as well as an independent centre of value and an originator of projects that demand respect. These counter-hegemonic virtues, she claims, help us to resist the reductionism of the dominant mechanistic conceptions of the non-human world, and to revise both our narrow epistemic objectives of prediction and control and our denial of non-human others as active presences and ecological collaborators in our lives.

Notes

1 *Feminism and the Mastery of Nature*, p. 194.
2 See Routley, 'Is There a Need for a New, an Environmental Ethic?', pp. 205–10; Plumwood's 'Critical Notice of Passmore's *Man's Responsibility For Nature*', pp. 171–85.
3 Routley and Plumwood, 'Human Chauvinism and Environmental Ethics', in D. Mannison, M. McRobbie and R. Routley (eds), *Environmental Philosophy*, Canberra: Philosophy Department, Australian National University, pp. 96–190, 1979; 'Against the Inevitability of Human Chauvinism', in Goodpaster and Sayre (eds), pp. 36–59.
4 Plumwood, 'Anthrocentrism and Androcentrism: Parallels and Politics', *Ethics and the Environment*, pp. 119–52 and 'Paths Beyond Human Centredness: Lessons from Liberation Struggles', in *An Invitation to Environmental Philosophy*, pp. 69–106.
5 Human vulnerability in the face of non-human agency was brought home to her dramatically in 1985 when she was attacked and nearly killed by a crocodile in Northern Australia. She has written memorably about the experience, now rare for humans, of being hunted for food in 'Being Prey', *Terra Nova*, pp. 33–44.
6 'Prospecting for Ecological Gold among the Platonic Forms', *Ethics and the Environment*, pp. 149–68.
7 See 'The Crisis of Reason, the Rationalist Market, and Global Ecology', *Millennium. Journal of International Studies*, pp. 903–26; 'Has Democracy Failed Ecology? An Ecofeminist Perspective', *Environmental Politics*, 4, pp. 134–68, 1995; 'Ecojustice, Inequality and Ecological Rationality', *Debating the Earth: The Environmental Politics Reader*, ed. J. Dryzek and D. Schlosberg, Oxford: Oxford University Press, 1998.

8   See 'Self-Realization or Man Apart? The Naess–Reed Debate', in N. Witoszek and A. Brennan (eds), pp. 206–12.
9   See 'Intentional Recognition and Reductive Rationality: A Response to John Andrews', *Environmental Values*, 17, pp. 397–421, 1998; *Feminism and the Mastery of Nature*.
10 In addition to *Feminism and the Mastery of Nature*, see two later essays, 'Self-Realization or Man Apart?' and 'Deep Ecology, Deep Pockets, and Deep Problems: A Feminist Eco-Socialist Analysis', in A. Light, E. Katz and D. Rothenburg (eds).

See also in this book

Lovelock, Passmore

Plumwood's major writings

'Critical Notice of Passmore's *Man's Responsibility For Nature*', *Australasian Journal of Philosophy*, 53, pp. 171–85, 1975.
'Against the Inevitability of Human Chauvinism', with R. Routley, in *Ethics and the Problems of the 21st Century*, ed. K.E. Goodpaster and K.M. Sayre, South Bend: Notre Dame University Press, 1978.
Feminism and the Mastery of Nature, London: Routledge, 1993.
'Anthrocentrism and Androcentrism: Parallels and Politics', *Ethics and the Environment*, 1, pp. 119–52, 1996.
'Being Prey', *Terra Nova*, 1, pp. 33–44, 1996; reprinted in *Terra Nova: New Environmental Writing*, Cambridge, MA: MIT Press, 1999.
'Prospecting for Ecological Gold among the Platonic Forms', *Ethics and the Environment*, 2, pp. 149–68, 1997.
'The Crisis of Reason, the Rationalist Market, and Global Ecology', *Millennium. Journal of International Studies*, 27, pp. 903–26, 1998.
'Paths Beyond Human Centredness: Lessons from Liberation Struggles', *An Invitation to Environmental Philosophy*, ed. A. Weston, Oxford: Oxford University Press, pp. 69–106, 1998.
'Self-Realization or Man Apart? The Naess–Reed Debate', *Festschrift for Arne Naess*, ed. N. Witoszek and A. Brennan, Savage, MD: Rowman & Littlefield, pp. 206–12, 1998.
'Deep Ecology, Deep Pockets, and Deep Problems: A Feminist Eco-Socialist Analysis', *Beneath the Surface: Critical Essays on Deep Ecology*, ed. A. Light, E. Katz and D. Rothenburg, Cambridge, MA: MIT Press, 1999.
Environmental Culture, London: Routledge, forthcoming.
The Eye of the Crocodile, Durham, NC: Duke University Press, forthcoming.

Further reading

Reviews of *Feminism and the Mastery of Nature*: by P. Hallen, *Australasian Journal of Philosophy*, 73, pp. 645–6, 1995; by L.M.B Harrington, *Journal of Rural Studies*, 12, pp. 319–20, 1996; by J. Cook, *Environmental Values*, 6, pp. 245–6, 1997.

Hawkins, Ronnie, 'Ecofeminism and Nonhumans: Continuity, Difference, Dualism, and Domination', *Hypatia*, 13, pp. 158–97, 1998.
Routley, Richard, 'Is There a Need for a New, an Environmental Ethic?', *Proceedings of the XVth World Congress in Philosophy*, Varna, i, pp. 205–10, 1973.

NICHOLAS GRIFFIN

J. BAIRD CALLICOTT 1941–

> There is no survival value in pessimism. A desperate optimism is the only attitude that a practical environmental philosopher can assume.[1]

For the past three decades John Baird Callicott has argued that philosophy and ethics lie at the root of our global environmental problems. He has steadfastly clung to a 'desperate optimism' that philosophy and ethics can both elucidate and help resolve these problems:

> Although an ethic, whether environmental or social, is never perfectly realized in practice, it nonetheless exerts a very real force on practice. Ideals do measurably influence behaviour. In envisioning, inculcating, and striving to attain moral ideals, we make some progress both individually and collectively, and gain some ground.[2]

Baird Callicott was born in Memphis, Tennessee, on 9 May 1941. He was educated at Rhodes College and Syracuse University, receiving his Ph.D. in philosophy (where he specialized in the philosophy of Plato) from Syracuse in 1971. He has taught and lectured at a vast number of universities in the United States and abroad and is currently a Professor of Philosophy at the University of North Texas. His contribution to the field of environmental ethics has been immeasurable. He was there at the beginning: teaching the very first university course in the world in environmental ethics in 1971 at the University of Wisconsin – Stevens Point, publishing in the very first issue of the original journal in the field in 1979, and establishing himself as one of the founders of the field.[3] One author recently referred to Callicott as 'the man who practically invented environmental ethics'.[4]

Although Baird Callicott did not begin publishing until age 38, his reputation for insightful and creative argument, lucid and engaging prose, and provocative thought have earned him the highest merit from countless contemporary environmental thinkers. One can hardly pick up an issue of *Environmental Ethics, Environmental Values* or any other journal in the field (and many related fields) without encountering numerous references to and comments upon Callicott's work. As the editor of *Environmental Values* recently wrote: 'Sustained critical interest in the work of J. Baird Callicott ... just won't lie down.'[5] Dave Foreman has commented that 'in scholarship, sincerity, and openness ... Callicott stands head and shoulders above his academic colleagues'.[6] A recent introduction at a wilderness conference in Montana invoked the words of Henry Miller to comment on the status of Callicott's contribution:

'Only a very few souls, at any time in man's history have been privileged to battle with the great problems, the problems of man.' Baird Callicott is just such a soul.

And always, Callicott's efforts have included a progressive attempt to bring philosophy out of the ivory tower of academia and apply it not only to real-world environmental problems but to other disciplines as well. He has written for many non-philosophical journals, various encyclopedias, textbooks in conservation biology, and has served on natural resource advisory boards.

Callicott's interest in environmental ethics grew out of his serious commitment to the discipline of philosophy. It has remained philosophically grounded ever since: 'My work has always been connected to philosophy; I see environmental ethics both as philosophy and as something that is challenging and transforming philosophy'.[7] His sense is that in the years to come the progress made by environmental philosophers on this front will be positively acknowledged:

I've bet my life on the belief that environmental philosophy will be regarded by future historians as the bellwether of a twenty-first-century intellectual effort to think through the philosophical implications of the profound paradigm shifts that occurred in the sciences during the twentieth century.[8]

If so, Baird Callicott will deserve much of the credit.
Callicott is most notably recognized as the leading interpreter of

the philosophical legacy of Aldo Leopold. Leopold's recognition that evolution and ecology altered our fundamental assumptions about ourselves and the world around us marks him as an early environmental philosopher. However, Leopold was not a philosopher in the formal sense. His ideas required unpacking. Baird Callicott provided the conceptual and philosophical foundations upon which to ground Leopold's metaphysical and ethical assumptions. Just as it is hard to see where the ideas of Socrates leave off and those of Plato (his student and scribe) begin, it is difficult at times to tell where Leopold's thoughts end and Callicott's emerge. However, both Leopold and Callicott view the evolutionary/ecological world-view as a dismissal of the modern mechanistic paradigm which, until quite recently, has been taken as a given. Denying the sharp divisions between self and nature and forcing a re-thinking of an atomistic and mechanical world in terms of an organic and systematically related world, Leopold and Callicott assert that such scientific paradigm shifts cannot be viewed in isolation – they have profound metaphysical and ethical implications, they challenge and change both. Building upon the work of biologist Charles Darwin and philosophers David Hume and Adam Smith, Leopold and Callicott point out that one's sense of ethical inclusiveness corresponds with one's sense of a shared community. And, since evolution and ecology portray the 'soils and waters, plants, and animals, or collectively: the land'[9] and human beings as part and parcel of a shared social community, Leopold and Callicott have argued that the ethical duties that we admittedly owe to one another can be and ought to be prompted and extended to this land community as well. Leopold refers to this set of ethical obligations as the 'land ethic', and Callicott's most acknowledged role has been that of defender of the land ethic. The power of the work of Leopold and Callicott, then, is that they portray the world as significantly more morally fertile than previously perceived.

Within the larger debate surrounding the extension of moral obligations to encompass the land, Callicott has argued that the land possesses intrinsic value: said to be value in and of itself as opposed to value as a means to some other end, or value in addition to merely instrumental value. Such a move designates Callicott as an ecocentrist – or one who attributes direct moral standing to such things as species, ecosystems, watersheds, biotic communities and the biosphere as a whole, not to mention those individuals which constitute those biological collectives[10] – and places him in the company of other environmental philosophers such as Holmes

Rolston III and Arne Naess. The debate surrounding the ascription of intrinsic value to environmental parts or wholes, and Callicott's contribution to this debate, has remained at the centre of environmental ethics from its inception.

Of course, anyone familiar with Baird Callicott's work realizes that he has made deep contributions in a multiplicity of other areas as well. Such sundry topics as environmental education, aesthetics, the distinction between animal welfare ethics and environmental ethics, Judeo-Christian stewardship, hunting ethics, farming, health and wellness, and environmental activism, have all garnered his attention.

Callicott possesses an uncanny knack for the provocative. If arguing that nature possessed intrinsic value and that we owe moral obligations to the land was not enough, a number of other topics he has taken up over the past thirty years have launched him into the centre of, sometimes, controversial debates.

Callicott became one of the earliest theorists on the environmental attitudes and ethics expressed by the overlapping world-views of North American Indian societies. He argues that an examination of the cosmology of American Indian tribes displays an environmental ethic worthy of notice – and one, interestingly, that shares strong affinities with Leopold's land ethic. As he once wrote:

> The implicit overall metaphysic of American Indian cultures locates human beings in a larger social, as well as physical, environment. People belong not only to a human community, but to a community of all nature as well. Existence in this larger society, just as existence in a family and tribal context, places people in an environment in which reciprocal responsibilities and mutual obligations are taken for granted and assumed without question or reflection.[11]

This line of thought later developed into commentary on the environmental attitudes and values expressed in a wide range of world cultures and religious traditions – from Christian to Islamic, from Pagan to Australian Aboriginal – which was published as his critically acclaimed book *Earth's Insights*.

Most recently, Callicott has been at the centre of the highly contentious debate over the concept of wilderness. Along with historian William Cronon, Callicott has argued that the concept of wilderness is a product of social construction; a product desperately in need of reconstruction. Callicott 'believes that the received wilderness idea has been mortally wounded by the withering critique to which it has been lately subjected'.[12] However, although this point is often misunderstood or ignored, he is no enemy of wilderness, but rather a friendly critic:

> I am as ardent an advocate of those patches of the planet called 'wilderness areas' as any other environmentalist. My discomfort is with an idea, the received concept of wilderness, not with the ecosystems so called.[13]

Callicott also emphasizes that through a conceptual re-thinking of wilderness those areas we refer to as wilderness will be better protected.

A journey through the writing and thoughts of J. Baird Callicott is always insightful, always challenging, always instructive, and always a lesson in the power of sound reasoning and good writing. And at all times in his work there is a sense of an empowering optimism, an affirmation that a successful ethical relationship between humans and the non-human world can and will be forged.

Notes

1 From 'Benevolent Symbiosis: The Philosophy of Conservation Reconstructed', in J. Baird Callicott and Fernando J.R. da Rocha (eds), *Earth Summit Ethics*, p. 157.

2 *Earth's Insights*, p. 3.

3 He also established one of the first environmental studies programmes in the United States at the University of Wisconsin – Stevens Point.

4 Arthur Herman, *Community, Violence, and Peace: Aldo Leopold, Mohandas K. Gandhi, Martin Luther King, Jr, and Gautama the Buddha in the Twenty-first Century*, Albany, NY: State University of New York Press, p. 234, 1999.

5 Alan Holland, 'Editorial', *Environmental Values*, 9/1, p. 1, 2000.

6 Callicott, 'The Ever-robust Wilderness Idea and Ernie Dickerman', *Wild Earth*, 8/3l, p. 1, 1998.

7 Personal communication, March 1999.

8 'Introduction: Compass Points in Environmental Philosophy', in *Beyond the Land Ethic*, p. 4.

9 Leopold, *A Sand County Almanac*, p. 204.

10 Callicott defines an ecocentric environmental ethic as 'An environmental ethic that takes into account the direct impact of human actions on nonhuman natural entities and nature as a whole', *Earth's Insights*, p. 10.

11 'Traditional American Indian and Western European Attitudes Toward Nature: An Overview', in *In Defense of the Land Ethic*, pp. 189–90.

12 'Introduction', in J. Baird Callicott and Michael P. Nelson (eds), *The Great New Wilderness Debate*, p. 12.
13 'The Wilderness Idea Revisited', in J. Baird Callicott and Michael P. Nelson (eds), op. cit., p. 339.

See also in this book

Darwin, Leopold, Naess, Rolston

Callicott's major writings

In Defense of the Land Ethic: Essays in Environmental Philosophy, Albany, NY: State University of New York Press, 1989.
Earth's Insights: A Multicultural Survey of Ecological Ethics from the Mediterranean Basin to the Australian Outback, Berkeley, CA: University of California Press, 1994.
Beyond the Land Ethic: More Essays in Environmental Philosophy, Albany, NY: State University of New York Press, 1999.

Further reading

Callicott, J. Baird (ed.), *Companion to 'A Sand County Almanac': Interpretive and Critical Essays*, Madison, WI: University of Wisconsin Press, 1987.
Callicott, J. Baird and Ames, Roger T. (eds), *Nature in Asian Traditions of Thought: Essays in Environmental Philosophy*, Albany, NY: State University of New York Press, 1989.
Callicott, J. Baird and Flader, Susan L. (eds), *The River of the Mother of God and Other Essays by Aldo Leopold*, Madison, WI: The University of Wisconsin Press, 1991.
Callicott, J. Baird and Nelson, Michael P. (eds), *The Great New Wilderness Debate*, Athens, GA: The University of Georgia Press, 1998.
Callicott, J. Baird and Rocha, Fernando J.R. da (eds), *Earth Summit Ethics: Toward a Reconstructive Postmodern Philosophy of Environmental Education*, Albany, NY: State University of New York Press, 1996.
Freyfogle, Eric T., *Bounded People, Unbounded Land: Envisioning a New Land Ethic*, Washington, DC: Island Press/Shearwater Books, 1998.
Hargrove, Eugene C., *Foundations of Environmental Ethics*, Englewood Cliffs, NJ: Prentice-Hall, 1989.
Leopold, Aldo, *A Sand County Almanac: And Sketches Here and There*, New York: Oxford University Press, 1949.
Rolston III, Holmes, *Environmental Ethics: Duties To and Values in the Natural World*, Philadelphia, PA: Temple University Press, 1988.

MICHAEL P. NELSON

SUSAN GRIFFIN 1943–

We know ourselves to be made from this earth. We know

> this earth is made from our bodies. For we see ourselves.
> And we are nature. We are nature seeing nature. We are
> nature with a concept of nature ... Nature speaking of
> nature to nature.[1]

Contemporary feminist poet Susan Griffin, who began writing at the age of 14, has published more than fifteen books of poetry, drama, fiction and non-fiction on subjects ranging from rape and pornography to war, *eros* and illness. She has received many book awards and honours, including the Ina Coolbirth Prize for Poetry, an Emmy Award, a National Endowment for the Arts grant, the Malvina Reynolds Award for cultural achievement, a Schumacher Fellowship, a Commonwealth Medal, a Women's Foundation Award, a MacArthur Foundation grant, and nominations for the Pulitzer Prize and National Book Critics Circle Award. Although nature is a concern throughout Griffin's body of work, her long prose-poem *Woman and Nature: The Roaring Inside Her* (1978) stands out as a key text of environmental thought and a germinative work of ecofeminism, a movement that originated in the 1970s and that has become an influential voice in environmental discourse of the twenty-first century.

Griffin's background is pertinent to understanding her work, for, concurring with the feminist insight that 'the personal is political', Griffin at times interweaves autobiography with cultural critique, a literary form she calls 'social autobiography'. Born in California in 1943, Griffin came of age during the Cold War, years marked by nuclear testing, anti-communist propaganda and social conformity, resistance to which caused her to identify herself as a radical. Griffin's parents divorced when she was 6, and she and her older sister were separated and sent to live with different relatives, their mother's alcoholism rendering her unable to care for either child. This early experience of abandonment and separation struck deeply into the psyche of Griffin, whose later writing would probe these wounds to provide insight into Western culture and whose years of therapy pre-disposed her to view the collective mind of Western civilization from a psychological perspective. Griffin was raised by conservative Republican grandparents near Hollywood, California, where she grew up loving movies and becoming a fan of Eisenstein, whose film montages and juxtaposition of images almost certainly inform the associative collage technique of much of her writing. As a teenager, Griffin lived with a close friend's Jewish family, where her consciousness was raised about the historical treatment of Jews; in later years the Holocaust became an important image and racism a recurrent theme in her work.

Attending the University of California, Berkeley, during the student unrest of the 1960s, Griffin became involved in the Free Speech movement, the Civil Rights movement and protest against the Vietnam War. She transferred to what is now San Francisco State University, where she graduated *cum laude* in English in 1965, received her M.A. in 1973 and worked for the radical magazine *Ramparts* as an editorial assistant, becoming troubled by the sexist attitudes of the staff. During the late 1960s and 1970s, Griffin married, became a feminist, gave birth to a daughter, divorced and became a lesbian. Simultaneously, she taught writing and developed her own writing career with several volumes of poetry, short stories and an award-winning play, *Voices* (1975), which were published by feminist presses and reflected her experience as a woman in society. Early exposure to the diverse worlds of gentile and Jew, conservatism and radicalism, heterosexuality and homosexuality, marriage, motherhood and divorce, allowed Griffin to become what she calls a 'bridge figure', someone who straddles boundaries rather than reinforcing them.[2] 'As a writer', Griffin says, 'I have always felt myself to be a kind of crucible, my mind a medium in which the many voices, spoken and unspoken, belonging to our age, are melted, mixed and transformed.'[3]

As Griffin reflects on her classic work *Woman and Nature* she notes that two voices (each set in a different type face) engage in an extended dialogue, 'one the chorus of women and nature, an emotional, animal, embodied voice, and the other a solo part, cool, professorial, pretending to objectivity, carrying the weight of cultural authority'.[4] The book opens with a stunning, heavily researched and annotated, chronologically arranged compendium of statements, or, rather, parodies of statements from Plato through Einstein – magisterial voices of science, philosophy, and religion from Western civilization – proclaiming parallel 'truths' about nature and women. For example, referring to Aristotle, Griffin writes, 'It is decided that matter is passive and inert, and that all motion originates from outside matter ... It is decided that the nature of woman is passive, that she is a vessel waiting to be filled.'[5] Later, paraphrasing Lamarck on evolution and the Marquis de Sade on women, Griffin writes: 'It is declared ... [t]hat "the stronger and the better equipped ... eat the weaker and ... the larger species devour the smaller"'. And it is stated that if women were not meant to be dominated by men, they would not have been created weaker.'[6] Only occasionally, in this first

SUSAN GRIFFIN

section, do the voices of women and nature speak, anguished cries such as '*Our voices diminish ... We become less ... And they say that muteness is natural in us*'.[7]

The next section of the book focuses on the tandem mistreatment women and nature have received at the hands of a patriarchal culture dominated by the mind-set chronicled above. In 'Timber', for example, drawing from her reading of forestry manuals and office management textbooks, Griffin juxtaposes the economics of timber harvest (where trees are referred to as so many 'board feet') with the efficient supervision of stenographers. Other chapters compare factory farming with modern childbirth, horse training and dressage with facelifts and breast implant surgery, nuclear waste disposal with hiding the body of a murdered woman, and strip mining with rape. Needless to say, these pages are deeply disturbing. Casting her extensive research into a poetic form, Griffin's goal is to evoke feeling even as she awakens consciousness. These pages convince the reader that the comparisons Griffin makes point beyond metaphorical similarities to systemic unity; namely, that these various cruelties are all part of the same system, founded on, in Griffin's words, 'a philosophy that is also a submerged psychology'.[8]

In *Woman and Nature* and throughout her later work, Griffin develops a diagnosis of the illness of the Western mind; we are suffering from a form of insanity that lies at the heart of our destruction of the environment. According to Griffin in essays such as 'Split Culture' and 'Ideologies of Madness' (the latter collected in *The Eros of Everyday Life*), we live in a culture of fear. We are afraid of physical pain, illness, change and death, and we are likewise terrified by the power of nature over our lives. In the face of such terror, we resort to denial and domination. We deny our physical natures, imagining instead that we are our minds, that we possess an immortal soul. We attempt to control nature, to master it, subdue it, shape it to our desires. However, argues Griffin, the repressed always returns to haunt us in our dreams. So, we project onto an 'other' the parts of ourselves that we wish to disown. In our culture, white men have been in power; thus, men have defined themselves as above matter, and they have construed women as closer to nature. Man's domination of women, of racial 'others', and of nature can be understood as part of his ongoing efforts to be in control of himself, to triumph over the body and to deny death. 'In a culture of delusion', Griffin writes, 'women symbolize a denied self who experiences what it is to be human, to be in and of nature. This self knows that we die, this self feels, suffers pain, knows love without boundary, grieves loss, knows the world through sensation, through the body, accepts that we are sometimes powerless before the powerful circumstances of this earth.'[9]

Why have women come to be allied so closely with nature in men's psyche? Griffin, paralleling the work of Dorothy Dinnerstein, reasons that mothers are a child's first experience of nature: she has the power to feed and to comfort; likewise, she has the awful power to withhold food and to abandon the child. It is this early feeling of helpless dependence on mother/nature that causes the grown man to strive for independence, creating a culture built upon fear of connection and alienation from nature. Paradoxically, though, in dominating nature, we threaten the very grounds of our continued existence. Griffin writes: '[W]e belong to a civilization which is bent upon suicide, which is secretly committed to destroying Nature and destroying the self that is Nature.'[10]

What is the way out of this madness? In psychotherapy, the first step in healing is naming. We must become conscious of the problem. Griffin conceives of herself as a witness, someone who is 'able to speak the unspeakable, to break the silence'.[11] She explains, '[B]ecause the assumptions that belong to a culture are often invisible in their fullest dimensions and consequences, one must make them visible before discerning change. The very process of seeing the structure of thought *is* itself a crucial kind of change and genesis.'[12] She contends that the root of the problem is our culture's construction of masculinity. Griffin notes differences between socially masculine and feminine values: 'The roles society [has] given to men and women [have] produced different thinking and different ways of being in us ... [M]en, valuing power, produce nations, conflict and wars, and ... women, valuing life, produce relationship, continuity and peace.'[13] '[T]here are lots of reasons why males are violent', she observes, 'and they have more to do with tradition than testosterone.'[14] What is needed, according to Griffin, is 'a deep transformation of consciousness'.[15] Hallmarks of this shift will be the celebration of sensual knowledge, respect for a multiplicity of views rather than a single perspective, a view of the earth as being imbued with intelligence and intrinsic meaning, and, most important, the reunification of body and spirit and nature and culture in our conception of humanity. Griffin hopes, 'If human consciousness can be rejoined not only with the human body but with the body of earth, what seems incipient in the reunion is the recovery of meaning within existence that will infuse every kind of meeting between self and the universe, even in the most daily acts, with an eros, a palpable love, that is also sacred.'[16]

In her analysis of patriarchy and articulation of a healthier cultural alternative, Griffin, along with mutually influential writers Carolyn Merchant and Adrienne Rich, is widely regarded as a leading ecofeminist thinker. *Woman and Nature* has been called 'fundamental to an ecofeminist library', a 'cultural feminist classic', and a 'touchstone text' for 'virtually all ecofeminists.'[17] Ecofeminism joins feminist thought with ecological thought, insisting that one cannot fully understand the oppression of women without understanding how Western civilization has regarded nature, and, conversely, one cannot adequately understand our civilization's abuse of nature without taking into account how our culture conceptualizes women. As Griffin explains: '[T]he social construction (exploitation, destruction) of nature is implicit in and inseparable from the social construction of gender.'[18] Although scholars have noted that there are different varieties of ecofeminism, all versions seem to agree that patriarchy rests upon a conceptual foundation of hierarchical dualism in which reality is categorized by oppositional pairs (such as spirit/matter, intellect/emotion, mind/body, man/woman, culture/nature), in which the first term of the pair is accorded greater worth, privilege and power than the second. In this system, man is allied with culture, spirit and intellect, while woman is identified with nature, the body and emotion. While some feminists seek to liberate women from the sphere of the natural and some separatist ecofeminists wish to bar men from that sphere, Griffin and the majority of ecofeminists celebrate the woman–nature bond and urge that men likewise cultivate a closer relationship with nature and their own material bodies. In general, ecofeminists aspire to move beyond dualistic thinking and to establish relationships based not on hierarchy and domination, but on caring, respect and awareness of interconnection.

Griffin's writing, which since 1976 has been published by major trade presses, reveals an exceptionally broad understanding of interconnection. Her studies of rape and pornography reveal motivations and mechanisms of domination that also explain our relationship to nature. Her poetry is intimately related to her prose, which itself is highly poetic, reflecting her conviction that poetry is 'a powerful way of knowledge' that arises out of bodily experience and 'teaches political theory imagination'.[19] Griffin's *A Chorus of Stones* draws connections between the private psyche formed in childhood and public acts of violence in war, showing conversely how war creates violence in private life. Her recent *What Her Body Thought* connects the story of the flamboyant nineteenth-century courtesan featured in the movie *Camille* with Griffin's own illness from Chronic Fatigue Immune Dysfunction Syndrome, which in turn, 'like canaries in a mine', becomes 'a signal of the sickness of the planet'.[20] Revealing the economics of illness, Griffin indicts society for failing to support those in need. As one supporter has noted, 'By refusing to respect the "commonsense" distinctions among historical, social and personal issues, Griffin creates a kind of network of meaning in which everything illuminates everything else.'[21] In the context of environmental thought, Griffin's profound insight that gender issues and ecological issues are interconnected has been responsible for transforming both feminism and environmental thought.

Notes

1 *Woman and Nature*, p. 226.
2 Griffin, quoted in 'Susan Griffin', *Utne Visionaries: People Who Could Change Your Life, 1995 Profiles*, http://www.utne.com/visionaries/95profiles2.html.
3 *Made from this Earth*, p. 3.
4 Ibid., p. 82.
5 *Woman and Nature*, p. 5.
6 Ibid., p. 27.
7 Ibid., p. 26, original emphasis.
8 'Ecofeminism and Meaning', p. 216.
9 *Made from this Earth*, p. 18.
10 'Split Culture', p. 199.
11 *The Eros of Everyday Life*, p. 12.
12 Ibid., p. 6.
13 *Made from this Earth*, pp. 14–15.
14 Griffin, quoted in 'Susan Griffin', *Utne Visionaries*.
15 *The Eros of Everyday Life*, p. 20.
16 Ibid., p. 9.
17 Judith Plant (ed.), *Healing the Wounds: The Promise of Ecofeminism*, Philadelphia, PA: New Society Publishers, p. 255, 1989; Ynestra King, 'Healing the Wounds: Feminism, Ecology, and Nature/Culture Dualism', in *Gender/Body/Knowledge: Feminist Reconstructions of Being and Knowing*, ed. Alison M. Jaggar and Susan R. Bordo, New Brunswick: Rutgers University Press, p. 124, 1989 ; Patrick D. Murphy, *Literature, Nature, and Other: Ecofeminist Critiques*, Albany, NY: State University of New York Press, p. 40, 1995.
18 'Ecofeminism and Meaning', pp. 219–20.
19 *Made from this Earth*, pp. 16, 242.
20 Griffin, 'The Internal Athlete', Ms. v. 2.6, p. 38, 1992.
21 'Susan Griffin', *Utne Visionaries*.

See also in this book

Aristotle, Lovelock, Plumwood, Schumacher

Griffin's major writings

Woman and Nature: The Roaring Inside Her, New York: Harper & Row, 1978; new edn, Sierra Club Books, San Francisco, 2000.

Pornography and Silence: Culture's Revenge Against Nature, New York: Harper & Row, 1981.

Made from this Earth: An Anthology of Writings, London: Women's Press, 1982; New York: Harper & Row, 1983.

'Split Culture', *The Schumacher Lectures*, vol. 2, ed. Satish Kumar, London: Blond & Briggs, pp. 175–200, 1984.

The Eros of Everyday Life: Essays on Ecology, Gender and Society, New York: Doubleday, 1995.

'Ecofeminism and Meaning', *Ecofeminism: Women, Culture, Nature*, ed. Karen J. Warren, Bloomington, IN: Indiana University Press, pp. 213–26, 1997.

Bending Home: Selected & New Poems, 1967–1998, Port Townsend, Washington, DC: Copper Canyon Press, 1998.

Further reading

Adams, B., 'Susan Griffin', *Contemporary Lesbian Writers of the United States: A Bio-Bibliographical Critical Sourcebook*, ed. Sandra Pollack and Denise D. Knight, Westport, CT: Greenwood Press, pp. 244–51, 1993.

Macauley, D., 'On Women, Animals and Nature: An Interview with Ecofeminist Susan Griffin', *American Philosophical Association Newsletter on Feminism and Philosophy*, 90, 3 (Fall), pp. 116–27, 1991.

Merchant, C., 'Earthcare: Women and the Environmental Movement', *Environment*, 23, 5 (June), pp. 6–13, 38–40, 1981.

'Susan Griffin', *Contemporary Authors, New Revision Series*, vol. 50, ed. Pamela S. Dear, Detroit, MI: Gale Research, pp. 169–72, 1996.

'Susan Griffin', dialogue with Nannerl Keohane in 1980, collected in *Women Writers of the West Coast Speaking of Their Lives and Careers*, ed. Marilyn Yalom, Santa Barbara, CA: Capra Press, pp. 40–55, 1983.

CHERYLL GLOTFELTY

CHICO MENDES 1944–88

> My dream is to see this entire forest conserved because we know that it can guarantee the future of all the people who live in it ... If a messenger from heaven came down and guaranteed me that my death would help to strengthen our struggle it would even be worth it. But experience teaches us the opposite ... I want to live.[1]

Francisco 'Chico' Alves Mendes Filho, man of courage, words and deeds, hero of the rubber tappers of the Amazon, played a major role in the transformation of the landscape of the Brazilian rainforest. Chico Mendes was born on 15 December 1944 on a rubber estate in Xapuri, Acre, in northwest Brazil. Forty-four years later, on 22 December 1988, he was brutally assassinated, leaving wife Ilzamar G. Bezerra Mendes and their two children, Helenira aged 4 and Sandino aged 2. Mendes' short life was devoted to leading the rubber tappers' fight to defend the Amazon Forest and its fragile eco-system against exploitation by powerful and wealthy land speculators and ranchers.

Mendes was born into poverty. His parents had come from the northeast during the Second World War, having been sent to cut rubber for the Allied war cause. He received no formal education and became a *seringueiro*, a rubber tapper, at the age of 9. He learned to read and write around the age of 20.

> My life began just like that of all rubber tappers as a virtual slave bound to do the bidding of the master. I started work at nine years old, and like my father before me, instead of learning my ABC I learned how to extract latex from a rubber tree ... schools were forbidden on any rubber estate in the Amazon. The estate owners wouldn't allow it ... If a rubber tapper's children went to school they would learn to read, and write, and add up, and would discover to what extent they were being exploited.[2]

Ruthless exploitation from a variety of sources was to become the dominating force in the rubber tappers' existence, and resistance to this the focus of Mendes' life. Traditionally rubber tappers were at the mercy of a system of debt bondage, but during the 1960s and 1970s this system faced collapse in Xapuri. Ranchers from southern Brazil began to buy up rubber estates and clear vast areas of the forest for cattle grazing. Many tappers were forcefully, often brutally, evicted. Others retreated deeper into the forest to continue their work, only to be exploited by local merchants.

Chico Mendes knew that the future of the forests and of the rubber tappers were inextricably linked; that in order to secure a future for the people, the forests had to be protected and managed by those who understood the eco-system and how to live in it sustainably. From his endeavours emerged the concept of 'extractive reserves', which are legally protected forest areas that are held in trust for people who live and work on the land in a sustainable manner.

Early in the 1970s, the Xapuri Rural Workers' Union was founded, and Mendes was elected its president. As exploitation and conflict intensified, the Union developed the technique of the *'empate'* or 'stand-off'. During the dry season ranchers hire labourers to clear the forest for pasture. Just before the rains come in September the cleared areas are fired. Faced with eviction the rubber tappers assembled at sites about to be cleared, preventing the clearing and persuading the labourers to lay down their chainsaws. During the months of June, July and August in the 1970s and 1980s the forests of the upper Acre valley were the scene of numerous *empates*.[3] In 1985, Mendes and other leaders founded the National Council of Rubber Tappers (CNS) and gained increasing international support for their cause and passive resistance demonstrations. The movement was recognized as a force not only for social justice, but also against environmental destruction. The rubber tappers were able to propose a socially equitable and environmentally sustainable development policy for the region based on securing and improving their way of life, rather than official investments in ranching and colonization projects that would lead to disaster both for them and for the forest.[4]

Chico Mendes played a crucial role in negotiating with governments, with the World Bank and the Inter-American Development Bank. For example, in 1987 he visited the USA at the invitation of the Environmental Defense Fund and the National Wildlife Federation in order to discuss an Inter-American Development Bank-funded road paving project in Acre. Chico's message of caution was that the project would be disastrous if environmental conditions in the loan were not fulfilled. The loan was later suspended.

In addition to a great deal of respect and support, Mendes won two international prizes for his efforts. He was awarded the Ted Turner's Better World Society Prize and the United Nations Global 500 Environmental Prize. In 1988, responding to ever-increasing international pressure and support for the cause, the Brazilian government established the first ever extractive reserve. Yet as rewards and support increased, so too did risk to the rubber tappers, and, as their leader, to Chico Mendes in particular. Despite the creation of the CNS and the increasing level of organization of the tappers, the political power of the landowners was formidable. Their movement, the UDR, was enormously influential throughout the country and in Congress. It had successfully defeated land reform proposals in the Constituent Assembly.

Here in Xapuri, the UDR is beginning to make its presence felt. Since April 1988, when it formally set itself up in Acre, the number of hired gunmen in Xapuri has increased, as have the number of assassinations and attempted assassinations of workers. These gunmen are in effect the armed wing of the UDR and we are the targets.[5]

Chico Mendes was well aware of the threat to his own life; perhaps he foresaw his death. The quotation at the opening of this account is taken from a letter he had written shortly before his assassination by the son of local cattle rancher Darli Alves da Silva.[6]

Perhaps the most significant element of the legacy of Mendes is the enhanced power and voice of the organizations associated with him and the rubber tappers' cause – the National Council of Rubber Tappers and the Amazon Work Group from whose membership emerged a new generation of environmental leaders and activists. In Acre, Mendes' co-campaigners won important elective offices. For example, Marina Silva, co-founder with Mendes of the union movement and the Workers' Party in Acre, was elected to the Federal Senate in 1994; colleague Jorge Viana was elected mayor of Acre state capital in Rio Branco in 1992 and governor in 1999; and environmentalist Jo o Alberto Capiberibe was re-elected in 1999 as governor of Amapa. Such political successes for the Mendes cause have transformed national debate in Brazil on the Amazon region. The new generation of environmentalists have a major task ahead – the environmental, ecological and social crisis of Amazonia remains critically serious. Yet the political conditions for potential change have never been better as state and federal policies which promote and support sustainability are framed.

At the time of writing, a total of twenty-one extractive reserves and extractive settlements have been established in the seven states in Brazil, covering an area of 3.3 million hectares, together with a number of state reserves. By law, residents of the reserves must prepare a management plan for their area in order to obtain long-term rights of use. Both local communities and government have rights and responsibilities which encompass principles of ecological sustainability. Beyond Brazil, international agreements enforce protection of rainforest ecosystems. Yet federal extractive reserves account for some 1.5 per cent of the Amazon area. Deforestation rates are as high as ever in many regions, land degradation becomes an increasingly significant issue as time goes by, fires are more frequent and harder to control, and illegal logging practices continue to strip hardwoods from within protected areas. Furthermore, rubber prices have fallen so low that the extractive reserves are not producing the income to support even the basic needs of some communities. The poverty, degradation and destruction of Amazonia are amongst the greatest socio-environmental challenges of the present day; challenges brought onto the world's political stage and the agendas of NGOs as a result of various significant influences. The charismatic and courageous leader of the Brazilian rubber tappers' union must surely be one of the most significant of all.

In early 1989, in the aftermath of Mendes' death, which made a great impact not only in Brazil but world-wide, the Second National Congress of Rubber Tappers was held in Rio Branco. Rubber tappers of Brazil were joined there in force by tappers from Bolivia, by indigenous communities from Acre and elsewhere, and by representatives of government, human rights groups, the Church and political organizations. The meeting published twenty-seven demands concerning environmental protection, social development and human rights protection. It also published the *Declaration of the Peoples of the Forest* in memory of Chico Mendes and in the hope of the fulfilment of his vision for the future of the Amazon:

> The traditional peoples who today trace on the Amazonian sky the rainbow of the Alliance of the Peoples of the Forest declare their wish to see their regions preserved. They know that the development of the potential of their people and of the regions they inhabit is to be found in the future economy of their communities, and must be preserved for the whole Brazilian nation as part of its identity and self-esteem. This alliance of the Peoples of the Forest, bringing together Indians, rubber tappers and riverbank communities, and founded here in Acre, embraces all efforts to protect and preserve this immense, but fragile life-system that involves our forests, lakes, rivers and springs, the source of our wealth and the basis of our cultures and traditions.[7]

Notes

1   *Fight for the Forest, Chico Mendes in His Own Words*, p. 6.
2   Ibid., p. 15.
3   T. Gross in *Fight for the Forest*, p. 2.
4   Ibid.
5   *Fight for the Forest*, p. 80.
6   Ibid., p. 6.

7   National Council of Rubber Tappers, Union of Indigenous Nations,
    Rio Branco, Acre, March 1989. In *Fight for the Forest*, p. 85.

Mendes' major writings

Mendes, C., *Fight for the Forest, Chico Mendes in His Own Words*, London:
Latin American Bureau (research and action) Ltd, 1989.

Further reading

Branford, S. and Glock, O., *The Last Frontier: Fighting Over Land in the
Amazon*, London: Zed Books, 1985.
Caulfield, C., *In the Rainforest*, London: Heinemann, 1985.
Hyman, R., 'Rise of the Rubber Tappers', *International Wildlife*, 18 (5),
pp. 24–8, 1988.
Revkin, A., *The Burning Season: The Murder of Chico Mendes and the Fight
for the Amazon Rain Forest*, New York: Plume, 1994.

JOY A. PALMER

PETER SINGER 1946–

> If it is in our power to prevent something very bad from
> happening, without thereby sacrificing anything of com-
> parable moral significance, we ought to do it.[1]

Peter Singer has been described as having more positive influence on
the world than any other living philosopher.[2] His book *Animal
Liberation* has been translated into fifteen languages, sold half a
million copies, and is known as the Bible of the Animal Liberation
Movement. *Practical Ethics*, published in eight languages, was
named one of the world's one hundred most significant philosophical texts.[3] He is currently De Camp Professor of Bioethics at
Princeton, having previously held a Chair at Monash University,
Melbourne, from 1977 to 1999, and stood for the Australian Senate
as a candidate for the Australian Green Party.

Singer was born on 6 July 1946, in Melbourne. His parents had
arrived in Australia eight years earlier, fleeing from their native
Vienna to escape the persecution of the Jews shortly after the
Anschluss, the political union of Germany and Austria. Singer,
however, did not learn German until he began high school. He then
studied Law, History and Philosophy at the University of
Melbourne, where he met his wife Renata, with whom he has had
three daughters. While at the University of Melbourne he

participated in the movement against the Vietnam War. Later, when he was continuing his studies at the University of Oxford, this experience inspired his first book *Democracy and Disobedience*. It was in Oxford that he first learned about the conditions in which animals are kept in laboratories and factory farms, when he met the vegetarians to whom *Animal Liberation* is dedicated.

Singer was also deeply influenced by his Oxford supervisor, Professor R. M. Hare, one of the leading philosophical advocates of utilitarianism, a moral outlook developed in the last century by Jeremy Bentham and John Stuart Mill. Like other consequentialist moral theories, utilitarianism claims that our ultimate moral aim should be to achieve outcomes which are best when the interests of all those affected are considered impartially. Utilitarianism also claims that the best outcome is one which contains the greatest sum of *utility*, a term which usually refers to pleasure minus pain, preference satisfaction minus frustration, or simply happiness.

A number of environmentally important implications follow from utilitarianism. Given the significance the theory attaches to certain mental states, it implies that we must take into consideration all those sentient creatures capable of possessing those states. The condition of animals must, therefore, be included in our calculations, and an outcome which involves their suffering is morally worse than one which does not. Moreover, even if we have little personal concern for individuals in distant countries or future generations, they too can thrive, or suffer, as a result of our actions, and so must also be taken into consideration. From the impartial perspective of morality, their interests are no less important than ours. Thus, as Singer shows, a welfarist theory like utilitarianism can prove very supportive of certain environmentalist concerns.

Singer also believes that welfarist premises are modest, and provide a sound basis for an environmental ethic. It is therefore unnecessary to appeal to more extravagant metaphysical assumptions about the 'inherent worth of all life', or 'the intrinsic value of species and ecosystems'. As Singer explains, it is difficult to see how 'a species or an ecosystem can be considered as the sort of individual that can have interests, or a "self" to be realised', let alone that 'the survival or realisation of that kind of self has moral value, independently of the value it has because of its importance in sustaining conscious life'.[4] By contrast 'an ethic based on the interests of sentient creatures is on familiar ground. Sentient creatures have wants and desires. The question: "what is it like to be a possum drowning?" at least makes sense, even if it is impossible

for us to give a more precise answer than "it must be horrible" ...
But there is *nothing* that corresponds to what it is like to be a tree
dying because its roots have been flooded.'[5]

A moral theory is stronger when it does not require faith in
controversial metaphysical assumptions, and has wide appeal based
on clearly intelligible and relatively weak premises, to which
everybody can relate. For this reason, although Singer openly
embraces utilitarianism, his arguments are often constructed in ways
which can appeal to people from a variety of moral positions. His
three best-known contributions to moral philosophy include his
work on animal ethics, famine relief and bioethics. His work on
abortion, euthanasia, reproductive technologies and other areas of
bioethics is extensive, and has generated considerable controversy.
The first two areas, however, which are more directly relevant to
environmentalism, have had an even greater impact on the
development of modern moral philosophy. They are not just
exercises in applied ethics, but path-breaking ideas about the nature,
scope and demandingness of morality, which raise issues of
fundamental theoretical importance.

Again, however, Singer's arguments are based on very simple and
apparently uncontroversial premises, such as 'pain is bad', and 'it is
wrong to cause intense pain unnecessarily or fail to relieve it when
we could do so at little cost to ourselves'. The following is probably
one of the best know passages making this point:

> The path from the library at my university to the humanities
> lecture theatre passes a shallow ornamental pond. Suppose
> that on my way to give a lecture I notice that a small child
> has fallen in and is in danger of drowning. Would anyone
> deny that I ought to wade in and pull the child out? This will
> mean getting my clothes muddy and either cancelling my
> lecture or delaying it until I can find something dry to
> change into; but compared with the avoidable death of a
> child this is insignificant.
>
> A plausible principle that would support the judgement
> that I ought to pull the child out is this: if it is in our power
> to prevent something very bad from happening, without
> thereby sacrificing anything of comparable moral signifi-
> cance, we ought to do it.'[6]

Relying on these modest premises, Singer argues that it is wrong to
cause millions of animals the most terrible suffering in factory farms

for the sake of a trivial difference in taste to our meals, and that it is wrong to allow people in poor countries to die of starvation, when we could prevent their deaths by making donations which do not represent unbearable costs to ourselves. In so doing, he develops a number of arguments which conclude that most of us should, like him, become vegetarians and donate – with some flexibility depending on our circumstances – at least 10 per cent of our incomes to charities like Oxfam that assist the world's poorest people.

Some people finds the case for vegetarianism more convincing because we actively cause animals to suffer and die while in the second case we may merely be passively allowing people to suffer and die. Others, by contrast, find the second case more persuasive because it concerns human beings rather than animals. Singer challenges the alleged moral importance of both of these distinctions: between killing and letting die (when they have the same consequences) and between causing a certain amount and type of pain to a human and to a non-human animal (which increases to the same degree the suffering in the world). Other consequentialists, as well as Singer, have scrutinized the first distinction at length.[7] Singer is particularly famous for opposing the second distinction, which he argues relies on *speciesism*, a discriminatory prejudice comparable to racism. Thus, like racists, speciesists do not base their decisions on the merits of an individual case but instead on group membership.

Singer's opposition to speciesism is often misinterpreted. First, anti-speciesists can accept that humans and animals are, in fact, different. Similarly anti-racists, and feminists, may accept the existence of racial or sexual differences since they need only deny that any such differences justify giving less importance to the interests of racial minorities or women. Second, anti-speciesists do not claim that killing any animal is as bad as killing a person. They may accept that taking the life of a self-conscious creature – with memories, expectations, plans and long-lasting friendships – like a chimpanzee is worse than killing a creature, like a fish or a worm, which lacks any of these capacities. By taking the chimpanzee's life, we would be depriving it of much more; a worm, however, has only a worm's life to lose.

Singer's view of the wrongness of killing – discussed in *Practical Ethics* and subsequent works – also fits with the view expressed in *Animal Liberation*, that whenever pain is of the same type and intensity it is not, in itself, morally worse if it occurs in the course of a human rather than an animal life. The views can be reconciled because the interest in avoiding suffering is very different to the interest in avoiding death. For example, when doctors cannot spare both a mother and her foetus a certain amount of pain, mothers tend to prefer to suffer quite considerably, to save their future baby from a smaller pain. However, when the conclusion the doctors reach is that they cannot save both lives, it is the mother that is generally saved, even when there are no relatives that could be affected by her death. This difference provides the plot of a good number of war movies. During the siege, all the sedatives, or the cognac, are given to the most gravely injured soldier. But when at the end of the movie they cannot all be rescued – for example, because someone must remain to detonate the explosives – the volunteer is standardly the most gravely injured soldier, who has least to lose. While an individual's interest in preserving their life depends on what sort of life it is going to be, the interest in avoiding suffering is universal, and when it really is of the same character and intensity, it has the same moral importance independently of who suffers.

This idea changed the life of Henry Spira, one of Singer's former students. Having participated in some of the century's key progressive struggles, for Civil Rights in the American South and for Trade Union reform in the US Labour Movement, amongst others, Spira devoted his last two decades to the fight for animal rights. Singer wrote a biography, and filmed a documentary about Spira because his life expressed so well what the philosopher has attempted to say: that there is a natural progression from human liberation to animal liberation. The same compassion, the same sense of justice, the same opposition to cruelty and exploitation which made us reject slavery and, later, so many other forms of oppression and abuse, have to make us react against the systematic and prolonged torture of millions of sentient creatures crammed in laboratory cages and factory farms. Singer also claims that by 'ceasing to rear and kill animals for food, we can make so much extra food available for humans that, properly distributed, it would eliminate starvation and malnutrition from this planet. Animal liberation is human liberation too.'[8] Furthermore, as well as contending that vegetarianism is required by interests of animals and the poorest human beings, Singer argues that the meat industry is so environmentally damaging that it cannot be part of a sustainable lifestyle, and must also be rejected on grounds of intergenerational justice.[9]

This is just one example of the integrity and depth of Singer's work. In an academic and political context favouring the development of novel and sophisticated theories which ultimately lead only to the same old reformist conclusions, it is refreshing to see a theorist begin with some modest premises, clear arguments and common sense, and inspire people to radically re-examine their lives, reach into their pockets, and take to the streets.

Notes

1 *Practical Ethics*, p. 229.
2 See, e.g., *The Philosophers Magazine*, 4, 1989, and R. Posner, 'Reply to Critics', *Harvard Law Review*, 111.7, p. 1816, 1998.
3 K. Worsley, 'Heartless Animal or Rational Beast?', *Times Higher Education Supplement*, 29 May 1998, p. 17.
4 *Practical Ethics*, p. 283.
5 Ibid., p. 277.
6 Ibid., p. 229. See also, 'Famine, Affluence and Morality', *Philosophy and Public Affairs*, 1, pp. 229–43, 1972; 'Reconsidering the Famine Relief Argument', in P. Brown and H. Shue (eds), *Food Policy: US Responsibility in the Life and Death Choices*, New York: The Free Press, 1977; *Practical Ethics*, chs 8 and 9; *The New York Times*, 5 September 1999.
7 But see, e.g., *Practical Ethics*, pp. 206–13, 218, 222–9, 309 and D. Jamieson, *Singer and His Critics*, pp. 311ff.
8 End of the 1975 Prologue to *Animal Liberation*.
9 See *Practical Ethics*, pp. 287–8 and *How Are We to Live?*, pp. 44ff.

See also in this book

Wang Yang-ming

Singer's major writings

Democracy and Disobedience, Oxford: Clarendon Press, 1973.
Animal Liberation, New York: New York Review/Random House, 1975, 2nd edn 1990.
Animal Rights and Human Obligations, ed. with Tom Regan, Englewood Cliffs, NJ: Prentice-Hall, 1976.
Practical Ethics, Cambridge: Cambridge University Press, 1979, 2nd edn 1993.
Marx, Oxford: Oxford University Press, 1980.
The Expanding Circle, Oxford: Oxford University Press, 1981.
Hegel, Oxford: Oxford University Press, 1982.
Should the Baby Live?, with H. Kuhse, Oxford: Oxford University Press, 1984.
Applied Ethics, ed., Oxford: Oxford University Press, 1986.
In Defence of Animals, ed., Oxford: Blackwell, 1986.
A Companion to Ethics, ed., Oxford: Blackwell, 1991.
How Are We to Live?, ed., Melbourne: Text Publishing, 1993.
The Great Ape Project, ed. with P. Cavalieri, London: Fourth Estate, 1993.
Ethics, Oxford: Oxford University Press, 1994.
Rethinking Life and Death, Melbourne: Text Publishing, 1994.
The Greens, with B. Brown, Melbourne: Text Publishing, 1996.

Ethics into Action: Henry Spira and the Animal Rights Movement, Lanham, MD: Rowman & Littlefield, 1998.

Further reading

Jamieson, D., *Singer and His Critics*, Oxford: Blackwell, 1999.

PAULA CASAL

VANDANA SHIVA 1952–

> Biotechnology, as the hand-maiden of capital in the post-industrial era, makes it possible to colonise and control that which is autonomous, free and self-regenerative. Through reductionist science, capital goes where it has never been before. The fragmentation of reductionism opens up areas for exploitation and invasion ... It is in this sense that the seed and women's bodies as sites of regenerative power are, in the eyes of capitalist patriarchy, among the last colonies.[1]

Born in the green valley of Dehradun on 5 November 1952, Vandana Shiva received her first lessons on the environment in the lap of the Himalayas from her mother, a farmer with a deep love of nature, and her father, a Conservator of Forests. Armed with her childhood aspirations of becoming a scientist, she was educated at St Mary's School in Nainital, and subsequently at the Convent of Jesus and Mary in Dehradun, from whence particle physics gave rise to her life-long passionate affair with the environment. She trained as a physicist, was awarded a Doctorate in Philosophy at the University of Western Ontario for her thesis 'Hidden Variables and Non-locality in Quantum Theory', and then entered into inter-disciplinary research in science, technology and environmental policy at the Indian Institute of Science and the Indian Institute of Management in Bangalore.

By 1982, Dr Shiva was a leading theoretical physicist in the ecology movement, but she then set aside her professional career to devote her next ten years to environmental activism. Her first step was to found an independent body to address the emerging ecological and social issues in partnership with local communities and social movements. This she named the Research Foundation for Science Technology and Ecology, and it was here that she began her campaign by highlighting the parallels between the world's major industrial revolutions: the first based on the mechanization of work chiefly within the textile trade; the second on the chemicalization of agriculture; and the emerging third, endeavouring to engineer biological process. With the frankness which has been the hallmark of her career, Shiva named 'poverty and underdevelopment' as integral to all three industrializations. By 1993 she had won the alternative Nobel Prize, also known as the Right Livelihood Award, the first of many awards acclaiming her research into the environmental and social injustice which underpins corporate solutions to the earth's declining renewable resources.

It can be said that when Shiva writes, the world reads. Certainly, she has been a prolific author, with each of her thirteen volumes on Biodiversity, Biopiracy, Biopolitics, Biotechnology, Ecofeminism, Globalization, and Food Security reflecting a profound multi-disciplinary scholarship. Her 1988 debut volume, *Staying Alive*, created common ground for feminists and environmentalists, providing an exemplary insight into the plight of women throughout developing regions. Drawing on historical evidence of the feminized poverty resulting from colonial rule, she identified 'modern development' as a product of Western patriarchy which further eroded women's productivity by removing land, water and forests from their management, while simultaneously impairing ecological productivity and sustainability via the destruction of soil, rivers and vegetation. Central to her argument was the theft of the natural biodiversity and food security which women had safeguarded over centuries, by a eurocentric science and economics which reshaped the seed and the earth to fit with the latest in patriarchal delusions. Shiva deemed the post-colonial development paradigm to be mal-development, a process subjugating women and nature while creating twinned social and environmental injustice. She called for a turn-around in the development mind-set, asking that the feminine principle be applied to substitute the *sanctity of life* for the *sanctified development concept rooted in patriarchy.*

In the early 1970s, India's Punjab was one of the fastest growing agricultural economies in the world and a showpiece of the Green Revolution, but Shiva's 1991 volume, *The Violence of the Green Revolution*, challenged the accepted gospel that Norman Bourlag's hybridized semi-dwarf, high-yielding wheat seeds had transformed the region's austerity into prosperity, while simultaneously correcting the widely held perception that the contemporary bloodshed which saw 15,000 Punjabis lose their lives in the 1980s was due to religious fundamentalism. Shiva exposed the truth of Bourlag's Nobel Prize winning miracle; nature's soil turned into waterlogged expanses or salinated deserts; and with the environmental destruction came community violence which was hardest felt by women and children. In effect, with control over both the environment and people essential to the centralized and centralizing tactics of the Green Revolution which integrated Third World farmers into the global market of fertilizers, pesticides and seeds, the ecological collapse, together with the political disruption of society, were predictable outcomes of a paradigm which had disconnected nature from society. To Shiva, it was already obvious that the seed was being colonized through a political process which removed control over biological diversity from the hands of peasant farmers and passed it over to corporate interests. New biotechnologies altered the role of farmer and of ecological procedure. Paraphrasing Shiva, the corporate seed had divested peasants, robbed them of their livelihoods, and was the very instrument of their underdevelopment and poverty. Whether farmers owned or leased their land, biotechnology's genetically programmed seed was corporate property. Unlike any before it, the progress and needs of the new age seed were met by a corporate computer. Similarly, the lifespan of the new age seed was regulated by corporations rather than by farmers guided by generations of traditional wisdom.

From the beginning, Vandana Shiva's research had synchronized with that of other feminist activists and multi-disciplinary academics representing every world region. Included were biologists, sociologists, engineers, political scientists and a consortium of experts in bioethics, development, economics, environment, law, medicine, nuclear hazards, and science and technology. In 1991, aware that the world's land, forests, rivers, oceans and the atmosphere were either colonized, eroded and/or polluted, and that global capitalism was seeking new territories – plants, animals and women's bodies – to invade and exploit in the quest for further wealth, Shiva convened a seminar on 'Women, Health and the Environment' in the southern Indian city of Bangalore. Sponsored by her own Research Foundation on Science Technology and Ecology, in partnership with the Swedish Dag Hammarskjöld Foundation, the meeting gathered feminists from an international circle, each committed to reconstructing the links with nature which a patriarchal and technocratic environmental science had incrementally destroyed. Led by Shiva, these women did not see 'environment' from an external or hypothetical perspective. Rather, 'environment' was the place where they lived, and therefore translated into everything which affected their lives. Their faith in the earth body–human body continuum meant that environmental hazards were health hazards. Their human rights and health ethics were at odds with the population control racism and misogyny advocated by neo-Malthusians to salvage the earth's resources, and for them the South was not the source of most, if not all, environmental problems, anymore than the North, for all its technology and capital, was the source of all environmental solutions. Shiva and her allies moved beyond the existing patriarchal dichotomies, to examine the manner in which the activity/passivity and culture/nature divisions became instruments for biotechnology to colonize the regeneration of plants and humans. Together, they launched ecofeminism into an extraordinary political movement; an alliance which was destined to become the strongest opponent of environmental degradation, economic exploitation, cultural globalization, and institutionalized gender and indigenous discrimination, and which by the end of the twentieth century had grown to influence global policy on environmental and social justice, human/women's rights, indigenous knowledge, human/women's health, world trade regulations and the multiple paradigms of development and economics.

Shiva saw that the second Green Revolution paved the way for human rights, including the right to a livelihood, to be exchanged for property rights protecting the processes of biotechnology. She laid bare the loopholes in the General Agreement on Tariffs and Trade (GATT) which allowed transnational corporations (TNCs) to market agricultural commodities without restriction, regulation or responsibility.[2] Encouraging the free trade of agricultural components, and aided by World Bank/International Monetary Fund Structural Adjustment programmes, GATT destroyed local food markets, converting subsistence Third World food production into a lucrative emporium for corporations. Small producers, most of whom are women, were destined for displacement by GATT.

By 1998, when the Intellectual Property Rights shaped by the World Trade Organization (WTO) were ordained to deny the world's poorest farmers both free access to their own seed and the liberty to exchange their own seeds between themselves, Shiva gave the opening keynote address at the First Grass Roots Gathering on Biodevastation: Genetic Engineering in the US city of St Louis. Interviewed afterwards, she was asked to further explain her 'Third World perspective'. She answered that following European colonization, the Third World was left with only its biodiversity, and a solitary renewable resource, the seed, to meet health and nutritional needs, and to retain a semblance of agricultural viability. Consequently, in the Third World, where the majority are totally dependent on agriculture for survival, 'You can't have a consumer society with poor people and therefore what you will have is deprivation, destitution, disease, hunger, epidemics, hunger, malnutrition, famine and civil war. What is being sown is the greed of the corporations in stealing the last resources of the poor.'[3]

While the Green Revolution invaded the seed to become a source of ecological disruption, biotechnology went further, colonizing the seed at two levels; first by robbing the seed of its fertility and self-regenerating capacity; and second via GATT patent protection (Trade Related Intellectual Property Rights or TRIPs), transferring the ownership of laboratory-spliced and/or relocated genes, none of which amount to newly created genes, to the seed's 'genetic tailors', most of whom were US-based TNCs and institutions. Shiva argued that TRIPs denied Third World farmers both their intellect and their rights, and calculated that the resulting transfer of funds from poor to rich countries had the potential to exacerbate Third World debt ten times over. She also emphasized that biotechnology, in addition to devaluing the seed 'from a living renewable resource into a mere raw material', demeaned women in the same fashion. Neither patriarchy's passive construction of nature, nor its politics of separation and fragmentation, was deemed an acceptable reason by Shiva, and as the health and ecology crises of the 1990s raised serious doubts about 'man's ability to totally engineer the world, including seeds and women's bodies', she further embraced the partnership which women shaped with nature in their everyday lives as the sustainable paradigm for 'dynamic and diverse' regeneration.

It can also be said that when Shiva speaks, the world listens. She is without doubt the most prized speaker on the global conference circuit, captivating audiences with her eloquence, her passion and her unquestionable logic. But to say that Shiva is an environmentalist, or an ecofeminist, is to sell her short. Vandana Shiva is a fearless campaigner for the environment, women, India, the Third World and planet Earth, tireless in her efforts to spread the word on existing and impending injustice within and beyond the institutional halls of government and academy. Readers of the print media are privy to her research on a regular basis. In 1997, writing in the London *Guardian*,[4] she drew European attention to the double standards of TNCs seeking to abduct global food security, and exposed the folly of biotechnology's answer to famine: 'The introduction of herbicide-resistant crops destroys biodiversity and rural livelihoods, which are supported by the full variety of nature. Herbicide use in societies where people collect "weeds" for vegetables and fodder can destroy nutrition and women's work. In India women gather more than 130 species of greens, or weeds – the most important source of vitamin A in rural areas. The irresponsible spread of herbicides through herbicide-resistant crops will aggravate malnutrition in poor communities.'

In 1999, Shiva used India's national print media to warn of Monsanto's impending agenda to monopolize global water supplies. To enter the water business, Monsanto had acquired an equity stake in Water Health International as part of a joint venture with Tata/ Eureka Forbes, but as Shiva wrote, 'The joint venture route has been chosen so that Monsanto can achieve management control over local operations but not have legal consequences due to local issues.'[5] To Shiva, it was clear that Monsanto's water initiative, like its seed and aquaculture trade, was designed for the express purpose of expanding its monopolies over the basic ingredients of life. In this instance, Monsanto's plan was to invent a market economy for water, with the company's investment underwritten by public finances. For Shiva, 'A more efficient conversion of public goods into private profit would be difficult to find. Water is, however, too basic for life and survival and the right to it is the right to life. Privatisation and commodification of water are a threat to the right to life.'[6]

At that time, India already had major water management movements, the pani panchayat and conservation movement in Maharashtra, and the Tarun Bharat Sangh in Alwar, each designed to regenerate and equitably share water as a commons, giving all the right to water and none the right to abuse or waste water. For Shiva too, water was a commons, and had to be managed as such, rather than 'controlled and sold by a life sciences corporation (i.e. *Monsanto*) that peddles in death'.[7] Her 1999 alert to Monsanto's water agenda proved prophetic. Ten months later, as Medha Patkar and Arundhati Roy took the Narmada Bachao Andolan struggle against displacement and human rights violations by the Sardar Sarovar large dam project to the Second World Water Conference in The Hague, so too the World Water Commission for the Twenty-First Century put forward its report, *A Water Secure Future: Vision for Water, Life and the Environment*. In this the Commission forwarded the notion that water security could only be implemented by the wholesale privatization of water supply and sanitation services across the world. Within hours, the report was condemned by various NGOs, women's groups and individual ecofeminists, all of whom shared Shiva's view that corporate control over water would end any concept of a universal right to water and sanitation, replacing yet another human right with a free market concept commodifying water.

While the popular catchphrase of the 1990s calls to 'think global, act local', Shiva thinks and acts at every level. At home in India, she has successfully filed Public Interest writs in the Supreme Court on a variety of environmental and trade-related issues, and in 1991 she founded Navdanya, a national movement designed to protect the diversity and integrity of living resources from corporate appropriation. *Navdanya* means nine seeds, and the programme bearing the same name has made seed-saving both a celebration of diversity and a mode of resistance. Demystifying GATT, working with farmers to explain TRIPs and the Agreement of Agriculture, Shiva has made significant contributions to fundamental intellectual debate and grassroots campaigns, including the mobilization of 500,000 farmers against GATT in 1993. Also at home, she has played a pioneering role in linking TRIPs to Biodiversity and Indigenous Knowledge and to the Convention on Biological Diversity; launched the idea of collective rights to defend indigenous knowledge; and was the first to suggest that the *sui generis* option in TRIPs should be based on community rights and farmers' rights. Less locally, in mid-1998, Shiva openly reminded Professor Mohammad Yunus from neighbouring Bangladesh that the Grameen Bank's impending partnership with Monsanto was a betrayal of the very women for whom his microcredit scheme promised self-reliance. Yunus listened, and a month later abandoned the agreement brokered between the Grameen Bank and Monsanto.

At an international level, Dr Shiva initiated the women's movement on food, agriculture, patents and biotechnology. Formally launched in Bratislava, Slovakia, in May 1998, as 'Diverse Women for Diversity', she led the movement to the WTO Ministerial meeting in Seattle to protest against international trade regulations which discriminated against the environment, women and the Third World. As Shiva described it, the successful Seattle protest 'demonstrated that globalisation is not an inevitable phenomenon which must be accepted at all costs but a political project which can be responded to politically'.[8] Paraphrasing Shiva, the rebellion both on the streets and within the WTO negotiations indicated the start of a new democracy movement; one where citizens from across the world and the governments of the South refused to be bullied and excluded from decisions in which they have a rightful share. WTO's dictatorial anti-people, anti-nature decisions, enabling corporations to steal the world's harvests via secretive, undemocratic structures and process earned itself names such as the World Tyranny Organization. The intolerance of democratic dissent, the hallmark of dictatorship, was unleashed in full force in Seattle. While the trees and stores were lit for Christmas festivity, the streets were barricaded by police. In retaliation, labour joined hands with environmentalists, as farmers from the North and farmers from the South made a common commitment to say 'no' to genetically engineered crops, acting not out of special interest, but in defence of the common interests and common rights of all. Shiva continues:

> A broad-based citizen's campaign stopped a new Millennium Round of WTO from being launched in Seattle, but launched its own millennium round of democratization of the global economy. The rights of all species and the rights of all people must come before the rights of corporations to make limitless profits through limitless destruction. Free trade is not leading to freedom. It is leading to slavery. Diverse life forms are being enslaved through patents on life, farmers are being enslaved into high-tech slavery, and countries are being enslaved into debt and dependence and destruction of their domestic economies. The future is possible for humans and other species only if the principles of competition, organised greed, commodification of all life, monocultures, monopolies and centralised global corporate control of our daily lives enshrined in the WTO are replaced by the principles of protection of people and nature, the obligation of giving and sharing diversity, and the decentralisation and self-organisation enshrined in our diverse cultures and national constitutions.[9]

On the eve of the Third Millennium, having overstayed her term as an activist by almost an entire decade, Vandana Shiva is the key environmental voice on the global stage. Her presence is one of authority, carrying weight within government and non-government, academic and non-academic, feminist and non-feminist, rural and urban, and national and international circles. She stands as a beacon of hope for the global Green movement and the world's expanding underclass of poverty-stricken farmers, most found in developing regions, and the majority of women, prey for transnational corporations carrying the global order's imprimatur to turn life resources into money-making commodities in the third millennium.

Notes

1   Vandana Shiva, 'The Seed and the Earth', in *Minding Our Lives: Women from the South and North Reconnect Ecology and Health*, New Delhi: Kali for Women, pp. 128–43, 1991.
2   Mies and Shiva, pp. 231–46.
3   Nic Paget-Clarke, 1998. An interview with Vandana Shiva in *MotionMagazine*.
4   Vandana Shiva, 'Genetic seeds of hope and despair', *Guardian*, 17 December, 1997.
5   Vandana Shiva, 'Monsanto's expanding monopolies', The Hindu, THE HINDU, 1 May, 1999.
6   Ibid.
7   Ibid.
8   Vandana Shiva, *The Historic Significance of Seattle*. Research Foundation on Science, Technology and Ecology, 12 December 1999.
9   Ibid.

Shiva's major writings

Staying Alive, New Delhi: Kali for Women, 1988.
The Violence of the GREEN REVOLUTION: Third World Agriculture, Ecology and Politics, London: Zed Books and Penang: Third World Network, 1991.
Biopiracy: The Plunder of Nature and Knowledge, Boston, MA: South End Press, 1997.
Stolen Harvest: The Hijacking of the Global Food Supply, Boston, MA: South End Press, 2000.

Further reading

Bandarage, Asoka, *Women, Population and Global Crisis*, London and Atlantic Highlands, NJ: Zed Books, 1997.
Hartmann, Betsy, *Reproductive Rights & Wrongs: The Global Politics of Population Control*, rev. edn, Boston, MA: South End Press, 1995.
Mies, Maria and Shiva, Vandana, *Ecofeminism*, London and New Jersey: Zed Books, 1993.
Roy, Arundhati, *The Cost of Living*, London: Flamingo, 1999.
Salleh, Ariel, *Ecofeminism as Politics: Nature, Marx and the Postmodern*, London and New York: Zed Books, 1997.
Shiva, Vandana and Moser, Ingunn (eds), *Biopolitics: A Feminist and Ecological Reader on Biotechnology*, London and New York: Zed Books and Penang: Third World Network, 1995.

LYNETTE J. DUMBLE